AF745592

Integrative Advances in Rice Research

Edited by Min Huang

Published in London, United Kingdom

IntechOpen

Supporting open minds since 2005

Integrative Advances in Rice Research
http://dx.doi.org/10.5772/intechopen.95205
Edited by Min Huang

Contributors
Raghu Shivappa, Devanna B. Navadagi, Mathew S. Baite, Manoj Kumar Yadav, Prabhukarthikeyan S. Rathinam, Keerthana Umapathy, Prajna Pati, Prakash Chandra Rath, Manasi Dash, Abinash Mishra, Pedro S. Sampaio, Carla M. Brites, Keiko Kataoka, Aarti Sharma, Sandhaya, Akanksha Srivastava, Snehanshu Singh, Subhash Mishra, Shiva Mohan, Chhavi, Akanksha Singh, Avinash Kumar Singh, Hemant Kumar Jaiswal, Noren Singh Konjengbam, Mayurakshee Mahanta, Andrean Lyngdoh, Md. Rashidul Islam, Md. Mahfujur Rahman, Md. Mostafa Masud, Md. Zahangir Alam, Md. Mamunur Rashid, Mohammad Ashik Iqbal Khan, Md. Abdul Latif, Krishna Pada Halder, Al-Imran Dianga, Ruth N. Musila, Kamau W. Joseph, Manonmani Swaminathan, Prasannakumar Mothukapalli Krishnareddy, Deepak Chikkaballi Annegowda, Hirehally Basavarajegowda Mahesh, Chetana Bangera Siddabasappa, Pramesh Devanna, Haniyambadi Basavegowda Manoj Kumar, Sahana Nagaraj Banakar, Siddegowda Rajendra Prasad, Manoj Kumar, Sandhya Kulhari, Pawan Kumar, Aravind Kumar Jukanti, Akash Gaurav Singh, Sathees Nagarajan, Nirubana Varatharajan, Deepika Chandra Sekaran, Karthikeyan Murugan, Vanniarajan Chockalingam, Adrian C. Dubock, Justus Wesseler, Robert M. Russell, Chen Chen, David Zilberman, Renganathan Vellaichamy Gandhimeyyan, Mahendra Kumar Mohanty, Nur-E-Tajkia Islam, Muhammad Iqbal Hossain

© The Editor(s) and the Author(s) 2022
The rights of the editor(s) and the author(s) have been asserted in accordance with the Copyright, Designs and Patents Act 1988. All rights to the book as a whole are reserved by INTECHOPEN LIMITED. The book as a whole (compilation) cannot be reproduced, distributed or used for commercial or non-commercial purposes without INTECHOPEN LIMITED's written permission. Enquiries concerning the use of the book should be directed to INTECHOPEN LIMITED rights and permissions department (permissions@intechopen.com).
Violations are liable to prosecution under the governing Copyright Law.

(cc) BY

Individual chapters of this publication are distributed under the terms of the Creative Commons Attribution 3.0 Unported License which permits commercial use, distribution and reproduction of the individual chapters, provided the original author(s) and source publication are appropriately acknowledged. If so indicated, certain images may not be included under the Creative Commons license. In such cases users will need to obtain permission from the license holder to reproduce the material. More details and guidelines concerning content reuse and adaptation can be found at http://www.intechopen.com/copyright-policy.html.

Notice
Statements and opinions expressed in the chapters are these of the individual contributors and not necessarily those of the editors or publisher. No responsibility is accepted for the accuracy of information contained in the published chapters. The publisher assumes no responsibility for any damage or injury to persons or property arising out of the use of any materials, instructions, methods or ideas contained in the book.

First published in London, United Kingdom, 2022 by IntechOpen
IntechOpen is the global imprint of INTECHOPEN LIMITED, registered in England and Wales, registration number: 11086078, 5 Princes Gate Court, London, SW7 2QJ, United Kingdom
Printed in Croatia

British Library Cataloguing-in-Publication Data
A catalogue record for this book is available from the British Library

Additional hard and PDF copies can be obtained from orders@intechopen.com

Integrative Advances in Rice Research
Edited by Min Huang
p. cm.
Print ISBN 978-1-83969-599-5
Online ISBN 978-1-83969-600-8
eBook (PDF) ISBN 978-1-83969-601-5

For EU product safety concerns:
IN TECH d.o.o., Prolaz Marije Krucifikse Kozulić 3, 51000 Rijeka, Croatia, info@intechopen.com or visit our website at intechopen.com.

We are IntechOpen, the world's leading publisher of Open Access books

Built by scientists, for scientists

5,600+
Open access books available

138,000+
International authors and editors

175M+
Downloads

156
Countries delivered to

Our authors are among the
Top 1%
most cited scientists

12.2%
Contributors from top 500 universities

WEB OF SCIENCE™

Selection of our books indexed in the Book Citation Index (BKCI) in Web of Science Core Collection™

Interested in publishing with us?
Contact book.department@intechopen.com

Numbers displayed above are based on latest data collected.
For more information visit www.intechopen.com

Meet the editor

Dr. Min Huang received his bachelor's degree in Biological Science from Zhejiang Normal University, China in 2005, and obtained his master's degree and Ph.D. in Crop Cultivation and Farming System from Guangxi University (GXU) and Hunan Agricultural University (HNAU), China in 2008 and 2011, respectively. He successively held the positions of assistant and associate professor in the Department of Agronomy at GXU from 2012 to 2014. He joined the faculty in the Department of Agronomy at HNAU as an associate professor in 2015 and was promoted to the position of professor in 2017. He served as a visiting fellow at the International Programs-College of Agriculture and Life Sciences, Cornell University, USA in 2017 and 2018.

Contents

Preface

Rice is a staple food for about half of the world's population. Since 1960s, rice yield has more than doubled in most parts of the world and has even tripled in certain countries due to considerable progress in the development of high-yield rice varieties and the improvement of crop management practices. In the next three decades to 2050, rice yield must increase by more than 1% annually to meet the growing demand for food that will result from population growth and economic development. This is not an easy task because several new challenges have been brought about by changes in socioeconomic and physical environments related to rice production. In particular, rapid urban expansion has resulted in a labor shortage and an increase in wages for agricultural production; over-use of chemicals (e.g., pesticides) has caused substantial soil degradation and environmental problems; and global climatic change has led to an increase in the appearance of abiotic stress events such as heat, chilling, drought, and flooding. Moreover, as living standards improve, the demand for high-quality rice (e.g., aromatic rice) and healthy food increases. To meet these challenges and changes, greater efforts are required to develop new rice technologies in areas of breeding, production, and processing.

This book describes some recent advances in rice research in terms of crop improvement and breeding (Section 1), crop production and protection (Section 2), and crop quality control and food processing (Section 3). In brief, the first section introduces the progress and prospects of two-line rice breeding in India (Chapter 1), the types and breeding strategies of aromatic rice in India as well as the biochemical, genetic, and molecular aspects and the extraction and quantification methods for aroma in rice (Chapters 2 and 3), the physiological responses to abiotic stresses and the biotechnological and breeding approaches to overcome abiotic stresses in rice (Chapters 4 and 5), and the breeding strategies for developing dual-purpose rice that is bioenergy efficient without compromising grain yield (Chapter 6). The second section introduces rice cultivation in the Northern Eastern Hill Region of India (Chapter 7), the constraints and prospects of rainfed rice production in Kenya (Chapter 8), the status and challenges of rice blast disease and emerging minor diseases of rice in India (Chapters 9 and 10), and the potential role of phylloplane and rhizospheric bacteria in controlling bacterial blight of rice (Chapter 11). The third section introduces the use of near-infrared spectroscopy and machine learning in quality control of rice (Chapter 12), the characteristics of fermented brown rice (Chapter 13), and what we can learn from the history of vitamin A deficiency and how golden rice has the potential to significantly contribute to public health in these 'Covid-times' (Chapter 14).

I am grateful to Dr. Zhaowen Mo at the South China Agricultural University, Dr. Yanhua Zeng at the Jiangxi Agricultural University, Dr. Dahui Huang at the

Guangxi Academy of Agricultural Sciences, and Dr. Huabin Zheng, Dr. Yue Wang, Dr. Xumeng Li, Dr. Jiwai He and Dr. Shuying Yang at the Hunan Agricultural University for their help with reviewing chapters.

Dr. Min Huang
Rice and Product Ecophysiology,
Key Laboratory of Ministry of Education
for Crop Physiology and Molecular Biology,
Hunan Agricultural University,
Changsha, China

Section 1

Crop Improvement and Breeding

Chapter 1

Progress and Prospects of Two Line Rice Breeding in India

Manonmani Swaminathan

Abstract

Increasing the yield potential through hybrid rice technology was very well proved in nearby countries. Three line hybrid rice technology is encounted with some of the difficulties in seed production. Identification of Environmentally influenced male sterility overcomes the problem encountered in seed production since it is regulated by the temperature it is called temperature regulated male sterility and hybrids produced using this line is called two line rice hybrids. Types of male sterility and methods to identify the new TGMS lines and breeding methods employed for producing the tgms lines were described. Characterization of tgms lines by both conventional method and molecular tools has been enumerated. Seed multiplication of tgms under fertility inducing environment and seed production of two line hybrids has been explained. Seed production location was exclusively identified for seed multiplication of tgms lines. Heterotic potential of two line rice hybrids has been studied. Future prospectes in tgms research and two line heterotic potential was explained for increasing the yield potential in rice.

Keywords: rice, TGMS lines, characterization, new line development, two line rice hybrids, yield

1. Introduction

Rice is the staple food for about half of the world's population. The global population is expected to reach 9 billion people by 2050 [1]. This demands for significant efforts to increase grain production, and it is expected to add 44 million tons staple crops per year for ensuring sufficient food production for such huge population [2]. Although rice production has significantly increased from 34.5 million tonnes in 1960–1961 to 117.5 million tonnes in 2020–2021, this significant increase over years was achieved by introduction of semi dwarf varieties, through improved crop management, adoption of hybrid varieties and improved plant protection practices. Rice production needed to be increased 42% by 2050 to feed the demands of an ever-increasing human population globally. Due to the potential of hybrid rice in increasing both rice production and productivity, many countries are focusing on exploiting the benefits of this technology. Success of hybrid rice technology and commercial exploitation was proved in China by the late 1970s marked the second major landmark in the history of rice breeding. It showed commendable breakthroughs in rice production and productivity, made other countries to revive its interest in hybrid rice breeding.

The use of hybrid rice has proved to be an effective and economical way to increase rice production. In rice, the phenomenon of heterosis was reported earlier

IntechOpen

by Jones [3] and Ramaiah [4]. However, several problems experienced in the production of hybrid seeds discouraged the commercial exploitation of heterosis in India. Later in the year 1976, it was accepted that large scale production of hybrid rice could be achieved through utilization of male sterility (MS) systems [5]. China initiated hybrid rice production in the year 1964 followed by India in the year 1989. Presently around 40 countries are actively involved in commercial hybrid rice production [6].

The over dependence on a single source of cytoplasmic male sterility (CMS) via WA (Wild abortive) and the difficulties in seed production and parental line development warrant the identification of alternate approaches to exploit hybrid vigor in rice. Two line breeding is one such possibility that emerged following the chance discovery of a photoperiod-sensitive genic male sterile plant called Nongken 58S, in the japonica variety Nongken 58 by Prof. Shi Ming Song of China [7–9] which was found to be sterile under longer photoperiods (> 14 hr) and fertile under shorter photo periods (13 hr) subsequently. Temperature sensitive genic male sterility (TGMS) line was identified by Chinese and Japanese scientists which was completely sterile under high temperature (> 32°C) and under low temperature (24°C) it was fertile [10, 11]. Using the PGMS system Yuan [12] put forth a new strategy of hybrid rice breeding which did not involve a maintainer, as the maintenance is taken care off by the shorter photoperiod (13 hr), hence it was called as two line method. During the sterile phase, EGMS plants can be used as a female parent to produce hybrid seed through self-fertilization without the use of a maintainer line as required in the CMS system. Since only two lines are required for the maintenance and multiplication of male sterile lines and production of hybrid seed, the system using this type of male sterility is known as the two-line system of hybrid breeding.

2. Advantages of two-line over three-line system of heterosis breeding

- More number of heterotic hybrids can be developed because of wide choice of parental lines.
- Simpler and more efficient seed Production system.
- Large scale use of single source of cytoplasm and the risk of outbreak of epidemics as well as the negative effects of sterility inducing cytoplasm are avoided.
- In Rice, two-line system is specifically useful for developing aromatic and inter racial hybrids.
- Two-line hybrids are having magnitude of heterosis 5 to 10% higher than in three line hybrids.

3. Types of environmental sensitive genic male sterility (EGMS)

Comprises mainly of four types namely:

3.1 Photoperiod-sensitive genic male sterility (PGMS)

This type of male sterility was discovered in rice by Professor Shi Ming Song in Hubei Province of China in 1973. Several male sterile plants were noticed in a late japonica cultivar Nongken 58, when exposed to photoperiod of more than 14 hours. The same sterile plants when grown in photoperiod of less than 13 hours and 45 minutes, turned to fertility. Subsequently detailed investigations were carried out and the findings have been reported by Shi [7, 8]. The male sterile mutant was originally designated as Hubie Photoperiod-Sensitive Genic Male Sterile Rice (HPGMSR). Subsequently the mutant was named as Nongken 58 S. Pioneering and extensive work has been done on this mutant at Wuhan in Hubei Province and at other centers in China. PGMS trait from Nongken 58 S has been transferred to several elite japonica and indica cultivars through backcrossing in China.

PGMS system is useful and can be deployed in temperate countries where the day length differs considerably during different seasons.

3.2 Thermo-sensitive genic male sterility (TGMS)

This type of male sterility which is controlled by the temperature prevailing at sensitive stage of the crop, was discovered in China [13, 14].

In most of the TGMS mutants reported so far, such as Annong – 1S from China, Norin PL-12 from Japan, IR 32364 TGMS from IRRI, Philippines and several mutants reported from India and Vietnam, the sterility is caused by higher temperatures (generally above 30°C) at the sensitive stage whereas at lower temperatures (generally below 24°C) fertility is observed. However, in few cases, sterility is observed at lower temperatures and fertility is observed at higher temperatures. Such type of male sterility is referred to as 'Reverse TGMS type'. Examples of reverse TGMS type reported are mutant Diaxin 1A and IV A and a mutant in variety 26 Zhaizao from China [15–17], JP-38S from India.

In tropical and sub-tropical countries, where there are large temperature differences across locations, regions, seasons and at different attitudes TGMS system can be utilized. India is one of the country with various regions and seasons and with attitude ranging from sea level to several thousand meters in hilly areas, is highly suitable for exploiting TGMS system for deployment and development of two-line hybrids.

3.3 Photo-thermo-sensitive genic male sterility (PTGMS)

Several japonica and indica male sterile lines have been developed, utilizing the Nongken 58S mutant. All such lines developed utilizing the PTGM mutant, were found to interact both with photoperiod as well as temperature. Such type of male sterility is called photo thermosensitive genic male sterility (PTGMS). In such a system, photoperiod is effective between critical fertility point (CFT) and critical sterility point (CSP). This range of temperature is called as Temperature Range of Photo sensitivity.

3.4 Micronutrient-deficiency induced male sterility

Deficiencies of copper, Boron and some other micronutrients are reported to cause male sterility in wheat and some other crops. High genetic variability has been reported in sensitivity to deficiency of these micronutrients. Very sensitive types are completely male sterile under micronutrient deficient conditions. It has been suggested that these sensitive genotypes can be used under deficient conditions as females and tolerant genotypes as males for producing F_1 hybrid seed. The sensitive types can be multiplied by growing them under micro nutrient sufficient conditions.

The PGMS, TGMS and PTGMS lines are governed by 2–3 recessive genes, hence they can be easily transferred through backcrossing to known elite, good combining varieties. TGMS system can be utilized for tropical countries like India where low and high temperature prevails in high altitudes and in plains respectively, whereas PGMS system for the temperate countries like China and Japan where the daylength variation is significant. PTGMS can be utilized in both the tropical and temperate conditions. The sensitive stage to temperature, photoperiod or both is generally stage IV (stamen and pistil primordia), to stage VI (Meiosis) of the developmental stages of the rice plant.

On the basis of the critical sterility point (CSP) the temperature at which complete sterility is induced and critical fertility point (CFP) the temperature at which maximum fertility is achieved, it can be classified into four types.

3.4.1 Type 1: High CSP (>32°C) Low CFP (<24°C)

This type is recognized by Chinese as ideal, as it is safe for both hybrid seed production and multiplication of PTGMS. According to Yuan [18] such an ideal type still, remains to be identified. Although no one is certain as to where to draw the lines for high CSP and Low CFP, based on the prevalent temperature and photoperiod regimes in a region, a narrow range can be determined. Spontaneous Mutant lines SM3 and SM5 fall under this category. SM5 with a CSP of 32.3°C is just on the border line [19].

3.4.2 Type 2: High CSP (>32°C) High CFP (>24°C)

Chinese have reported several of the EGMS lines identified by them to fall under this category. Zhang *et al.,* [20] classified this category as 7001 S type. Under this category TGMS like 8902S and W7415S have been listed. It is not suited to Chinese condition, as it would introduce risk in hybrid seed production. Zhang *et al.,* [20] suggested their usefulness in tropics. JP 2 falls under this category [19].

3.4.3 Type 3: Low CSP (<32°C) – Low CFP (<24°C)

By virtue of its stable sterility duration over a large region in China, it can be used in hybrid seed production without any problem. However in this type of EGMS lines, their seed multiplication becomes difficult and hence limits their wide utilization in China [20]. Nevertheless this type (Pei ai 64S type) was preferred until ideal lines were bred [18, 20]. In subtropical countries like India however this type would be most suited, as only sterile phase is required to be more stable in such situation. Most of the TGMS Lines viz., TNAU 45S, TNAU 60S, TNAU 95S etc., developed from Department of Rice, TNAU, India fall under this category.

3.4.4 Type 4: Low CSP (< 32°C) - High CFP (>24°C)

Some of the lines *viz.*, GDR 38S, GDR 39S and GDR40S developed from Hybrid Rice Evaluation Centre recorded this type of category.

4. Detection and identification

A detailed procedure for identifying Thermo-sensitive genic male sterile (TGMS) lines under field conditions, from germplasm and mutagenised populations in rice, has been given by Virmani *et al* [21]. These populations are critically observed for occurrence of male sterility when the crop gets exposed under natural conditions after panicle initiation stage to temperatures of above 30–35^0 Male sterility can be easily identified in the field, by the presence of partially filled hanging panicles and completely sterile erect green panicles in the same plant. Those showing complete lack of seed formation in self pollinated crop or partial seed set in cross pollinated crops under sterility inducing phase and partial seed set both in self and cross pollinated crops under fertility inducing phase are suspected to be EGMS. Such suspected EGMS plants are them studied critically in phytotrons or growth chambers under appropriate environmental conditions to confirm the presence of EGMS. If such facilities are not available, then periodical sowings over a period of time can be resorted to with change in temperatures under natural conditions, to observe the transformation from fertility to sterility vice-versa.

5. Characterization of EGMS

Characterization of EGMS lines, essentially involves precise determination of sensitive stage and in case of TGMS lines, determination of critical sterility point (CSP) and critical fertility point (CFP). In case of PGMS lines, critical light length is determined.

5.1 Characterizing TGMS lines under field conditions

- Meteorological Data of 10–15 years can be collected on minimum and maximum temperature, day length, humidity, etc., of the location where the lines are to be characterized.

- During the year, Identify 3–4 distinct periods of high and low temperatures.

- Select the sowing season with the period of 15–25 days before heading (5–15 after PI) coincides with the high temperature. Such plants which remain sterile at high temperature will be selected.

- Note the temperature data pertaining to 15–25 days before heading, this is the critical sterility point of a given line.

- Allow and multiply the plants (selected in #3) by ratooning and subject them to lower temperature regimes at the same growth stage. Plants showing partial fertility or become fully fertile will be identified.

- Record the temperatures which prevailed during the period 15–25 days prior to heading. This is the critical fertility point of a TGMS line.

6. Inheritance of EGMS

In most of the studies inheritance of EGMS has been reported as monogenic recessive. However, there are few studies, where it has been reported as digenic and in one case as dominant, depending upon the crosses in which inheritance has been studied. In PGMS mutant Nongken 58S, Shi [8] reported the inheritance of the character as monogenic recessive, whereas reported it as digenic. Oard and Hu [22] in PGMS mutant M-201S, reported this trait to be controlled by one to three recessive genes. Huang and Zhang [17] in the mutant CIS-25-10 S, reported a single dominant gene. However, Xue and Deng [23] reported that the PGMS trait was quantitatively inherited.

TGMS trait is reported to be monogenic recessive in 5460 S [14], R 59 TS [24] H89–1 [25, 26] and IR 32364 TGMS [25]. However, this trait was reported to be controlled by two recessive genes in Annong S-1 [27] and UPRI 95–140.

TGMS genes in 5460S and H 89–1 (later renamed as Norin PL-12) were designaed as tms_1, and tms_2 respectively Virmani and Borkakati [25] found TGMS gene in IR 32364 TGMS to best non-allelic with the gene in Norin PL-12 and tentatively designated it as tms_3. For lack of accessibility to Chinese TGMS mutant 5460S, allelic test with tms_1, could not be carried out. Subsequently, Reddy *et al.* [28] and Dong *et al.* [29] reported tms_4(t) gene in their studies. Recently Wang *et al.* [30, 31] reported tms_5 gene in Annong S-1 mutant.

Ashraf et al., [32] reported that pollen-mother-cell (PMC) formation, as well as meiosis stages, are induction detection sites for TGMS because at high-temperature wrinkled or abortive pollen grains were produced due to abnormal meiosis in microspore-mother-cells (MMC). Zhou et al. [33] quoted that, other TGMS-lines were also reported from Japan, The Philippines, India, and Vietnam [26, 34–36]. Mostly, reported TGMS-lines or mutants induce male sterility at high temperatures and male fertility at low temperatures [37–39]. The stated TGMS genes/lines are tms1, tms2, tms3, tms4, tms5, tms6, tms7(t), tms8, tms9, tms9–1, and tms10 [30, 31, 40–47] and Zao25S, Lu18S, N28S, 95,850 ms, XianS, Zhu1S, Meixiang851S, and HD9802S [48–53], that provide useful material for two-line HR production. Intriguingly, the reverse phenomena were also observed such as male sterility induced at low temperature and fertility restored at high temperature. Such kinds of TGMS rice-lines are termed as reverse TGMS (rTGMS) lines. Herein, the reported rTGMS genes/lines are rtms1, Diaxin-1A, and IVA and the mutant of Indica-rice variety 26-Zhaizao from China and JP-38S from India [16, 54–57]. The tms5 is an important factor that regulates thermosensitive sterility among many tgms lines.

6.1 Linkage with molecular markers

Linkage with morphological or molecular markers facilitates transfer of EGMS genes to desirable agronomical backgrounds, since mutants are rarely suitable for direct utilization in plant breeding programs. Linkage with morphological markers are rare. Secondly, this character of sensitivity to environmental factors is expressed only under certain specific ranges or conditions of these factors. Under such situation molecular markers are very handy and useful.

A summary of the molecular markers linked to the EGMS genes and the chromosomes on which they are located is given in **Table 1**.

The molecular mechanism underlying the TGMS Lines are studied by Pan *et al.*, [61]. They reported the usefulness of thermosensitive genic male sterile (TGMS) lines and photoperiod-sensitive genic male sterile (PGMS) lines to improve rice yields. The male sterility in recently developed TGMS CO 27 is based on co-suppression of a UDP-glucose pyrophosphorylase gene (Ugp1). They studied Microarray-based transcriptome profiling by growing the TGMS-Co27 line and wild-type Hejiang 19 (H1493) line at high and low temperatures. A total of 8303 genes were differentially expressed in the two lines, under the two conditions, or both. Global gene expression was strongly affected by environmental factors. Some genes were strongly repressed in TGMS-Co27 at high temperature were important for pollen development. Notably, series-cluster analysis of differentially expressed genes (DEGs) between TGMS-Co27 plants grown under the two conditions showed that low temperature induced the expression of a gene cluster. This cluster was found to be essential for sterility transition. Many meiosis stage-related genes were included that are probably important for thermosensitive male sterility in TGMS-Co27. Temperature plays a major role in global gene expression and may be the common regulator of fertility in PGMS/TGMS rice lines.

EGMS gene	Linked molecular markers	Chromosomal location	Reference
PGMS Pms_1	RG 477 and RG 511	Chro-7	Zhang et al. [58]
PGMS Pms_2	RG 191 and RG 348	Chro-3	Zhang *et al* [58]
TGMS tms_1	RAP marker 1.2 TGMS	Chro-8	Wang *et al.* [45]
TGMS tms_2	RFLP Marker R 643 A and R 1440	Chro-7	Yamaguchi *et al.* [46]
TGMS tms_3	RAPD Markers OPF 18_{2600} $OPAC3_{640}$ $OPAA7_{550}$ $OPM19_{750}$	Chro-6	Subudhi *et al. [59]*
TGMS tms_4	RM-27 RM-257	Chro-2 and Chro-9	Dong *et al.* [29] Reddy *et al.* [28]
TGMS tms_5	STS Marker C 365–1	Chro-2	Wang *et al.* [30, 31]
TGMS tms_6	RM 3476	Chro 5	Robin *et al.[60]*

Table 1.
Molecular markers for EGMS genes.

7. Breeding of TGMS Lines

Procedures for breeding TGMS lines are similar to conventional breeding procedures, with one major difference. The trait to be selected is male sterility, which requires a particular set of conditions for it's expression and another set of conditions for multiplication of selected segregants. Sterile plants are selected under appropriate conditions in F_2 generation. Such plants are ratooned and grown under fertility inducing conditions to obtain seed of the selected segregants. Since TGMS is a recessive character, and if it is controlled by a single gene in selected segregants, there will be no further segregation for sterility/fertility in F_3 and subsequent generations, though there may be segregation for other plant characters. Thus F_3 generation onwards, the selected segregants can be grown in fertility inducing conditions for selection and forwarding the generations. By F_6 generation, stabilized elite TMGS lines can be developed.

8. Breeding procedures for TGMS Lines

8.1 Germplasm search

Make a detailed and systematic study on germplasm or any stabilized breeding material and look out for spontaneous sterile mutants which may revert to fertility under low temperature. Robin *et al.*, [62] developed the new TGMS line (TNAU 60S) and was identified as spontaneous mutant from the rice variety PMK 3 with desirable floral characteristics and stable sterility. This TGMS line has the duration of 125 days with semi dwarf plant type. The panicle exertion percentage is 76.9% with wide angle of glume opening which makes the line with higher out crossing potential highly amenable for commercial exploitation. The grain quality of the TGMS line is highly preferable. TNAU 60 S has been used in hybridization and many heterotic hybrids were developed. This line was registered with NPBGR for its unique TGMS trait as IC 0622805 and INGRI 17028.

8.2 Pedigree breeding

Agronomically adopted line or variety will be crossed with TGMS donor and F1 will be studied under normal Environments and selfed seeds of F1 will be raised under sterility inducing Environment. From this F2 population Sterile plants will be are selected under high temperature *ie* sterility inducing Environment. Such plants are ratooned and grown under fertility inducing conditions to obtain seed of the selected segregants. Since EGMS is a recessive character, and if it is controlled by a single gene in selected segregants, there will be no further segregation for sterility/fertility in F_3 and subsequent generations, though there may be segregation for other plant characters. Thus F_3 generation onwards, the selected segregants can be grown in fertility inducing conditions for selection and forwarding the generations. By F_6 generation, stabilized elite TMGS lines can be developed. Kavithamani et al. [63] developed the many new tgms lines by crossing CO 49 an agronomically adopted variety with TS 29 (promising TGMS line) as a donor line. The lines shows superiority over their parents in sterility with good floral traits and grain yield. The details were given in **Figure 1**.

One of the line were characterized by [64]. TNAU 135S (TS 29/CO 49) with a long duration line showed 100% sterility during January month sowing at Department of Rice, Coimbatore and its complete sterility is under maximum temperature of 31.2°C, and minimum temperature of 21. ^{0}C and critical stage for expression of pollen sterility is 85–104 days. Other lines TNAU 137S developed from TNAU TNAU4S-1-2 /CB 06–564 with CSP of 20.3 also showed complete sterility whereas another line with same female parent TNAU 4S with BPT 5204 showed partial sterility indicated that it requires different month of sowing.

8.3 Molecular tagging of TGMS gene and development of new lines

The TGMS parental line, TS 29 has a stable sterile and fertile phases with substantially low critical temperature. Inheritance studies of the F_2 population revealed that the thermo-sensitive genic male sterility in TS 29 was under the control of single recessive gene. Molecular tagging of TGMS genes in the F_2 mapping population was done by using SSR markers. Out of 50 primer pairs (putatively linked to the six reported TGMS genes in rice) assayed for studying polymorphism, 19 primer pairs produced polymorphic alleles between the parents. The SSR markers revealed 38 percentage of polymorphism between TS 29 and CO(R) 49, the recipient fine grain parent. The identified 19 primer pairs were used for bulked segregant analysis A total of 400 F_3 progenies were raised during summer (fertility

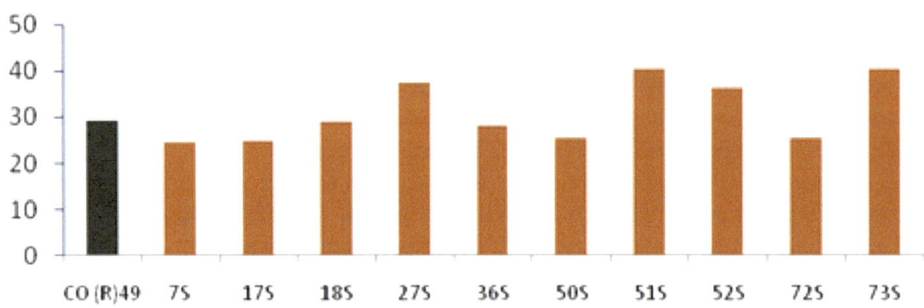

Figure 1.
Promising TGMS lines in comparison with its parents.

limiting season). The DNA of F2 plant which contributed the sterile/ fertile F3 plants were identified and bulked. The study identified one SSR marker, RM 3476 which co-segregated with the phenotypic observations recorded under the field condition. The marker, RM 3476 has already been identified as located adjoining *tms* 6 gene in chromosome 5 of rice. Population advancement for fixing stable segregants with TGMS trait and diversifying the resistance through other crosses and MAS are in progress. However the stability and heterotic potential of TGMS segregant lines were assessed in F4 generation by crossing with tester parents. The F_1s were evaluated along with check hybrids few promising hybrids *viz.*, TNAU 61 S/DE 2, TNAU 31S/ JGL 385 were identified as superior in grain yield and grain quality. Robin *et al.*, [60].

8.4 Mutation breeding

Any breeding materials can be mutated and the progenies will be screened for the presence of TGMS gene. Strict observations are to be made in M2 generation planted under high temperature region, as the trait is governed by a recessive gene. To improve the floral traits favoring out crossing two stable TGMS lines *viz.*, TS 6 and CBTS 0282 were subjected to gamma rays (300 and 350 Gy). In the M_2 generation, a total of 469 progeny rows with 1,28, 975 plants of CBTS 0282 and 854 progeny rows with 1,28,100 of TS 6 were raised. 361 sterile plants with good stigma exertion and wide angle of glume opening were selected and stubble planted at Hybrid Rice Evaluation Centre, Gudalur, a low temperature region for inducing pollen fertility and further seed increase. The M_3 and M_4 generations of the selected plants with desirable floral traits were evaluated at high temperature conditions at Coimbatore and seed increase was done at Gudalur. Finally a total of 11 TGMS lines were developed with good floral traits *viz.*, better stigma exertion percentage, wider angle of glume opening and better panicle exertion than the control were identified and are being utilized for two line hybrid rice development [65]. Out of these 11 lines two lines TNAU 84S (TS 29 150Gy) and TNAU 139S(TS 29 100 Gy-3) were characterized by [64] and found that TNAU 139S was completely sterile in all five staggering during January sowing but TNAU 84S did not showed complete pollen fertility but it showed complete spikelet sterility. TNAU 139S recorded CSP of 20.3. Apart from using already available TGMS lines released varieties were also used for developing new TGMS lines.

Mutation studies were further followed with released rice varieties by [66]. Two rice varieties viz., ADT 39 and CR 1009 were utilized to generate genetic variability by exposing them to gamma rays at 50, 100, 150, 200, 250 and 300 Gy. The main focus of this study is to identify TGMS mutants which could help in hybrid breeding programme. Chlorophyll mutants were observed in both the varieties in M_2 generation. The male sterile plants were identified in M_2 generation under high temperature condition (Coimbatore) and the reverted lines in the low temperature region (Gudalur) were planted again in the high temperature condition to confirm their TGMS nature. All the plants expressed complete sterility. Seven plants (comprising five plants from ADT 39 and two plants from CR 1009) isolated from M_3 generation recorded 100 percent pollen and spikelet sterility under high temperature condition and more than 60 percent spikelet fertility under low temperature condition. These lines were further advanced for attaining homozygosity and out of seven lines one of the promising line TNAU 100S was Isolated from the ADT 39 100 Gy. The line was characterized and found that 72–91 days were critical stages for expression of pollen sterility and CSP was 20.5°C [64]. This line also showed complete sterility with wider sterility period. This line is having good grain quality *ie* medium slender grain type which is highly preferred by south Indian people can be exploited for commercial two line hybrid development.

8.5 Backcross programme

Transferring the genes from already available sources to elite genotypes or lines with high combining ability. Tanee et al., [67] utilized this metod for developing new TGMS lines. To transfer *tgms* gene(s) controlling TGMS to Thai rice cultivars by backcross breeding method, Thai rice cultivars ChaiNat 1, PathumThani 1, and Suphan Buri 1 were used as recurrent parents and a male sterile line was used as a donor parent. The BC_2F_2 lines were developed from backcrossing and selfing. An individual plants were evaluated for tmsX gene by conventional breeding and 28 plants were selected from the total plants 78. A total of 18 SSR markers covering the 12 rice chromosomes were employed to select the outstanding genetic background for backcrossing in order to maintain genetic background of the recurrent parent. Selected 60 plants were screened for tmsX gene by phenotyping, subsequently 32 selected plants were screened by added more 18 SSR markers for genetic background in BC2F1 generation. In order to increase the recurrent parent genetic background, MAS was applied at BC2F1 generation by using more number of loci for providing opportunity to find individual plants with the highest genetic background of the recurrent parent. Two plants were isolated from BC2F1 generation and were found to carry 97.22% genetic background. These plants had genetic background higher than BC2F1 generation. Selfed seeds from selected BC2F1 plants were planted at a temperature higher than 30°C and phenotypic selection was employed at flowering stage for selection of the sterile plants. In this way they successfully introgressed the *tgms* gene into Thai rice cultivars.

9. Characterization of the TGMS lines

To study the critical fertility (CFT) and critical sterility (CST) point for TS 29, staggered sowing of seeds in weekly interval was taken up in large cement pots. Pollen fertility was observed. Fertility variation in comparison with maximum temperature prevailed during that month was compared. Complete sterility was observed at a temperature of more than 32/19°C and it was fertile below this temperature. But occasionally sterility was observed above and below this temperature that may be due to combined influence of the other weather parameters. The results of correlation analysis between pollen sterility and weather factors revealed that maximum and mean temperature were the primary factors influencing fertility transition. In this result the negative association was observed between pollen sterility and relative humidity. During the sterile phase relative humidity was low (< 85 per cent) and during fertile phase relative humidity was high (> 90 per cent). Minimum temperature was also observed a significant association with pollen sterility in the latter phase of panicle developments. Sunshine hours had lower level of influence over pollen sterility. Negative significant association was observed between relative humidity and pollen sterility percentage (**Figures 2–4**) [60].

A study was carried out an experiment at the Paddy Breeding Station, TNAU, Coimbatore. The materials comprised, 60 suspected TGMS lines from different populations viz., for screening the sterility /fertility expression. All the population of TGMS lines initiated panicle development during April when the maximum/

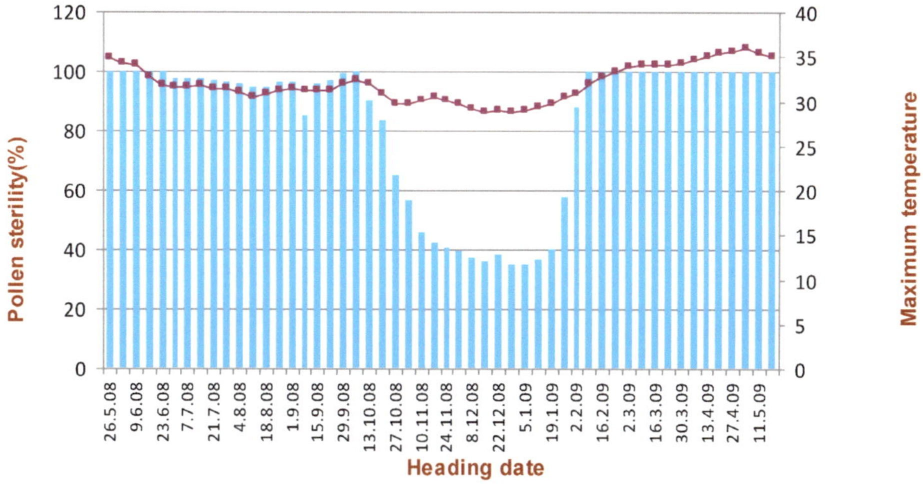

Figure 2.
Fertility behavior of TS 29 for maximum temperature.

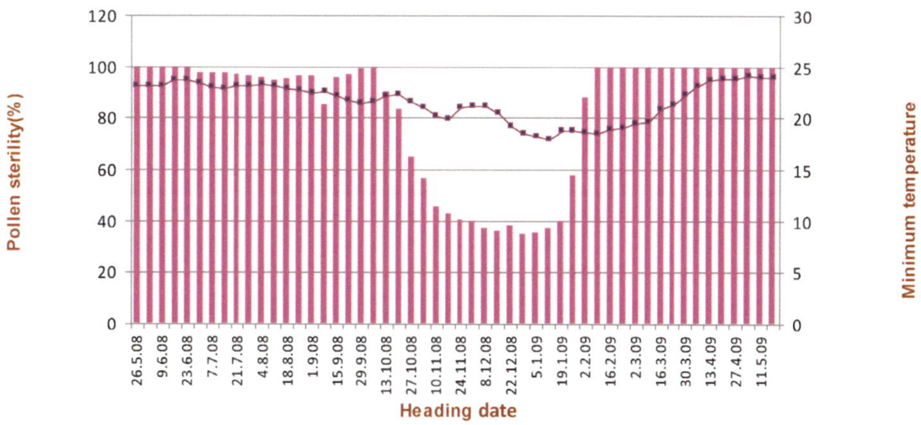

Figure 3.
Fertility behavior of TS 29 for minimum temperature.

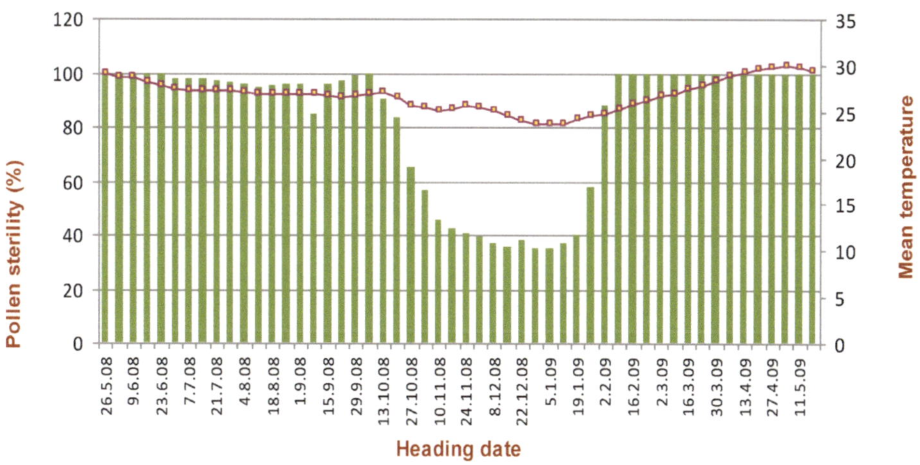

Figure 4.
Fertility behavior of TS 29 for mean temperature.

minimum temperature (day/night) was 30.5–37.8 0 C/22.0–26.4°C. The suspected TGMS lines were evaluated for pollen fertility by using 1% Iodine Potassium Iodide (I-KI) solution. Pollen grains from three randomly chosen fields were evaluated and pollen fertility was expressed in percentage. Five panicles per plant were evaluated for spikelet fertility. Sterile plants identified from promising TGMS lines were ratooned for self multiplication of seeds to confirming fertility transformation during kharif. Pollen fertility/spikelet fertility observation were recorded for ratooned plants. Seeds were collected from each ratooned promising TGMS lines and raised for evaluation during rabi. Pollen/spikelet fertility was assessed for all TGMS lines. Promising TGMS lines identified during summer. Out of 60 population evaluated for TGMS expression, 175 sterile plants were identified based on pollen/spikelet sterility. The sterile plants consisted of 27 F4'S, 40 Fs's, 20 DH's and 88 GD No's. The results showed that 25 lines found promising for stable TGMS expression along with good floral traits. Among the TGMS lines CBTS 0280, CBTS 0283, CBDHTS 025, GD 98014, GD 98028 had early flowering. All the twenty five promising TGMS lines exhibited 100 per cent pollen/spikelet sterility during summer season when the maximum, minimum temperature was 30.5–37.8/22.0–26.4 (day/night). The TGMS lines viz., CBTS 0268 and CBTS 0272 are found to be possessing long slender grain, purple tip, well exerted purple stigma. The TGMS line viz., CBTS 0252 and CBTS 0254 were developed from Indica/Japonica crosses which showed 100% pollen sterility, medium slander grain with purple stigma. The results revealed that the TGMS line viz., CBTS 0252 and CBTS 0254 could be useful to produce two line hybrids with high heterosis for yield with good plant type [68].

Newly developed 66 tgms lines were screened under sterility favoring environment for tgms gene expression during summer and sterility limiting environment during winter at Paddy Breeding Station, Coimbatore for the past five years. Out of 66 TGMS lines, 15 lines showed stable performance and seven TGMS lines *viz.*, COTGMS 02, COTGMS 07, COTGMS 10, COTGMS 11, COTGMS 12, COTGMS 13 and COTGMS 15 were completely pollen sterile through out the summer period. It shows that, these lines are having wider sterility expression period. These lines also recorded very good floral traits *viz.*, higher pollen sterility per cent, panicle and stigma exsertion per cent, wider glume opening favorable for enhancing out-crossing rate and seed set percentage during seed production [69].

A total of 21 TGMS lines which showed complete sterility were raised during kharif for seed multiplication in the fertile phase [70]. Among the 21 TGMS lines, 12 TGMS lines viz., TNAU 9S, TNAU 14S, TNAU 15S, TNAU 28S, TNAU 30S, TNAU 32S, TNAU 63S, TNAU 64S, TNAU 67S, TNAU 69S, CBTS 0282-27-1 and TS 6–182-1 were found to be highly fertile and uniform in plant and grain type. Utilizing these lines Crossing block with 7 TGMS lines and 120 male parents were raised for developing new two line hybrid combinations and 23 hybrids were synthesized. All these hybrids were evaluated with already existing hybrids and promising hybrids were forwarded. A total of 9 two line rice hybrids were raised in the advanced yield trial to assess the yield performance along with the check varieties / hybrids viz., CO 48, CORH 3, CORH 2 and ADTRH 1. Six hybrids viz., TNTRH 1, TNTRH 2, TNTRH 5, TNTRH 8, TNTRH 10 and TNTRH 12 performed well over the checks for which mini seed production has been taken up for further evaluation. First time in India these two line hybrids were developed and evaluated under yield trial. The hybrid combination viz., TNTRH 5 has been nominated for Initial hybrid rice trial (IHRT-Medium) during Kharif 2007. This hybrid recorded 6893 kg/ha of grain yield. The promising medium duration hybrid TNTRH 19 recorded the grain yield of 8210 kg/ha which is 28.12% increased yield over CO (R) 49 and was tested under MLT 2010 hybrid Rice – Medium Trial.

Salgotra et al. [71] characterized eight TGMS lines, DDR 1S, DDR 18S, DDR 19S, DDR 20S, DDR 23S, DDR 27S, DRR 28S and DDR 29, showed complete sterility at low altitude and satisfactory seed-set percentage at high altitude. Characterization of floral traits and sterility-sensitive stage were determined by the tracking method. At low altitude, with an average air temperature of 35.4°C, TGMS lines DRR 19S, DRR 20S and DRR 29S displayed a sterility-sensitive stage at 21 days prior to normal heading. For complete sterility the TGMS line DRR 1S requires a temperature of 36.6°C at 17 days prior to normal heading. The temperature for complete sterility ranged from 33.9°C to 35.8°C at low altitude in the remaining seven lines. A significant positive correlation with opening duration of lemma and palea and with size of stigma and angle of opened lemma and palea.

9.1 Out crossing potential of TGMS lines

The newly developed TGMS Lines are to be assessed for its outcrossing potential then only it can be successfully exploited for hybrid breeding programme. [72] studied the outcrossing potential of TGMS lines. In this study average style length of the TGMS lines was 1.87 mm. Maximum style length was recorded in TNAU 18 S with 2.01 mm while TS-29–150 GY and TNAU 60 S had the styles with 1.85 mm and 1.75 mm, respectively. TNAU 18 S had the maximum spikelet opening angle of 23.01^0 and the TS-29–150 GY and TNAU 60 S had the angles of 20.23^0 and $18.54^{0,}$ respectively. TNAU 60 S involved in five cross combination and the minimum of 60.8 cm and the maximum of 74.4 cm plant height was obtained. Similarly TS-29–150 GY had the minimum height of 62.2 cm and the maximum plant height of 66.80 cm involving in two cross combinations. It was noted that the tillering and flowering of TGMS lines were prolonged even after the completion of flowering in male parents. TNAU 18 S showed the 92.8 per cent with highest panicle exertion rate, TS 29 150 GY had the medium value of 66.6 per cent while TNAU 60 S had the lowest of 51.8 per cent. This parameter did not show significant difference among the lines tested. The degree of spikelet opening angle and the duration of panicle opening usually bear significant influence on seed setting percentage. The height difference also played a major role in seed setting percentage. The height difference between the male and the female plants was the maximum of 54 cm in TNAU 18 S x IET 27044. The medium height difference was observed with the cross involved in the crosses of TNAU 18 S as female and the minimum height difference of 22.0 cm was observed in the hybrid generated form TNAU 60 S and CB-09–106. The height differences between the parents of ten hybrids during the flowering period showed notable influence in out-crossing percentage. The medium differences between two parents (22.8 cm to 51.20 cm) had great influences in the seed productions. This is in accordance with the statement elaborated by Virmani *et al,* [73] as this may be attributed to the appropriate height differences which might ensure the contacts of female line's stigma with the maximum pollens at peak anthesis period of the parents concerned.

The seeding interval was determined by the growth duration between the two parental lines. The one with longer duration was sown earlier according to the number of days of difference between the two parents in terms of days to 50 percent flowering [73]. Synchronizing period, though it did not have much variation in this study, had a little effect on the out-crossing potentials of these TGMS by exhibiting varying seed yield in different cross combinations. Three of four cross combination involving TNAU 18 S had the highest out-crossing. Cross combinations with TS-29–150 GY revealed the second highest out-crossing rate.

Manonmani [74] studied Tgms gene introgressed 200 lines forpollen fertility in plains (lowaltitude300 MSL) at Coimbatore during the summer. During the critical period of the crop growth the average temperature was 25–29°C. Then the selected lines were stubbleplanted in high altitude (1500MSL) at Hybrid Rice Evaluation Centre, Gudalur during Khariff. Average temperature of less than 20°C was recorded. During flowering stage, on microscopic observation with potassium iodide stain, some of the sterile lines recorded pollen free anthers (GDR 33S, TNAU 84S & TNAU 86S) and also differences in size of the pollens (GDR 29S). Pollen sterility level observed was 0–98%.All the seventy sterile stubbles from coimbatore were planted at Gudalur, fertility reversion rate was studied and selfed seeds were collected. Based on the pollen fertility observation the70 lines were grouped into four categories. Thirty eight lines showed >90% reversion, 13 lines showed 50–90% and < 11 lines showed <50% reversion and 8 lines showed no reversion. The Selfed seeds from the revesed lines were collected. These lines will be further exploited for their stability and will be used for the development of the two line hybrid in Tamil Nadu.

TGMS lines were also characterized with molecular markers. [75] investigated to study the genetic relationship of thermosensitive genic male sterile lines developed at Tamil Nadu Agricultural University, Coimbatore using morphological traits and SSR markers. Wide genetic variation among TGMS lines were observed for morphological and floral traits. SSR markers survey using 100 SSR markers revealed that 27 were polymorphic, amplifying a total of 71 alleles with an average of 2.67 alleles. TNAU 18S exhibited better performance based on the morphological characters, for a number of tillers per plant, angle of glume opening and panicle length and TNAU 45S expressed good floral characters. Cluster analysis differentiated six TGMS lines into four clusters.

Two TGMS lines (TNAU 60S and TNAU 95S) showed 100 per cent pollen and spikelet sterility and the remaining lines are in the range of 97–98 per cent pollen and spikelet sterility. The stable pollen sterility showed by TNAU 95S was also reported by Srimathi et al. [76] and Kanimozhi et al. [77]. There is a narrow variation in the angle of glume opening which ranged from 20 to 23° among the lines. Panicle length was observed to be more in TNAU 18S (20.25 cm) and while less in TNAU 39S (11 cm). There is a narrow variation in 100 seed weight ranged from 1.97 to 2.44 g. TNAU 45S had the highest stigma length (0.27 mm), stigma breadth (0.08 mm), anther length (0.31 mm) and anther breadth (0.06 mm) among all TGMS lines and TNAU 95S (0.13) had the highest pollen volume compared to all TGMS lines. Euclidean distance values ranged from 4.464 to 6.558 indicating the presence of a wide range of genetic diversity among the six TGMS lines. The ED value was maximum (6.558) between the genotypes TNAU 39S and TNAU 95S, indicating that these genotypes are diversely related to each other. Meanwhile, the minimum ED value was observed between genotypes TNAU 95S and TNAU 60S (4.464) and followed by genotypes TNAU 14S and TNAU 39S (4.507) indicating that these genotypes were closely related to each other.

Identification of the new polyploid rice photoperiod - and thermo-sensitive genic male sterile lines will provide material for further research into polyploidy and hybrid vigor in rice and promote the exploitation of polyploid hybrid rice [78].

Pardeep et al., [79] investigated the eighteen TGMS lines and being used for molecular characterization by sixteen SSR markers and correlated with critical sterility temperature. Based on the data generated on 18 TGMS lines, the UPGMA dendrogram was constructed using Jaccard's similarity coefficients. A total of 47 alleles were amplified using 16 SSR primer pairs. All the lines except for marker RM499 were found to be polymorphic. The range of alleles was 2–5, while the average number of alleles per primer was 2.93. All the three clusters contained one or two fertile lines in each namely, cluster I (TGMS-6), cluster II (TGMS-9) and (TGMS-18) and cluster III (TGMS-1). These fertile lines separated to other sterile lines by three markers with unique bands. TGMS-6 and TGMS-9 showed 200 bp specific band by RM 324 marker, TGMS-1 and TGMS-18 showed 180 bp and 200 bp specific bands and in TGMS-1 also showed 180 bp specific band with RM 254 marker, it means that the specific bands 180 bp and 200 bp generated by different markers in different lines responsible for fertility.

Grouping of TGMA lines based on molecular markers were studied by Mengchen *et al.,* [80]. They studied 48 simple sequence repeat (SSR) markers, and genotyped a panel of 208 *indica* P/TGMS lines and confirmed three subgroups, named *indica*-I, *indica*-II and *indica*-III, in *indica* P/TGMS lines. Further diversity analysis indicated *indica*-II had the highest genetic diversity. The genetic differentiation between *indica*-II and *indica*-III was demonstrated as the largest among the three subgroups. Moreover, *indica/japonica* component identification was detected that five P/TGMS lines possess *indica* components less than 0.900. These results improve our knowledge on the genetic background for P/TGMS lines in China and will be beneficial for hybrid rice breeding programs.

TGMS Lines also introgressed with broad spectrum resistance for many diseases. Wang et al., [81] successfully bred the broad-spectrum resistance gene Xa23 through marker-assisted selection (MAS) combined with phenotypic selection in two novel inbred rice varieties and two photoperiod - and thermosensitive genic male sterility (P/TGMS) lines. All of the developed lines and derived hybrids exhibited enhanced resistance to BB with excellent yield performance.

9.2 Stability of TGMS lines under fertility inducing and sterility inducing environments

In Tamil Nadu, there is an exclusive centre for two line hybrid rice research under Tamil Nadu Agricultural university, Coimbatore. The centre was established during 1996 at Gudalur, Nilgiris District. The seed production procedure mentioned above is followed for evaluating tgms line at Gudalur and Coimbatore. For exploiting the two line breeding system a stable TGMS Lines are needed and seed production to be standartised based on the line to be used and prevailing weather parameters in the particular location.

Manonmani *et al.,* [82] studied the stability of new TGMS lines for sterility and standardized the seed multiplication of tgms lines at sterility and fertility favoring Environments. The experiments were conducted at Paddy Breeding Station, Coimbatore, Farmers field at Sathiyamangalam and Hybrid Rice Evaluation Centre, Gudalur during the *rabi and khariff* seasons in 2013 & 2014 to assess the pollen fertility expression under different temperature regimes in new generation temperature sensitive genic male sterile lines of TNAU. Weather parameters at Coimbatore and Gudalur during for the past fifteen years was analyzed for fixing the sowing season (**Figure 5** and **6**).

Tgms lines were evaluated under sterility inducing Environments *viz.*, Coimbatore and Sathiyamangalam during the month of December 2013 & 2014 (Rabi 2013 & 2014). The same lines were stubble planted and evaluated for pollen sterility under pollen fertility inducing Environment during the month of July 2014 at high altitude (1500 MSL) with cool climate at Hybrid Rice Evaluation Centre, Gudalur. The new TGMS lines developed at TNAU *viz.*, TNAU 45S, TNAU 60S, TNAU 95 S, TNAU 19S and TNAU 39S were evaluated for their stability of pollen sterility under different temperature regimes were given in the **Table 2**.

At sterility inducing Environments the lines showed 100% pollen sterility. These lines were seeded during December at Coimbatore and Sathiyamangalam to expose them to a sterility inducing temperature (>29°C /< 23°C day night) during panicle initiation to flowering stage to test their sterility behavior so that their critical stage

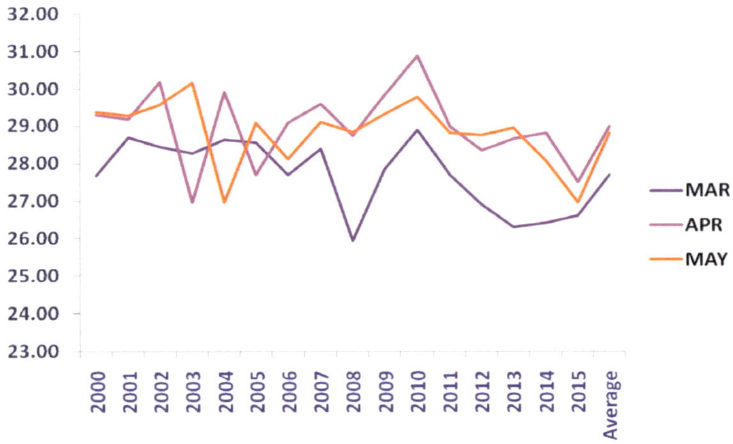

Figure 5.
Mean Weather data at Coimbatore Location.

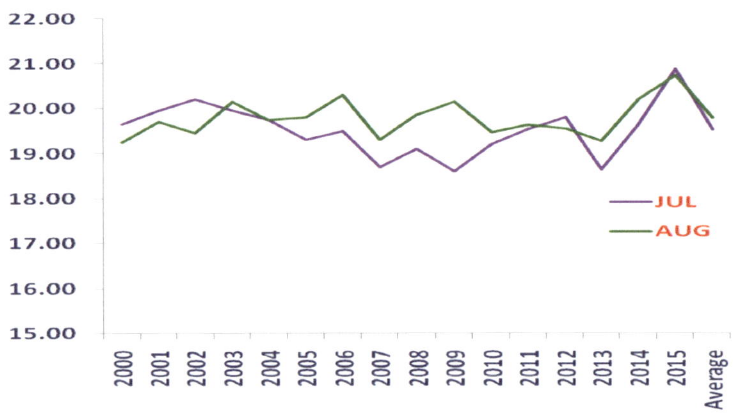

Figure 6.
Mean Weather data at Gudalur Location.

TGMS lines	Rabi 2013		Rabi 2014		Kharif 2013	Khariff 2014
	Coimbatore	Sathy	CBE	Sathy	Gudalur	Gudalur
TNAU 45S	100	100	100	100	5	4
TNAU60S	100	100	100	100	3	5
TNAU95S	100	100	100	100	6	5
TNAU 19S	100	100	100	100	7	9
TNAU39S	100	100	100	100	4	6

Table 2.
Pollen sterility of different TGMS lines in Rice.

of flowering coincides with more than 29°C. Mean weather data for both locations was provided in the **Figures 6** and 7 and it showed that the temperature recorded in both the places exceeded >25°C during the month of March, April and May. During the flowering stage all these lines showed 100% pollen sterility at both the locations for more than 60 days and was test verified for next year also.

Figure 7.
Mean Weather data at Coimbatore and Sathyamangalam.

The daily mean temperature of 24 to 26°C was found to be the critical temperature for fertility alteration. The sterile stubbles of these lines were planted at HREC, Gudalur during May to induce fertility for their seed multiplication and were exposed at critical stages to fertility inducing temperature (24°C / 18°C day / night). Maximum, minimum, and mean temperature significantly influenced the pollen and spikelet fertility in all five TGMS lines at high altitude. At Gudalur the temperature range during the month of July and August was less than 20°C. The appropriate sowing date of TGMS lines was fixed during June–July in such a way that the critical stages of panicle development would be exposed to the required temperature. The individual lines were maintained under isolation and genetically pure seeds were produced at Gudalur.

The above TGMS lines with wider pollen sterility period under plains can be very well exploited for developing two line rice hybrids during the period of December to April at Coimbatore. The same lines can be easily seed multiplied at Gudalur during July to November.

10. Two line hybrid rice seed production

The TGMS seed produced from high altitude can be brought to locations like Coimbatore as proved above, where stable high temperature prevails during its sensitive stage for nearly 30 days. TGMS plants are planted in 6 rows and sandwiched with two rows of non TGMS good combiner lines on either sides. TGMS line must be randomly checked for complete pollen sterility during its sterile phase at high temperature. Supplementary pollination techniques as applied to three line system must be adopted such as GA_3 spray, rope pulling, flag leaf clipping etc. to increase hybrid seed production. At the time of harvest care must be taken to harvest separately the non TGMS lines first. Seeds harvested from the TGMS line must be cleaned, packed, labeled, and sold to farmers as two line hybrid seed.

Hence Coimbatore and Gudalur locations were identified for the TGMS Seed multiplication and hybrid seed production in Tamil Nadu.

Based on the studied conducted at Department of Rice, Coimbatore and Hybrid Rice Evaluation Centre, Gudalur with Stability of TGMS lines compared with weather parameters over the ten years study resulted in the identification of season for seed multiplication and hybrid seed production as follows in the **Table 3**.

S. No.	Location	Sowing season	Suitable for
1.	Gudalur	June–July	Seed increase of TGMS lines
2.	Coimbatore	December–January	Hybrid seed production
3.	Coimbatore	September–October	Seed increase of TGMS lines
4.	Sathyamangalam	December–January	Hybrid seed production

Table 3.
Standardization of sowing season for the TGMS lines.

10.1 Multiplication of EGMS lines

EGMS lines, if multiplied continuously for several generations without any selection, may segregate for Critical Sterility Point, thereby causing major problems in maintaining purity of the hybrid seeds. Therefore, nucleus and breeder seed production must be taken up on a continual basis.

10.1.1 Method I

- Seeding of TGMS or PGMS lines is arranged in such a way that the sensitive stage occurs when the temperature or photoperiod is favorable for a higher seed set. Nucleus seed production of an EGMS (TGMS or PGMS) line begins in the fertility inducing environment.

- About 100 plants will be selected at the time of flowering, from the population of an EGMS (TGMS or PGMS) line and their panicles are bagged. Within a week selection process should be completed.

- After the harvest, 50 plants with higher spikelet fertility (above 30%) are selected.

- About 30 seeds are taken from each of the selected plants to grow single-row progenies and the remaining seeds are stored carefully. Progenies of the selected plants are grown in the sterility-inducing environment. The balance of the seeds of the progenies that are uniform and completely male sterile must be marked and bulked to form the nucleus seed.

- Nucleus seed of the EGMS line is used for producing breeder seed under strict isolation. Breeder seed for the EGMS line is produced in the fertility-inducing environment.

- The breeder seed produced under the direct supervision of the plant breeder has high genetic purity and is used for producing foundation seed of parental lines, which in turn will be used for producing hybrid seed.

10.1.2 Method II

- Under a sterility-inducing environment select a completely male sterile plant with typical characteristics of the original EGMS line.

- Ratoon the selected plant. Multiply the ratooned stubbles under a fertility-inducing environment. The nucleus seed will be harvested from the ratooned stubbles.

- The nucleus seed is used for producing breeder seed and the latter for producing foundation seed.

- Preserve the selected stubbles under favorable temperature conditions with good management. The new nucleus seed will be produced continuously.

10.2 Seed multiplication of TGMS Lines

Seed production potential in the TGMS lines were studied at Tamil Nadu. During fertility reversion phase can be enhanced by growing them under medium hill regions of Gudalur (1500 m MSL) in Nilgiris district, vTamil Nadu [83]. At Gudalur, the temperature range during the month of July and August was less than 20°C. The appropriate sowing date of TGMS lines was fixed during June–July in such a way that the critical stages of panicle development would be exposed to the required temperature. The individual lines were maintained under isolation and genetically pure seeds were produced at Gudalur. The TGMS line TNAU 60S was evaluated at different locations for their stability in sterility and it was proved that under high temperature (Coimbatore) it expressed 100% sterility and at low temperature it produced more than 90% seed set at Gudalur [74]. This line with wider pollen sterility period under plains can be very well exploited for developing two line rice hybrids during the period of December to April. The same lines can be easily seed multiplied at Gudalur during July to November.

Alternate to hilly areas fertility reversion was studied at plains also in the cooler months by [84]. They evaluated 255 TGMS derivatives generated at IARI-RBGRC, Aduthurai, for fertility phase changes during kharif season. The lines were grown during Apr – Aug at Aduthurai (elevation 19.5 m) conditions when ambient temperature was above 27°c (Season 1). Fully sterile plants identified were stubble planted during Aug – Dec at Gudalur (elevation 416 m) for testing the reversion phase (Season 2). Concurrently, a subset of 43 random fully sterile stubbles were retained at Aduthurai during the same period (Aug-Dec) and their fertility behavior was also observed. There was 69% of the lines showing SF of >45%, since 45% seed set or more is desirable for commercial seed production. In Season 2, the lines stubble planted at Aduthurai also showed fertility reversion between 12 and 65% with 35.6% of the population showing >45% seed set. Temperature of Gudalur ranged between 19 and 33°c during flowering (Nov), while at Aduthurai, it was between 21 and 34°c in the same season. Results indicate that minimum temperature is more crucial than the maximum temperature in fertility reversion behavior.

11. Heterotic potential of two line hybrids

The magnitude of heterosis in two line hybrid is also 5–10% higher than in three line hybrids as it does not have a cytoplasmic penalty. Reported that for most of the characters, the mean heterosis percent was in the order of indica/japonica F1 > Tropica japonica/indica F1 > indica/indica F1 > Tropical japonica/japonica F1.

A comparative studies on two - line, three - line and conventional hybrids of rice (l.) was made at TNAU by [85]. To compare the efficiency of the available systems in hybrid rice technology, a study was conducted to evaluate 120 hybrids belonging to four different group of combinations using cytoplasmic genic male sterile lines (CMS), temperature sensitive genic male sterile lines (TGMS), temperature sensitive genic male fertile lines and well adapted varieties as female parents for their genetic potential related to yield and yield components.

The two line hybrids, TNAU (TGMS)4 x BPT5204, TNAU (TGMS)4 x JGL1798 and TNAU (TGMS)4 x Karnataka Deluxe Ponni were the high yielders and in CMS based hybrids, the hybrid IR 79156 A x Karnataka Deluxe Ponni possessed high yield and good restoration capacity. Of the conventional hybrids, IR 79156B x BPT5204, IR80151B x PSBRC82, IR80151B x WGL32100 and IR80151B x Karnataka Deluxe Ponni have exhibited significant *sca* and standard heterosis for yield [86]. Among the three groups of hybrids the two line hybrids were superior for yield compared to the CMS based and conventional hybrids and its application has great potential to revolutionize rice production in breaking the yield plateau.

Highly heteotic two line hybrids were identified by [87]. They studied three TGMS lines and 20 testers were used to generate 60 two line rice hybrids in a LxT mating design. All the three TGMS lines *viz.,* TNAU 18 S, TNAU 60 S and TS-29–150 GY and nine testers, *viz.*, CB 493, CB 55, CB 508, CB 513, CB 004, CB 921, CB-09–104, CB-09–106, CB 306 showed desirable general combining ability values for two or more characters of which three testers *viz.*, CB 044, CB 009 and CB 306 had positive and significant general combining ability values for single plant yield. Thirty eight hybrids had desirable *sca* effects for at least one of the eight characters. Top hybrids selected with high standard heterosis were TNAU18 S X CB 55, TNAU 18 S X CB 508, TNAU 18 S X CB 044, TNAU18 S X CB 921, TNAU 60 S X CB-09–106, TNAU 60S X CB 493, TNAU 60S/CB 55, TNAU60 S X CB 513 and TS-29–150 GY X CB 306 (**Table 4**).

Considering both physical and cooking quality traits primarily with head rice recovery, the four hybrids namely TNAU 60 S X CB 009, TNAU 60 S X CB-09–106, TNAU 18 S X CB 921 and TNAU 18 S X CB 044 had acceptable grain quality traits with maximum phenotypic scores. These hybrids were studued for their adoptability under three environments. Comparing both Eberhart and Russell and AMMI models to all hybrids and checks, four two - line hybrids *viz.,* TNAU 60 S X IET 21009, TNAU 60 S X CB-09–106, TNAU18 S X CB 921 and TNAU 60 S X CB 513 gave above average values and outperformed the checks thus showing considerable stability in the tested locations. On comparison of the two line rice hybrids with their test entries of similar maturity groups belonging to CGMS system and HYVs, obviously the two line rice hybrids gave additional per day productivity ranging from 3.23 to 44.53 percent increase. The hybrid TNAU 60 S

SN	Hybrid combination	Single plant Yield (Kg/plant)		
		d_i	d_{ii}	d_{iii}
1	TNAU18SX CB 55	122.36**	113.3**	108.88 **
2	TNAU18S X CB 508	124.81**	122.5**	122.49 **
3	TNAU18SX CB 044	203.18**	202.7**	196.45 **
4	TNAU18SX CB 921	118.54**	77.26**	178.99 **
5	TNAU 60SXCB-09–106	80.48**	70.39*	66.86 **
6	TNAU60SX CB 493	70.34**	27.76	98.82 **
7	TNAU60SX CB 55	141.98**	125.7**	102.96 **
8	TNAU60SX CB 513	224.32**	151.00**	95.27 **
9	TS-29–150GYX CB 306	118.2**	100.5**	151.48 **

*,**Significant at 5% and 1%, respectively.*

Table 4.
Heterotic potential of Two line rice hybrids.

X CB 55 significantly differed from other entries for six traits in initial hybrid evaluation and notably recorded 34.30 g of single plant yield. This hybrid also had medium slender grain type, high head rice recovery with intermediate amylose content and desirable pasting properties leading to a score of 40 in IRRI quality scale. This cross combination yielded fairly good amount of hybrid seed in mini seed production plot and also the female parent TNAU 60 S had acceptable out-crossing percentage. Multi location trials revealed that this particular hybrid fell under medium maturity group with 131 days of maturity, highest panicle length and also the highest yielded grain yield of 13082 $kgha^{-1}$ with 99.6 $kgha^{-1}$ day^{-1} productivity.

The hybrid combination was given with name TNTRH 55. The new two line rice hybrid TNTRH 55 with a duration of 125 days was synthesized using tgms line TNAU 60S with CB 55. The hybrid seeds were produced with minimum staggering between the parental lines. The hybrid was tested with station trials at Coimbatore and also at HREC, Gudalur the exclusive station for tgms line multiplication situated at 1500 MSL in Tamil Nadu for three seasons (Khariff 2014, 2015 & 20 (**Tables 5** and **6**) [88]. All the biometrical traits along with blast reaction were also studied with check varieties. Per day productivity of this hybrid was 39.7 kg/day.

This hybrid showed 13–25% yield increase over the check variety. It showed resistance to blast with the score of 1. The hybrid produces medium slender grain type with Intermediate amylose content and Gel consistency. In Multilocation Evaluation Trial it recorded a grain yield of 6562 kg/ha which was 17.4% over ADT 39 and on par with medium duration check TNAU rice hybrid CO 4 (6578 kg/ha). At present the hybrid is under advanced stage of evaluation in Tamil Nadu. If this hybrid qualifies the criteria for release it will be the first two line rice hybrid ever released for cultivation in India. By the release of two line rice hybrid we can reduce the hybrid cost drastically as it involves only two parental lines for seed production [82].

Chandrasekhar [89] studied 1000 hybrids by crossing 500 germplasm lines (male) with one CGMS female line (IR79156A) and one TGMS female line (IR75589) and were evaluated in test cross nursery. Among the lines tested in test cross nursery on CGMS female, 60% lines are either partial restorers or partial maintainers. The maintainers proportion was 9% and restorer was 33%. TGMS female on an average across 161 combinations yielded 6.25 F_1 seed yield per plant in comparison to 4.95 g F_1 seed in CMS female. It was observed that TGMS females in general have higher seed production potential than the CMS female and TGMS female yielded 26% higher seed yield that CGMS female on an average across male lines in the study. For grain yield 187 hybrids recordedsignificant

Entries	DFF	Plant Ht (Cm)	Panicle length	Spikelet fertility	Duration	Grain yield (kg/ha)	Per day productivity (kg/day)
CO R 51	95	82.33	20.8	70.32	115	4373	**38.028**
CORH 3	95	70.66	17.46	76.28	115	4276	37.178
TNTRH 58	105	72.66	21.66	46.38	130	2478	19.063
TNTRH 55	**105**	**75.00**	**22.53**	**80.10**	**125**	**4956**	**39.650**
CORH4	109	96.00	25	66.95	139	3644	26.218
CO (R) 50	112	90.16	22.13	64.32	142	5102	35.930

Table 5.
Evaluation of two line rice hybrids at HREC, Gudalur.

Genotypes	Yield (kg/ha)			
	Kharif 2014	Kharif 2015	Kharif 2016	Average
TNTRH 55	4956	4194	6410	5186
ADT 39	3644	3065	5128	3945

Table 6.
Performance of two line rice hybrid over seasons.

Hybrid code	Combination	Days to 50% flowering	Grain Yield (t/a)	Standard heterosis
H137	IR75589TGMS x PLASD20	82	9.80	34.4**
H203	IR75589TGMS x PLIR547452231983	86	9.62	31.9**
H323	IR75589TGMS x PLSANTOSH	88	9.24	26.7**

*, ** *Significant at 5% and 1%, respectively.*

Table 7.
Heterotic potential of the hybrids.

positive heterosis over better parent and 128 hybrids over standard check. The top two line hybrids H137 (34.4%), H203 (31.9%) and H323 (26.7%) recorded the highest significant positive heterosis for grain yield (**Table 7**). TGMS hybrids exhibited higher average grain yield heterosis than the CGMS hybrids.

12. Status of two line rice hybrids in India

TGMS or two line system based rice hybrids are predominantly cultivated in Northern India (Punjab, Haryana) and some parts of Chhattisgarh. Commercially in India, Savannah Seeds Private Limited is the major player which supplies their TGMS hybrids in the brand name of "SMART Rice". Currently based on estimates the two line hybrids occupy area of around 400000–450000 acres in India. Some of the prominent hybrids in the market are Sava smart rice −127, Sava smart rice −134, Sava smart rice −200, Sava smart rice −300 etc. The two line system hybrid seed production requires stable weather parameters and currently in India major hybrid seed production in the northern states of Haryana, parts of Rajasthan and Punjab during Kharif season. This is quite opposite to 3 line system in which the major seed production area in southern states and in rabi season. Some of the other companies actively involved in the parental line and hybrids development in 2 line system are Syngenta, Corteva Agriscience, Monsanto-Bayer etc.

13. Future prospects

- Long lasting research in TNAU resulted in the development of short duration, medium duration and long duration TGMS lines with better out crossing percentage, good grain quality with better agronomic traits can be exploited for developing good grain quality two line rice hybrids. It can be achieved through the selection of good grain quality similar duration tgms lines subsequently by heterosis breeding approach.

- Development of TGMS lines with herbicide tolerance for making the seed production ease and maintaining the genetic purity enables to reduce the seed production cost. Markers are developed for this herbicide tolerant line. By adopting marker assisted back cross breeding method one can convert tgms line with herbicide tolerant trait.

- Application of Marker Assisted Breeding for introgression of biotic and abiotic stress tolerant gene in to the TGMS line and male parents. Promising donors are available with multiple stress tolerance that can be utilized.

- Exploitation of *japonica* germplasm with wide compatible gene for developing new TGMS lines.

- Using the conventional and molecular approach it can be achieved.

- Developing seed production packages for enhancing the row ratio and seed produce ability of two line hybrids by formulating the experiments with more number of female rows it can be standardized.

- Exploiting the ideal locations already identified places for seed production and seed multiplication. As it was indicated in the text one can exploit their own locality for seed production utilizing the past ten to fifteen years weather data.

- Employing Genome editing techniques for quick development of TGMS lines with proven or mega varieties. The success report was already published by employing the Crisper CAS 9 technique they developed the tgms line. Similar approach can be adopted.

- Molecular characterization of available TGMS lines and identification of mechanism involved in regulation of genes through omic technologies.

- Development and testing of large number of two line rice hybrids across the locations. The hybrids developed will be tested across the locations and stability models can be exploited for identifying the adoptable hybrids with higher yield potential.

Author details

Manonmani Swaminathan
Department of Plant Genetic Resources, Centre for Plant Breeding and Genetics, Tamil Nadu Agricultural University, Coimbatore, Tamil Nadu, India

*Address all correspondence to: manonmanitnau@gmail.com

IntechOpen

© 2021 The Author(s). Licensee IntechOpen. This chapter is distributed under the terms of the Creative Commons Attribution License (http://creativecommons.org/licenses/by/3.0), which permits unrestricted use, distribution, and reproduction in any medium, provided the original work is properly cited.

References

[1] Feng X, Wang C, Nan J, Zhang X, Wang R, Jiang G, Yuan Q, Lin S. 2017. Updating the elite rice variety Kongyu 131 by improving the *Gn1a* locus. Rice, **10**,: 35.

[2] Tester M. and Langridge 2010. Breeding technologies to increase crop production in a changing world. Science 12;327(5967):818-22 doi: 10.1126/science.1183700

[3] Jones, J.W. (1926). Hybrid vigour in rice. American Society of Agronomy, 18:423-428.

[4] Ramaiah, K. Inheritance of flowering duration in rice. Indian J. Agric. Sci. 1933, 3, 377-410.

[5] Lin S.C. and l.P.Yuan.1980.Hybrid rice breeding in China.In:Innovative approaches in rice breeding, IRRI, Manila, Phillipines:33-51.

[6] Changxiang Mao 2012. Hybrid Rice Development and outside China. In:International Symposium on 100 ears for rice science and looking beyond,Tamil Nadu Agricultural Univer sity,Coimbatore,India:183-188.

[7] Shi MS. Preliminary report of breeding and utilization of late japonica natural double-purpose line. J. Hubei Agricultural Sciences. 1981;7:1-3

[8] Shi MS. The discovery and the study of the photosensitive recessive male sterile rice (*O. sativa L*. Subsp. Japonica). Scientia-Agricultura-Sinica. 1985;**2**:44-48

[9] Shi MS, Deng JY. The discovery determination and utilization of the Hubei Photosensitive genic male sterile rice (*O. sativa L*. subsp. Japonica). Acta Genetica Sinica. 1986;**13**(2):107-112

[10] Maruyama, K., Araki, H and Kato, H. 1990. Thermosensitive genic male sterility induced irradiation. Rice Genetics II P.O. Box 933, Manila, Philippines, pp. 227-235.

[11] Tan ZC, Li YY, Chen LB, Zhoul GQ. Studies on ecological and adaptability of dual purpose line An-Nong S-1. Hybrid Rice. 1990;**3**:35-38

[12] Yuan LP. Strategy conception of hybrid rice breeding. Hybrid Rice. 1987;**1**:1-3

[13] Zhou, T.; Xiao, H.; Lei, D.; Duan, Q. The breeding of indica photosensitive male sterile line. Hunan Acad.Agric. Sci. 1988, 6, 16-18.

[14] Sun, Z. A temperature sensitive male-sterile line found in rice. Rice Genet. Newslett. 1989, 6, 116-117.

[15] Jiang YN. Studies on the effect of high temperature on fertility of the male sterile lines in Dian-type hybrid rice (In Chinese) J. Yunan Agric. Univ. 1988; **3**(2):99-107

[16] Shen Y, Gao M, Cai Q. A novel environment-induced genic male sterile (EGMS) mutant in indica rice. Euphytica. 1994;**76**:89-96

[17] Zhang ZG, Yuan SC, Zen HL, Li YZ, Li ZE, Wei CL. Preliminary observation of fertility changes in the new type temperature sensitive male sterile rice IVA. Hybrid Rice. 1991;**1**:31-34

[18] Yuan LP. The strategy of breeding rice PGMS and TGMS lines. Hybrid rice. 1992;**1**:1-4

[19] Ali, J. 1993. Studies on temperature sensitive genic male sterility and chemical induced sterility towards development of two line hybrids in rice (*Oryza sativa L.*) Ph.D. (Genetics) thesis, Indian Agricultural Research Institute, New Delhi pp. 138 (Thesis).

[20] Zhang ZG, Zeng HL, Yuan SC, Wang BX, Li YZ, Zhang DP. Restudies

on the model of photo thermo reaction of fertility alteration in photo sensitive genic male sterile rice. Journal of Haazhong Agricultural University. 1992;**11**(1):1-6

[21] Virmani, S.S., Viraktamath, B.C., Casal, C.L., Toledo, R.S., Lopez, M.T. and Manalo, J.O., 1997. Hybrid Rice Breeding Manual: International Rice Research Institute, Manila, Philippines.

[22] Oard JH, Hu J (1995) Inheritance and characterization of pollen fertility in photo periodically sensitive rice mutants. Euphytica 82:17-23

[23] Xue G, Deng JY (1991). Studies on the Hubei photosensitive genetic male-sterile rice (Oryza sativa L. ssp. japonica). Sci. Agr.sin 20:13-19

[24] Yang, R.C.; Liang, K.L.; Wang, N.Y.; Chen, S. A recessive gene in indica rice 5460S for thermosensitive genic male sterility. Rice Genet. Newslett. 1992, 9, 56-57.

[25] Borkakati RP, Virmani SS. Genetics of thermosensitive genic male sterility in rice. Euphytica. 1996;**88**:1-7

[26] Maruyama K, Araki H, Kato H. Thermo-sensitive genic male sterility induced by irradiation. In: Rice Genetics II. PP. Manila, Philippines: International Rice Research Institute; 1991. pp. 227-235

[27] Li, Y., Lin, P.L., Li, J., Wu, B.Z., Ji, H.P., Zheng, H.J., Li, S.X., Zhang, J.B. and Zhang, X.F. 1994. Study on the identification for resistance of rice varieties to bakanae disease. Heilongjiang Agric. Sci. 1 : 29-31.

[28] Reddy OUK, Siddiq EA, Sarma NP, Ali AJ, Hussain AJ, Nimmakayala P, et al. Genetic analysis of temperature-sensitive male sterility in rice. Theor. Appl. Genet. 2000;**100**:794-801

[29] Dong NV, Subudhi PK, Luong PN, Quang VD, Quy TD, Zhang HG, et al. Molecular mapping of a rice gene conditioning thermo-sensitive genic male sterility using AFLP, RFLP and SSR techniques. Theor. Appl. Genet. 2000;**100**:727-734

[30] Wang, X.G., Xing, Q. H., Deng, Q.Y., Liang, T.s., Yuna, L.P., Weng, M.L., and Wang B., 2003a. Fine mapping of the rice thermosensitive genic male sterile gene tms5. Theor. Appl. Genet. (Published on line June 25, 2003).

[31] Wang YG, Xing QH, Deng QY, Liang FS, Yuan LP, Weng ML, et al. Fine mapping of the rice thermo-sensitive genic male-sterile gene tms5. Theor. Appl. Genet. 2003b;**107**:917-921

[32] Ashraf MF, Peng G, Liu Z, Noman A, Alamri S, Hashem M, Qari SH, Zoubi O. M. 2020. Molecular Control and Application of Male Fertility for Two-Line Hybrid Rice Breeding. Int. J. Mol. Sci. 2020, 21, 7868; doi:10.3390/ijms21217868 www.mdpi.com/journal/ijms.

[33] Zhou H, Zhou M, Yang Y, Li J, Zhu L, Jiang D, et al. RNase Z(S1) processes UbL40 mRNAs and controls thermosensitive genic male sterility in rice. Nat. Commun. 2014;**5**:4884

[34] Ali J, Siddiq E, Zaman F, Abraham M, Ahmed I. Identification and characterization of temperature sensitive genic male sterile sources in rice (*O. sativa* L.). Indian J. Genet. Plant Breed. 1995;**55**:243-259

[35] Pandey M, Rongbai L, Singh J, Mani S, Singh H, Singh S. The identification and nature of a new thermosensitive genic male sterility source, UPRI 95-140 TGMS in rice. Cereal Res. Commun. 1998;**26**:265-269

[36] Virmani S, Voc P. Induction of photo-and thermo-sensitive male sterility in indica rice. Agron. Abstr. 1991;**119**

[37] Chuan-Gen L, Zou J-S, Ning H, Yao K-M. Plant temperature for sterile

alteration of a temperaturesensitive genic male sterile rice, Peiai64S. Agric. Sci. China. 2007;**6**:1283-1290

[38] Xu M, Zhou G, Chen L. Response of fertility of Peiˆ ai 64 S to temperature and photoperiod in rice. Zuo Wu Xue Bao. 1999;**25**:772-776

[39] Yuan L. Progress of two-line system hybrid rice breeding. Sci. Agric. 1990;**3**: 1-6

[40] Lang, N.T.; Subudhi, P.K.; Virmani, S.S.; Brar, D.S.; Khush, G.S.; Li, Z.; Huang, N. Development of PCR-based markers for thermosensitive genetic male sterility gene tms3 (t) in rice (*O. sativa* L.). Hereditas 1999, 131, 121-127.

[41] Lee DS, Chen LJ, Suh HS. Genetic characterization and fine mapping of a novel thermo-sensitive genic male-sterile gene tms6 in rice (*O. sativa* L.). Theor. Appl. Genet. 2005;**111**:1271-1277

[42] Liu X, Li X, Zhang X, Wang S. Genetic analysis and mapping of a thermosensitive genic male sterility gene, tms6 (t), in rice (*O. sativa* L.). Genome. 2010;**53**:119-124

[43] Qi Y, Liu Q, Zhang L, Mao B, Yan D, Jin Q, et al. Fine mapping and candidate gene analysis of the novel thermo-sensitive genic male sterility tms9-1 gene in rice. TAG. Theor. Appl. Genet. Theor. Und Angew. Genet. 2014;**127**:1173-1182

[44] Sheng ZH, Wei X, Shao G, Chen ML, Song J, Tang SQ, et al. Genetic analysis and fine mapping of tms9, a novel thermosensitive genic male-sterile gene in rice (*O. sativa* L.). Plant Breed. 2013;**132**:159-164

[45] Wang B, Xu WW, Wang JZ, Wu W, Zheng HG, Yang ZY, et al. Tagging and mapping the thermo-sensitive genic male-sterile gene in rice (*O. sativa* L.) with molecular markers. Theor. Appl. Genet. 1995;**91**:1111-1114.82

[46] Yamaguchi Y, Hirasawa H, Minami M, Ujihara A. Linkage analysis of thermosensitive genic male sterility gene, tms-2 in rice (*O. sativa* L.). Jpn. J. Breed. 1997;**47**:371-373

[47] Zhou G, Chen Y, Yao W, Zhang C, Xie W, Hua J, et al. Genetic composition of yield heterosis in an elite rice hybrid. Proc. Natl. Acad. Sci. USA. 2012;**109**: 15847-15852

[48] Guo G, Guo M, Yin M, Yin J, Meng W. Discovery and analysis of new PTGMS lines. Guangxi Agric. Sci. 2004;**35**:187-188

[49] Ku SJ, Yoon H, Suh HS, Chung YY. Male-sterility of thermosensitive genic male-sterile rice is associated with premature programmed cell death of the tapetum. Planta. 2003;**217**:559-565

[50] Peng H, Qiu Z, Chen X, Wan B, Zhang G, Lu Y. Pollen fertility and cytological observation of a thermo-sensitive genic male sterile line of non-pollen type Xian S in rice (*O. sativa* L.). Acta Ecol. Sin. 2006;**26**:2322-2327

[51] Peng H, Wan B, Zhang G, Lu Y, Zhou G, Chen X. Microstructure observations of pollenless abortion in thermo-and photoperiod-sensitive genic male sterile line N28S in rice (*O. sativa* L.). Acta Bot. Yunnanica. 2009;**31**:15-20

[52] Yang Y, Fu C, Hu X, Zhang Z, Zhou Y, Song Y. Discovery of thermosensitive genic male sterile gene in Zhu1S and study on super hybrid early rice breeding. J. China Rice. 2007;**6**:17-22

[53] Zhou Y, Ju C, Xu G, Huang Z, Zhao H, Xie P, et al. Breeding and utilization of fine quality indica TGMS line HD9802S in Rice. Hybrid Rice. 2008;**23**:7-10

[54] Ali AJ, Siddiq E. Isolation and characterization of a reverse temperature sensitive genic male sterile mutant in rice. Indian J. Genet. Plant Breed. 1999;**59**:423-428

[55] Jia JH, Zhang DS, Li CY, Qu XP, Wang SW, Chamarerk V, et al. Molecular mapping of the reverse thermo-sensitive genic male-sterile gene (rtms1) in rice. Theor. Appl. Genet. 2001;**103**:607-612

[56] Zhang Z. Preliminary observation of fertility changes in the new type temperature sensitive male sterile lines. IV. A. Hybrid Rice. 1991;**1**:31-34

[57] Yiming J (1988) Studies on the effect of high temperature on fertility of the male sterile lines inDian-type hybrid rice [J]. J Yunnan Agric Univ 2000

[58] Zhang QF, Shen BZ, Dai XK, Mei MH, SaghaiMaroof, MA, Li BZ (1994).Using bulked extremes and recessive class to map genes forphoto-period-sensitive genic male sterility in rice. Proc nat. Acad. Sci.91: 8675-8679.

[59] Subudhi PK, Borkakati RP, Virmani SS, Huang N. Molecular mapping of a thermosensitive genic male sterility gene in rice using bulked segregant analysis. Genome. 1997;**40**: 188-194

[60] Robin S, Kavithamani, Manonmani S, Mohana Sundaram K, Thiyagarajan K. 2010. Molecular tagging of thermo-sensitive genic male sterile (tgms) gene and Identification of new tgms lines in rice. Proceedings of the 3rd International Rice Congress held at Hanoi, Vietnam, 8-12,November, 2010 (Oral paper # 4208).

[61] Pan Y, Li Q, Wang Z, Wang Y, Ma R, Zhu L, He G, Chen R (2014) Genes associated with thermosensitivegenic male sterility in rice identified by comparative expression profiling. BMCGenomics 15(1):1114. https://doi.org/10.1186/1471-2164-15-1114

[62] Robin, S., S.Manonmani, R.Pushpam, SJ Arasakesary and K. Thiyagarajan. 2019. TNAU 60S (IC 0622805; INGRI7028), a TGMS Line of Rice (*Oryza sativa* L.). Indian J. Plant Genet. Resour. 32(2):239-297.

[63] Kavithamani D, Robin S, Manonmani S, Mohanasundaram K, Thiyagarajan K. Identification and Characterization of Promising TGMS Mutants with Good Floral Traits in Rice. The Madras Agricultural Journal. 2012;**99**(4-6):158-160

[64] Vinodhini M, Saraswathi R, Viswanathan PL, Raveendran M, Jeyakumar P. Studies on sterility behavior in thermo-sensitive genic male sterile lines of rice. A International Journal of Chemical Studies 2019. 2019;**7**(6):55-61

[65] Thiyagarajan K., S. Abirami, S. Robin, S.J. Jambhulkar and Manonmani, S, 2006. Development of TGMS Lines with Improved floral traits through mutation breeding in Rice. Trends in Research and Technologies in Agriculture and Food Sciences Abstracts, Mumbai, 18-20 December, 2006.

[66] Raveendran TS, Souframanien J, Manonmani S. Induction of Temperature Sensitive Male sterility in rice using Gamma Radiation. In: FAO/International Symposium on Induced Mutations I n Plants, 12-15th August 2008. Vienna: Austria; 2008

[67] Tanee Sreewongchai, Matthayatthaworn Weerachai, Phumichai Chalermpol, Sripichitt Prapa. 2014. Introgression of Gene for Non-Pollen Type Thermo-Sensitive Genic Male Sterility to Thai Rice Cultivars. Rice Science. 21 (2). pp. 123-126.

[68] Kalaiyarasi R, Thiyagarajan K, Manonmani S, Pushpam R, Asish K, Binodh Biju S. 2004. Developing new TGMS lines for exploitation of two line heterosis in rice (*O. sativa* L.) In abstracts of national seminar on hybrid breeding in crop plants, held at Annamalai University, Annamalai Nagar 3-4th March 2004. pp.: 31-33.

[69] Thiyagarajan K, ManonmaniS MD, RobinS PR, Mohana SK. Development of new TGMS lines with good floral

traits in rice. Electronic Journal of Plant Breeding. 2010;**1**(4):568-571

[70] Pushpam R., Thiyagarajan K., Manonmani S., Robin S., Ezhilarasi T and Umadevi M. 2012. Identification and characterization of new TGMS lines for hybrid rice breeding. International Symposium on 100 Years of Rice Science and Looking Beyond, January 9-12, 2012, TNAU, India. P. 210.

[71] Salgotra RK, Gupta BB, Ahmed MI. Characterization of thermo-sensitive genic male sterility (TGMS) rice genotypes ('*O. sativa* 'L.) at different altitudes. Australian Journal of Crop Science. 2012; 6(6): 957.

[72] Arasakesary S.J., S. Manonmani, R. Pushpam, S. Robin and K. Thiyagarajan. 2012. Out crossing potentials of new generation temperature sensitive geneic male sterile (TGMS) lines of rice. International Symposium on 100 Years of Rice Science and Looking Beyond, January 9-12, 2012, TNAU, India. P. 193.

[73] S.S. Virmani Z.X. Sun T.M. Mou A. Jauhar Ali C.X. Mao. 2003. Two-Line Hybrid Rice Breeding Manual. *International Rice Research Institute*. P. 1-88.

[74] Manonmani S. Pushpam R and S. Robin. 2016. Stability of TGMS Lines under different temperature regimes for pollen sterility. Journal of Rice Research, 9(1):17-20.

[75] Sai Rekha K, SaraswathiR KM, Manonmani S, Raveendran M, Robin S. Morphological and molecular diversity of thermosensitive genic male sterile lines in rice (*O. sativa* L.). Electronic Journal of Plant Breeding. 2020;**12**(1):177-182

[76] Srimathi K, Pillai MA, Aananthi N, Rajababu C. Genetic studies on TGMS lines for development of superior two-line rice hybrids. *Electronic Journal of Plant Breeding*. 2019;**10**:620-626

[77] Kanimozhi P, Pushpam R, Binodh AK, Kannan R, Pillai MA. Evaluation of TGMS lines for good floral and outcrossing related traits in rice. *Electronic Journal of Plant Breeding*. 2018;**4**:1497-1502

[78] Zhang, X., Zuo, B., Song, Wei Wang, Yuchi He, Yuhua Liu & Detian Cai .2017. Breeding and study of two new photoperiod- and thermo-sensitive genic male sterile lines of polyploid rice (*Oryza sativa* L.). *Sci Rep* 7, 14744 (2017). https://doi.org/10.1038/s41598-017-15241-8

[79] Pardeep Kumar , Nautiyal MK, Kumar P, Kumar K (2016) Morphological and Molecular based Characterization of different Thermo-sensitive Genetic Male Sterile (TGMS) lines in Rice (Oryza sativa L.). Vegetos 29:4. doi: 10.5958/2229-4473.2016.00101.4

[80] MengchenZhang, WangShan, YanJianfang, SunShuiyong, XuXin, XuQun, YuanXiaoping, WeiXinghua, YangYaolong. 2019. Classification and Identification of indica P/TGMS Lines in China. *Rice Science*. 26 (3). pp. 195 – 198

[81] Wang Shiguang, Wei Liu, Dongbai Lu, Zhanhua Lu, Xiaofei Wang, Jiao Xue, and Xiuying He.2020 Distribution of Bacterial Blight Resistance Genes in the Main Cultivars and Application of Xa23 in Rice Breeding. Front. Plant Sci., 02 March 2021. https://doi.org/10.3389/fpls.2021.629314

[82] Manonmani S., Pushpam R, Robin S. 2015. Stability of TGMS Lines under different Temperature Regimes for Pollen Sterility Oral Paper (IRS15 – OP550) presented during International Rice Congress, Hyderabad. Compendium of Abstracts P. 37.

[83] Arasakesary SJ, Manonmani S, Pushpam R, Robin S. New Temperature Sensitive Genic Male Sterile Lines with better Outcrossing Ability for production of Two-Line Hybrid Rice. Rice Science. 2015;**22**(1):49-52

[84] Nagarajan M., Vinod KK, Manonmani S, Gopalakrishnan S, Bhowmick PK, Singh AK. 2015.Fertility Reversion Behavior And Seed Set Evaluation Of Thermosensitive Genic Male Sterile Lines In Rice. Poster Paper (IRS15 – PP 097) presented during International Rice Congress, Hyderabad. Compendium of Abstracts P. 195.

[85] Van, Tran Thihi Thuy. 2008. A Genetic appraisal of two line, three line and conventional hybrids of Rice (*O. sativa* L.)Master thesis submitted to the Tamil Nadu Agricultural University, Tamil Nadu, India.

[86] Manonmani, S. 2004. Two line breeding system in Rice. In: Proceedings of Zonal Conference on Hybrid Rice: Problems and Prospects for increasing rice production and its Quality held at Indira Gandhi Agricultural University, Raipur, Chhattisgarh from 29th – 30th November, 2004,p. 70-78.

[87] Araskesary SJ, Robin S, Manonmani S, Pannerselvam S, Marimuthu N. Identification of stable two-line rice hybrids (*O. sativa* l.) Using additive main effects and Multiplicative interaction (ammi) model. *Annals of Sri Lanka Department of Agriculture*. 2013;*2013*(*15*):*115-126*

[88] Manonmani S, Pushpam R, Saraswathy R. Performance of two line rice hybrid at Hybrid Rice Evaluation Centre, Gudalur, Tamil Nadu, India. In: Oral paper presented in the International Hybrid Rice Symposium held on 27-28 feb to 1 st march. Yognakarta: Indonesia; 2018

[89] Chandrashekar S Somanagoudra 2021.Development and identification of superior CGMS and TGMS based rice hybrids for their grain yield and quality. Ph.D.Desertation submitted to the Department of Genetics and Plant Breeding,TNAU,Coimbatore,Tamil Nadu,India.

Chapter 2

Rice Aroma: Biochemical, Genetics and Molecular Aspects and Its Extraction and Quantification Methods

Nirubana Varatharajan, Deepika Chandra Sekaran, Karthikeyan Murugan and Vanniarajan Chockalingam

Abstract

Aroma in rice is unique and a superior grain quality trait, varieties especially Basmati and Jasmine-type are fetching a high export price in the International markets. Among the identified volatile aroma compounds, 2AP (2 acetyl-1-pyrroline) is believed to be the distinctive biochemical compound contributing the flavor in rice. Genetically, aroma in rice arises by the phenotypic expression of spontaneous recessive mutations of the *OsBadh2* gene (also known as *fgr/badh2 /osbadh2/os2AP* gene) which was mapped on chromosome 8. An 8-bp deletion in the exon 7 of this gene was reported to result in truncation of betaine aldehyde dehydrogenease enzyme whose loss-of-function lead to the accumulation of a major aromatic compound (2AP) in fragrant rice. Among the different sampling methods and analytical techniques for the extraction and quantification of scentedness, simultaneous distillation extraction (SDE) is traditional and normalized, whereas solid-phase micro extraction (SPME) and supercritical fluid extraction (SFE) are new, very simple, rapid, efficient and most importantly solvent-free methods. These methods are coupled with Gas Chromatography–Mass Spectrometry (GC–MS), Gas Chromatography-Flame Ionization Detector (GC-FID) and/or Gas chromatography olfactometry (GC-O) and also with sensory evaluation for readily examining 2AP compound found in rice. The major factor affecting the aroma in rice was their genetic makeup. However, the aroma quality may be differed due to different planting, pre-harvest and postharvest handling and storage. For a more extensive elucidation of all effective and fundamental factors contributing to fragrance, it is essential to explore target quantitative trait loci (QTLs) and their inheritance and locations.

Keywords: aromatic rice, 2-acetyl-1-pyrroline, *fgr*, *badh2*, evaluation methods, affecting factors

1. Introduction

Rice (*Oryza sativa* L.) is a dietary staple food crop and the grain being consumed by atleast 50 per cent of the world's population [1]. It had a decisive role in food, security and in improving the livelihood of people. With the continuous/marked

IntechOpen

improvement in standard of living of the people, the ethnic preference of rice is under conversion which increases the demand for superior fine quality rice. Most of the scented rice are inferior in agronomic performances and highly prone to environmental variations [2] yet it paves much attention for their first-rated aroma. The origin and evolution of the aroma gene betaine aldehyde dehydrogenase (*BADH2*) remains unclear but the haplotype analysis firmly establishes a distinct origin of the *badh2.1* allele within the Japonica varietal group [3] and the centre of origin is considered to be the Himalayan foothills in the Indian subcontinent from where it spreads to various parts of the world [4].

It was reported that rice aroma was controlled by single recessive nuclear gene in rice [5, 6]. The biochemical analysis of rice grain reveals the presence of numerous volatiles in fragrant rice revealing the major aromatic compound, 2-acetyl 1-pyrroline (2AP). Recent advances had also discovered the biochemical pathway for biosynthesis of 2-Acetyl-1-pyrroline (2AP) from different types of amino acids and polyamines [7]. Based on the rice genome sequence information [8], the *OsBadh2* gene (present on chromosome 8) is identified as a candidate gene for aroma, which is the most important aroma gene till now. However, several other genes and locus had been reported to be the contributor for aroma [9, 10].

Rice aroma quality evaluation is quite complex due to the complex interaction of numerous volatile compounds, and many affecting factors during planting and processing. With the rapid advancement in the instrumentation and sampling methods for the isolation and determination of 2AP concentration levels it is possible to analyze compounds even at very low concentrations (ppb levels) [11]. This chapter gives insights into the flavor chemistry, the progresses pertaining to the genetic and molecular understanding of fragrance, various extraction, quantification methods and their interaction with genetic and non-genetic factors in rice.

2. Fragrant rice germplasm and varieties

Scented rices had been grouped into small, medium and long grained types based on grain length and could be categorized by their scentedness as mild and strong aromatic types. Broadly, the aromatic rice germplasm are grouped into three categories *i.e.* the Basmati, jasmine, and non-basmati/jasmine typed scented rice. The word 'Basmati' has its origin from two Sanskrit roots (vas = aroma) and (mayup = ingrained or present from the beginning) making the word vasmati and it has been pronounced as Basmati in course of time. Of the largest aromatic germplasm maintained at IRRI, about 86 had its root word as Basmati irrespective of grain dimensions and intensity of aroma [12].

A number of non-Basmati scented rices had been considered superior to Basmati in one or more characteristics like flavor, texture, linear elongation ratio on cooking, taste etc. Moreover, many of them can be cultivated under conditions and in areas where Basmati cannot be. Small-grain Bindli, for example, is superior to Basmati in aroma, grain elongation, taste and digestibility (as perceived by the farmers) and it performs well under water-logged conditions. A few such potential candidates could be Kalanamak, Tilakchandan, Sakar-chini and Dhania (U.P.), Ambemohar (Maharastra), Badshahbhog (Bihar and West Bengal), Bindli (waterlogged conditions of U.P.), Chakhao (Manipur), Madhumalti and Mushkan (H.P.), Kon-Joha - 1, Raja Joha and Krishna Joha (Assam), Randhuni Pagal (W.B.), Vishnubhog and Dhubraj (H.P.), Katarani and Sonachur (Bihar) [13]. The small and medium grain aromatic rices could be explored further and improved by selecting short stature, better yielding and early maturing plant types in order to develop varieties to be cultivated in non-traditional areas of basmati cultivation.

The scented rices are mainly cultivated and consumed in India, Pakistan, Thailand, Bangladesh, Afghanistan, Indonesia, Iran and United States. Of these, the major exporter of fine-grained fragrant rices includes India, Pakistan and Thailand Major aromatic rices of different states of India were presented in **Table 1** [12].

States	Small grain	Medium grain	Long grain
Southern zone			
Andhra Pradesh	—	Jeeragasambha	—
Kerala	Jeerakasala, Gandhkasala	—	—
Karnataka	—	Kagasali	—
North eastern zone			
Assam	Bengoli Joha, Bhaboli Joha, Bhugui, Boga Joha, Bogamanikimadhuri, Boga Tulsi, Bogi Joha, Bokul Joha, Borjoha, Borsal, Cheniguti, Chufon, Goalporia Joha-1, Goalporia Joha-2, Govindbhog, Joha Bora, Kaljeera, Kamini Joha, Kataribhog, Khorika Joha, Kola Joha, Koli Joha, Kon Joha-1, Kon Joha-2, Krishna Joha, Kunkuni Joha, Manikimadhuri Joha, Ramphal Joha, Ranga Joha	—	—
Manipur	—	Chahao Amubi (black scented rice), Chahao Angangbi (pink/red scented rice)	—
Eastern zone			
Bihar	Badshahbhog, Deobhog, Karia Kamod, Katami, Shyam Jeera, Kanak Jeera, Kanakjeeri, Badshapasand, Mircha, Brahmabhusi, Ramjain, Kamina, Dewta Bhog, Tulsi Pasand, Chenaur, Sona Lari, Sataria, Tulsi Manjari.	Gopalbhog, Champaran Basmati (Lal), Champaran Basmati (Kali), Champaran Basmati (Bhuri), Bhilahi Basmati, Amod, Abdul, Bahami, Kalanamak, Kesar, Sonachur	Baikani
West Bengal	Badshabhog, Chinisakkar, Danaguri, Gandheshwari, Kalo Nunia, Kataribhog, Radhuni Pagal, Sitabhog, Tulai Panji, Tulsibhog	Kanakchur, Katanbhog	—
Northern zone			
Haryana	—	—	Basmati 370, Khalsa 7, Taraori Basmati, Pakistani Basmati
Punjab	—	—	Basmati 370, Basmati 385, Pakistani Basmati
Rajasthan	—	—	Basmati (local), Basmati 370

States	Small grain	Medium grain	Long grain
Himachal Pradesh	—	Achhu, Begmi, Panarsa local	Baldhar Basmati, Madhumalti, Chimbal Basmati, Mushkan, Seond Basmati
Central zone			
Madhya Pradesh	Chinore, Dubraj, Kalu Mooch, Vishnubhog, Tulsi Manjari, Badshabhog,	Chatri, Madhuri, Vishnu Parag	Laloo
Uttar Pradesh	Adamchini, Badshapasand, Bhanta Phool, Bindli, Chhoti Chinnawar, Dhania, Jeerabattis, Kanak Jeeri, Laungchoor, Moongphali, Rambhog, Ramjawain, Sakkarchini, Tinsukhia, Bengal Juhi, Thakurbhog, Yuvraj, Bhantaphool	Karmuhi, Kesar, Parsam, Sonachur, Tilak Chandan, Kesar, Kalanamak, Vishnuparag	Basmati 370, Dehraduni Basmati, Type 3, Hansraj, Nagina 12, Safeda, Vishun Parag, Kala Sukhdas, Lal Mati, Tapovan Basmati, T-9, Dubraj, Duniapat (T9), Ramjinwain (T1)
Western zone			
Maharastra	Ambemohor, Chinore,	Kagasali, Prabhavati, Sakoli-7	—

Table 1.
Zonal classification of scented Rices of different states of India [12].

3. Biochemical basis of fragrance

Generally, the aromatic rice cultivars are enriched with large volatile and semi volatile compounds *viz.*, alcohols, aliphatic aldehydes, alkane, alkene, aromatic aldehydes, aromatic hydrocarbon, carboxylic acid, ester, furan, ketone, N-heterocyclic, phenol, and terpenes [14–17].

3.1 Structure and chemistry of 2AP

Among the different volatile compounds, 2-acetyl-1-pyrroline (2AP) with popcorn-like aroma and lowest odor threshold is reported to be the potent biochemical compound to impart fragrance in rice [18]. The chemical structure of 2AP is an N-heterocyclic compound containing 1-pyrroline ring in which the hydrogen at position 2 is replaced by an acetyl group with a methyl ketone group. 2AP content in scented rice varieties include 0.04–0.09 ppm, whereas non-aromatic varieties have 10x less (<0.006–0.008 ppm) [19].

3.2 Biosynthetic pathway for 2AP

There are many contradictions and views regarding the biochemical pathway for 2AP synthesis and it is still being explored. It was reported that *L*-proline was the precursor for the production of 2AP [20]; and is involved in polyamine degradation pathway which is the main enzymatic pathway and there are some other non-enzymatic pathways reported having an influencing action on 2AP concentration. In the enzymatic polyamine pathway, arginine, ornithine, spermidine, putrescine, etc. are degraded into GAB-ald which spontaneously cyclises

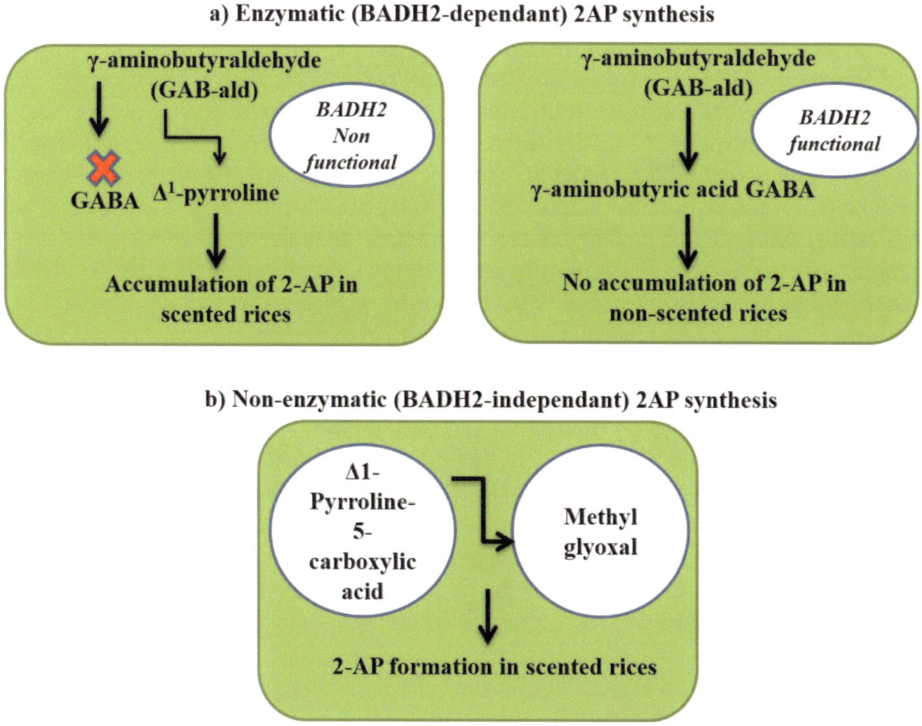

Figure 1.
2AP biosynthetic pathway in rice. (a) Enzymatic (BADH2-dependant) 2AP synthesis [21, 22] (b) non enzymatic (BADH2-independant) 2AP synthesis [23].

to Δ^1-pyrroline, an immediate precursor of 2AP biosynthesis [21]. The non-functional badh2 enzyme (encoded by *osbadh2* gene) inhibits the conversion of the γ-aminobutyraldehyde (GAB-ald) to γ-aminobutyric acid (GABA) thereby allowing the formation of Δ^1-pyrroline and ultimately the 2AP in scented rice whereas the reverse happens in non-scented rices (functional BADH2 enzyme coded by *OsBadh2* gene inhibits 2-AP formation) [22].

Some non-enzymatic direct pathways had also been described by many scientists and researchers. Glutamate produces proline and the proline accumulated during stress is converted to Δ^1-Pyrroline-5-carboxylic acid (P5C) by the enzyme Δ^1-Pyrroline-5- carboxylate synthase (P5CS). This P5C combines directly with methylglyoxal without involving any enzymes or might be converted to Δ^1-pyrroline and thereby enhancing the 2AP concentration [23]. In normal plants the methylglyoxal produced from glycolysis is detoxified by glyoxalase enzymes and their concentration was kept low. It was speculated that 2AP is a generative volatile compound to detoxify methylglyoxal in rice plant [24] (**Figure 1**).

2AP concentration differs in different plant parts of rice. The concentration is more in grains and flag leaf than in any parts of the plant [25–27]. Glutamate, ornithine and proline are important amino acids that serves as nitrogen (N_2) source in the ring of Δ^1-pyrroline [11]. The high aroma content in grains is mainly from the larger availability of N_2 from the soil. So, the aroma concentration may vary depending upon the nitrogen availability to the plants [28]. Advanced researches are essential in correlating the genetic and biochemical aspects of scented rice varieties, particularly with regard to the nitrogen and acetyl group donor in 2AP in order to reveal the key enzymes that are involved in the biosynthetic pathway of aroma in rice.

4. Genetic and molecular basis of fragrance

Inheritance of aroma is quite difficult to understand because it is controlled by number of unknown genes at different growth stages of rice and influenced by various concentrations of volatile and semi-volatile compounds. Although, plant breeders have reported the aroma inheritance by monogenic, digenic and polygenic pattern with recessive, dominant, complimentary and duplicate gene actions, indicating that complex genetic control of the trait. In majority of studies, the genetics of fragrance in rice is mainly due to single recessive gene [6, 9, 29–33] while other studies have also identified two, three or four genetic loci having influence on fragrance [9, 34–38]. Studies on the genetic control of aroma/fragrance/scent in rice have been presented in **Table 2**.

However, much of this conflicting information on the inheritance of aroma might have arisen due to (i) unreliable and cumbersome phenotyping methods used for fragrance determination [6], (ii) failure to consider the endosperm fragrance in rice seeds [29] and (iii) segregation distortion [9]. The nature of aroma inheritance appears to be cross/genotype specific due to the number of genes and the type of gene action varied with the genotype. However, the fragrance trait is a highly heritable as some of the lines derived from T142 (scented) x IR 20 (non-scented) cross, and some of the high yielding released aromatic rice varieties show strong scent.

The implementation of marker assisted selection is a significant supplement to traditional approaches, altering the selection process directly or indirectly from

S.No.	Gene action	References
1.	Monogenic dominant	[39, 40]
2.	Monogenic recessive	[29–32, 38, 41–63]
3.	Monogenic recessive with an inhibitor	[41, 64]
4.	Digenic or trigenic dominant	[34]
5.	Monogenic or digenic recessive	[37]
6.	Digenic recessive	[38, 65, 66]
7.	Three recessive genes	[67]
8.	Two dominant complimentary genes	[68]
9.	Three dominant complimentary genes	[69, 70]
10.	Four dominant complimentary genes	[35]
11.	Monogenic or digenic recessive or dominant, complimentary	[71]
12.	Monogenic or digenic dominant, duplicate	[72]
13.	Digenic dominant suppression epistasis interaction	[73]
14.	Polygenic	[9, 10, 74]

Table 2.
Inheritance pattern of aroma in rice.

phenotype to genes [75]. A novel compound namely 2AP (2-acetyl-1-pyrroline) plays a major role in most of the aromatic rice cultivars for the presence and absence of unique popcorn like characteristic aroma. Several attempts have been made at molecular level for genetic mapping the fragrance gene governing the 2AP synthesis in different aromatic rice varieties such as Della [30], Azucena [9, 76], Suyunuo [77, 78] and Wuxianjing [77]. Quantitative trait locus (QTL) mapping was also performed in indica aromatic rice KDML105 (Jasmine) [52, 79], Kyeema [80] and Wuxiangxian [77] (**Table 3**).

By using RFLP technique, a single recessive gene (*fgr*) that controls fragrance was mapped on chromosome 8 tightly linked with a single-copy marker RG28 and found that genetic distance between aroma gene and RG28 was 4.5 cM [30]. The close linkage between RG28 and *fgr* (5.8 cM) was confirmed by [9], also identified two quantitative trait loci for fragrance, one on chromosome 4 and the other on chromosome 12.

Further, a gene responsible for 2AP synthesis was mapped in a Jasmine rice variety KDML105 between the flanking regions of RG1 and RG28 [86]. The original region (1.13 Mb) flanking between RG1 and RG28 was narrowed down to 82.2 Kb in segregating population, within this region three KDML BACs were cloned and identified three new candidate genes. Among them, a single recessive gene (Os2AP) was identified which majorly contributing the 2AP synthesis in rice. The comparative analysis between aromatic KDML105 and Nipponbare for Os2AP gene sequences revealed two important mutational events within the exon 7 of Os2AP of KDML105, at positions 730 (A to T) and 732 (T to A), followed by the 8-bp deletion "GATTAGGC" starting at position 734 [87]. A similar mutational event was also reported by [79] within the flanking regions of RM515 and SSRJ07, a gene responsible for 2AP in Kyeema fragrant rice cultivar.

Number of Genes	Type of markers	Chromosome location	References
1	RFLP	8	[30]
1	RFLP	8	[81]
1	RAPD	—	[52]
1 major gene and 2 QTLs	RFLP, STS	8, 4 and 12	[9]
1	SSR	8	[56]
1	SSR	8	[32]
1	SNP	8	[33]
1	SSR	8	[80]
1	EST, SSR	8	[82]
1	SSR	8	[77]
1	SSR	—	[66]
1	SSR	8	[83]
3 QTLs	SSR	QTLs on 3, 4 and 8	[10]
1	SSR, RFLP	8	[58]
1	SSR	8	[60]
1	SSR	8	[84]
1	SSR	8	[85]
2	—	—	[73]

Table 3.
Molecular mapping of fragrance gene in rice.

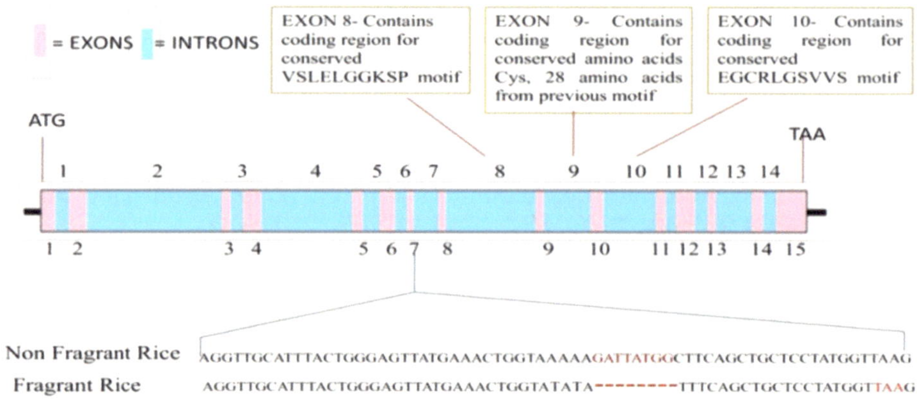

Figure 2.
Structure of the fgr gene showing ATG, initiation codon, exons (15), introns (14) and the termination site (TAA).

Using four BAC of Nipponbare spanning within a region of 386 bp from RM515 to SSRJ07, an in silico physical map was developed and suggested that one BAC clone (clone AP004463) as most likely to be having the gene. Further, resequencing of all 17 genes lying within the BACs helped in identification of a novel gene with 3 single nucleotide polymorphisms (SNPs) with the 8 bp deletion in the 7th exon of the gene, which resulted in a premature stop codon [10]. The newly identified gene was showing homolog with BAD1 (betaine aldehyde dehydrogenase 1) locus of chromosome 4 and hence named as BAD2 [79]. A comparative study between amino acids and sequences of Os2AP and BAD2 suggested them as one gene with two different names. Recent surveys of diverse fragrant germplasm support the association of *badh2* with fragrance [76, 78, 88], and transformation of a fragrant variety with the dominant non-fragrant allele has been proved to abolish aroma [21], confirming that *badh2* is the major and effective genetic determinant of aroma in rice (**Figure 2**).

The nucleotide sequences of 7 exon are shown for both the rice varieties. The fragrant variety shows a deletion of 8 bp with 3 SNPs that terminates prematurely, within this exon. Thus, fragrant varieties truncated protein might lack the conserved sequences which is encoded by 8, 9 and 10 exons and that are believed to be important for correct protein function [84].

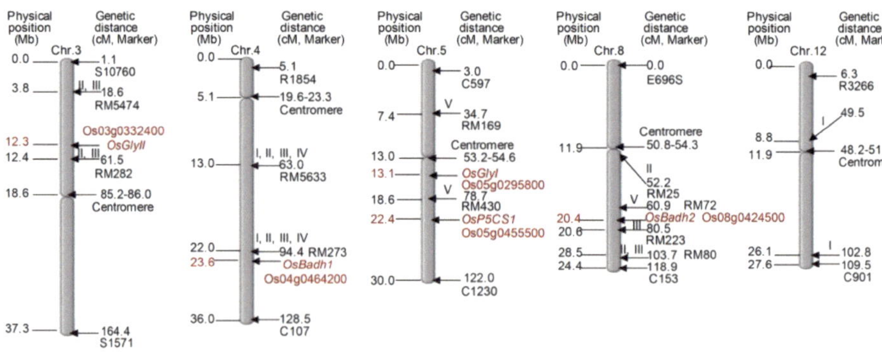

Figure 3.
QTLs for aroma with respective candidate genes in rice.

Since the *Badh2* gene was isolated and cloned, more than a dozen mutation sites have been found in *Badh2* [3, 89–91] and a series of molecular markers were designed for these loci, which could be used for the identification of gene responsible for aroma, selection of different aromatic rice varieties and cultivation of new varieties of aromatic rice.

4.1 QTLs for aroma

A number of QTLs for aroma have been identified on chromosomes 4, 8 and 12, at least three QTLs have been located on chromosomes 3, 4 and 8 in Pusa 1121 [9, 10, 92]. Recently, three QTLs were detected for rice grain aroma on chromosome 5 (one QTL) and chromosome 8 (two QTLs) [93]. However, until now only a few QTLs and associated markers have been confirmed (**Figure 3**).

5. Aroma extraction, identification and quantification methods

The extraction process is influenced by various criteria viz., type of matrix, volatility of the analyte, concentration of constituents in the sample and extraction conditions; therefore, efficient methods of extraction is needed [14]. However, so far there is no single method that will prove ideal for aroma extraction, identification, and quantification of rice. Although, several traditional and modern methods are available for extraction and isolation of rice aroma chemicals which are coupled with analytical methods (**Figure 4**).

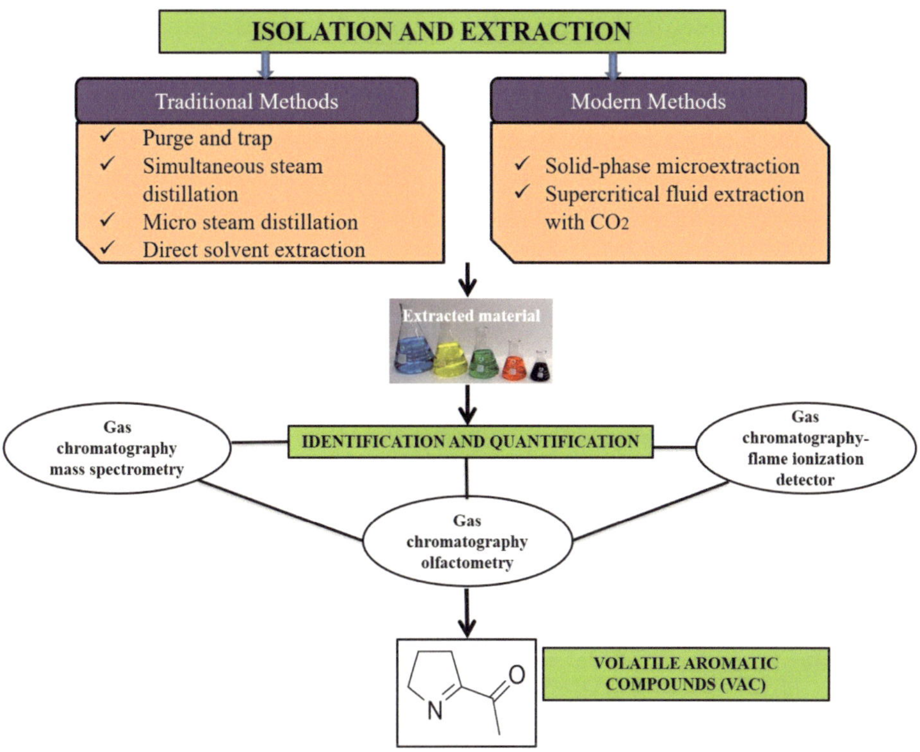

Figure 4.
Extraction and isolation methods of rice aroma.

5.1 Isolation and extraction methods

Several extraction methods are available for extraction of VACs (volatile aromatic compounds) in rice. While considering the extraction efficiency, it differs dramatically for each method. Based on that, the selection of extraction technique can efficiently extract volatile compounds viz., alcohols, aldehydes, hydrocarbons, carboxylic acids, esters, furans, ketones, N_2-containing, phenols and terpenes [94]. Most of the volatile compounds are insoluble in water so conventional extraction methods need non polar solvent as a medium.

The isolation techniques such as vacuum SDE apparatus; PTM, SDE, SEfbDI, SPME, SFE, HAS, and HSSE have their own distinguished characteristic feature. For extraction and isolation of 2-AP concentration from rice sample a desired efficient technique is a prerequisite. From the above methods, simultaneous distillation extraction (SDE), solid-phase microextraction (SFME) and supercritical fluid extraction (SFE) are the most widely used method for extraction of volatile compounds.

5.1.1 Simultaneous distillation extraction (SDE)

Simultaneous distillation extraction is a combination of vapor distillation and solvent extraction method for extraction of VACs [95]. SDE is also known as the Likens–Nickerson steam distillation, it is one of the most popular method for rice aroma chemical analysis. Solvent extract is the final product of these method. Many researchers have used this method for extraction of 2AP volatile compounds and it showed to be the most effective approach for a quantitative evaluation of 2AP. A laborious concentration step is still needed for traditional SDE procedure. Therefore, a modified version of so called micro-SDE device was proposed to overcome the problem [96].

The main advantage of this method is only a small amount of solvent is used for extraction and it also shortens the extraction time and improves the extraction efficiency. The solvents, such as: hexane, dichloromethane (DCM) and n-pentane, can be used and the amount of solvent required for SDE has been dramatically reduced compared to that of the conventional LLE. However, the major disadvantage of this method is the atmospheric pressure SDE was also obvious. Due to its high temperature, there is a possible occurrences of undesirable ester hydrolysis, Maillard reaction and sugar degradation [95].

5.1.2 Solid-phase microextraction (SPME)

This method was first introduced in early 1990s by Arthur and Pawliszyn [97]. Solid-phase microextraction is a newly emerging extraction technique for extraction of rice aroma compared with other method. It was applied in both laboratories as well as on-site [98]. Because of its persistence over other method of extraction, many results have been reported on 2AP aroma compound from rice grains [99].

Solid-phase microextraction has been used for the extraction of volatiles due to certain advantages, such as low-cost, simple, solvent-free, rapid and time-saving technique when compared with SDE method. This method can eliminate contamination, prolong the fiber lifetime and lead to reproducible results [98]. The chemistry of volatiles is decided with the help of desorption and adsorption behavior. If such extracted analytes are varied in their polarities that requires a different chemistry of SPME fiber [99].

5.1.3 Supercritical fluid extraction (SFE)

Supercritical fluid extraction is a separation and extraction process and it uses the supercritical fluids (SCFs) as the extraction solvent. It is a type of solvent which is clean and pure. The supercritical fluid is considered as 'Green Chemistry', because it is less toxic in compression compared to the organic solvents. The carbon dioxide (CO_2) is an extensively used SCF; sometimes it is modified by ethanol or methanol as such co-solvents [100].

Nowadays, Supercritical fluid extraction (SFE) is widely used for extraction of volatile compounds from rice and also in other plant samples such as vegetables, fruits and so on [101].

It is a quick technique and it can recover the majority of the VACs [102]. Within 10–60 minutes the whole process would be completed and it produces the pure extract by releasing the pressure. One of the main disadvantage of SFE is less effective, than solvent extraction [101].

5.2 Identification and quantification methods

The identification and quantification of volatile aromatic compounds from various types of rice is a tedious process. The research on aroma in rice is conducted by the researchers, scientists and industry groups for more than four centuries but still now it is not possible to identify all aroma compounds presented in rice. To determine the sensory quality of foods, need more concentration/efforts towards the application of modern and recently developed technologies.

There are several analytical methods for identification and quantification of rice aroma *viz.*, GC–MS, GC–MS-FID, GC–MS-AFID, GC–MS-FTD, GC–MS-SIM, Capillary GC–MS, GC-O, GC-FID, GC-PFPD, GC-TOF-MS, GLC, GLC-Capillary or GLC-Packed column. From the above methods, Gas Chromatography–Mass Spectrometry (GC–MS), Gas chromatography-olfactometry (GC-O), Gas Chromatography-Flame Ionization Detector (GC-FID) these three methods are widely used for identification and quantification of aroma in rice sample.

5.2.1 Gas chromatography-mass spectrometry (GC-MS)

Gas Chromatography–Mass Spectrometry (GC–MS) method was the most common instrumental analysis method for rice aroma analysis and effective method for analyzing volatiles, and widely used for qualitative and quantitative analysis of volatiles in rice [103]. In this method, the volatile compounds present in rice are determined and separated by GC and then identified by GC–MS [12]. This method can separate VACs having a molecular weight of less than 1,000 Dalton [104]. To date, the identification and quantification of volatile components from rice depends on advanced technologies and improved GC with multidimensional use. The qualitative and quantitative analysis of VACs is proved to be very sensitive in this method. The performance of MS is based on generated charged particle (ions) from molecules of analyses.

5.2.2 Gas chromatography-olfactometry (GC-O)

Gas chromatography-olfactometry (GC-O) is considered as one of the most advanced analytical method for the identification and quantification of VACs in sample matrix of rice. In GC-olfactometry (GC-O) system, human nose was applied to detect the odor intensity of volatiles. Two detectors which perceived the odor-active compounds (hexanal, longifolene, 2-methoxyphenol and so on) eluted from the chromatographic column [105]. Although this method was very much useful for identification of aroma-active compounds from food samples. However, it is not suitable method for quantitative and qualitative analysis of VACs. As a result, the GC–O analytical method is not only an instrumental but also a sensorial analysis [106].

5.2.3 Gas chromatography-flame ionization detector (GC-FID)

Gas Chromatography-Flame Ionization Detector is the combination of FID with GC. This method is considered as very effective and crucial GC method because of its excellency [97]. It enables the separation, identification, and quantification of volatile compounds with their existing levels of concentrations from different food sample [98]. By comparing the retention time (RT) in GC-FID, the identification of volatile aromatic compounds of rice is completed and the retention times are converted into system-independent constant known as Kovatx retention index [99].

6. Factors affecting rice aroma

6.1 Genetic factors

The genes controlling aroma was found to be a highly heritable and also relatively complex in nature. In chromosome 8 the main candidate gene was *fgr/badh2/Os2AP* homologous to betaine aldehyde dehydrogenase (BADH), whereas many other genes were also reported [103]. Deletion of 8 base pair in exon 7 or deletion of 7 base pair in exon 2 of BADH2 gene on chromosome 8 results in a loss of function of BADH2, which catalyzes the oxidation of 4-aminobutanal to 4-aminobutanoic acid. It was reported that 4-aminobutanal existed in solution equilibrium with its cyclic form 1-pyrroline which was a precursor for 2AP [107].

According to [88], the gene *fgr/badh2/Os2AP* was not the only aromatic gene in rice. The aromatic compound 2AP was identified in a number of rice varieties not carrying the 8-bp deletion. The aromatic landraces in Japan consists of six clades, none of which had the 8-bp deletion in exon 7 of BADH2 and also Japanese aromatic and non-aromatic landraces were found genetically different [108]. About 84 Subsp. indica rice landraces were investigated with respect to 8-bp deletion in BADH2 gene [109]. The results showed that aroma traits were genetically controlled by recessive monogenes, independent of cytoplasmic genes, however, aroma was also studied as a quantitative trait, and many genes were included in the expression [110].

6.2 Planting and harvesting factors

The maximum down regulation of BADH2 gene was reported in temperature of 25°C, highest 2AP content and excellent phenotypic aroma score, indicating the function of temperature on regulating phenotypic expression of aroma and final rice aroma quality. BADH2 gene expression is influenced by the temperature, phenotypic aroma score and 2AP content were investigated in three different temperatures (ambient or 28.29 ± 0.91°C, 25°C and 20°C) [2].

There is a significant positive effect on 2AP content in rice grains (Meixiangzhan and Nongxiang 18) by the application of manganese, which esults in probably improvement of enzyme activities involved in 2AP formation. Higher total soil nitrogen plays a major role in producing rice aroma. During flowering stage, it was found that Si contents in leaves were positively related with 2AP contents.

Thus, indicating that Si application to some amount will improve 2AP contents in grains [111].

An increase of 2AP content in grains with salinity was observed for three improved aromatic rice varieties and salinity was thought to have a positive effect on rice aroma quality [112]. NaCl stress enhanced aroma production in Tulaipanji, Radhunipagal and Gobindobhog rice varieties while weaken that in Kalonunia [113]. Shading treatments during grain filling significantly increased 2AP content in both Yuxiangyouzhan and Nongxiang rice varieties, and had a selective effect on the metabolism of other volatiles [114].

6.2.1 Effect of planting density on 2-acetyl-1-pyrroline content

2AP content decreases with an increase in planting density. The content of 2-AP in rice grains obtained during the early season will be stored for 6 months. However, other seed quality attributes at the exception of head rice yield and grain vitreosity were not affected by planting density [115].

6.2.2 Effect of harvesting time on 2-acetyl-1-pyrroline content

Reduction in 2AP was observed with increasing harvest date during the early season. During the late season, however, the concentration of 2AP is gradually decreased from 10 DAH and seemed to stabilize at 40 DAH, a reduction rate of 60%. However, it is well compensated for by the high level of 2AP in both brown and white rices, which remains significant even after a storage period of 3 months at ambient temperature [116].

6.3 Processing factors

6.3.1 Cooking

Presoaking is a traditional pretreatment before cooking. It would result in uniform cooking and less cooking time. Presoaking for 30 min before cooking resulted in significant increase in sewer/animal flavor and summed negative flavor attributes, and significant decrease in sweet taste and summed positive flavor attributes, mainly as a result of an increase in sulfur-containing free amino acids and their breakdown products [19].

According to [117], divided the cooking process into four stages and identified the major compounds of Japanese rice cultivar Akitakomachi. In stage I (25 min, from the start of heating to start of steam coming out of rice cooker) were aldehydes such as n-nonanal, n-decanal, and (E)-4-nonenal. The dominating compounds identified at cooking stage II (13 min, from the start of steam coming out of rice cooker to the end of steam coming out of rice cooker) were hexadecanoic acid and tetradecanoic acid. The major compounds identified at cooking stage III (10 min, from the end of steam coming out of rice cooker to automatic stop of heating) and IV (keeping the rice warm for another 30 min) were aldehydes and fatty acids.

6.3.2 High hydrostatic pressure and superheated processing (HHP)

High hydrostatic pressure (HHP) had stabilized effects on low molecular weight volatiles [118], and it is one of the effective processing to improve products flavor. HHP was thought to be a good pretreatment option to enhance aroma quality of cooked rice. HHP process enhanced the formation of aldehydes, alcohols and ketones in germinated brown rice [119]. The volatile compounds in rice were cooked by superheated steam rice cooking machine were compared -with those of ordinary cooked rice [105].

6.3.3 Roasting and parboiling

While roasting there is a change in volatiles by the Maillard and caramelization reactions, and consequently form unique flavor, and usually increase the popularity of consumers. Increases the content of heterocycle compounds and decreases the content of hydrocarbons and benzene derivatives by roasting process [120]. Parboiling cause concomitant changes in the physical, chemical, and nutritional properties of grains, and consequently greatly affect organoleptic and other qualities. Hydrothermal treatment during parboiling would inactivate lipases, and inhibit the development of off flavors [121]. Hence, it was a good method to keep rice aroma during storage.

6.3.4 Milling

Un-milled black rice contained significantly larger amounts of total volatiles than milled black rice [122]. That is, the volatile compounds were mainly distributed in the bran layer of black rice (624 ± 17.7 ng g – 1), and significantly decreased by milling, especially the contents of acids, esters, and alcohols. When milling aromatic rice (Cheonjihyang-1-se), hexan-3-one, benzene, 2-pentylfuran, and pentanal decreased to 79%, 70%, 54%, 78% with milling time from 10s to 140 s, while (E)-non-2-enal, pentadecanal, (5E)-6,10-dimethylundeca5,9-dien- 2-one, and menthol increased 252%, 185%, 172% and 159% [123].

6.4 Storage factors

It was reported that, proteins, lipids and carbohydrates were decomposed into volatiles contributing rice odor during storage [124]. Enzyme catalyzed reactions were drastically inhibited at low temperature. This was one reason for slower deterioration of rice aroma. In general, lower storage temperature and better packaging materials would be more appropriate for aromatic rice to better maintain desirable rice aroma. OPP/Al/LLDPE package was superior to Nylon/LLDPE package, and storage at lower temperature (15°C) was better than ambient temperature, since they better retarded the formation of lipid oxidation products and other characteristic odorants in organic red aromatic rice [125]. Some paper assumed that lower temperature during storage would minimize volatilization of 2AP from rice [126, 127].

6.4.1 Effect of storage time and temperature on 2-acetyl-1-pyrroline content

Fragrant rice harvested in June and kept for 6 months at – 4°C contained up to four times 2-AP in all forms (brown and white), compared to those kept at 30°C. High losses of 2-AP occurred under a very warm condition of 30°C. There were also significant differences in the concentration of 2-AP between samples collected in November with losses of 25 to 35% occurring after storage of 3 months at 20°C compared to 8°C [115].

Therefore, insights into extraction and quantification methods and various factors affecting the quality of aroma are essential, and also modern biotechnological advances like Transcription Activator Like Effector Nuclease (TALENs), Zinc Finger Nuclease (ZFNs) and Clustered Regularly Interspaced Short Palindromic

Repeats (CRISPR) associated endonuclease Cas9 (CRISPR/Cas9) are being entrusted in improving the rice aroma content and quality. Researchers succeeded in editing *Badh2* gene and generating high yielding fragrance rice varieties by using TALENs [128] and CRISPR/CAS9 [129–131] technologies, which led to increased accumulation of 2AP.

7. Conclusion and future prospects

Aroma in rice is a key quality trait determining its acceptability and marketability. 2AP has gained major importance among other volatiles as the primary compound for aroma. Aroma compound is encoded by betaine aldehyde dehydrogenase 2 (*badh2*) gene also called fragrance (*fgr*) gene which is located on chromosome 8 and the level of aroma depends on this gene caused by mutation in *badh2* of 8 bp deletion and 3 SNPs. Apart from Basmati genotypes possessing long slender grains, only few other medium/short slender grain rice varieties possessing aroma are in the market. Those short/medium slender aromatic genotypes are not high yielding and possess several other disadvantages. Development of aromatic rice varieties possessing superior grain qualities through conventional or molecular breeding approaches takes considerable number of years and in some cases retaining the superior grain qualities of elite genotypes still remains a challenge. In recent years, genome editing technologies like TALENs, ZFNs and CRISPR/Cas9 has been employed in developing superior quality rice grains and it has opened new avenues for accelerated improvement of rice varieties thereby gaining competitive advantage in improving economy on national and global scale. These new technologies seems to be an attractive strategy to overcome the number of years required for developing desired genotypes and also to overcome the problems due to linkage drag. It will accelerate the cultivation of new aromatic rice varieties with high quality, yield and multiple resistance.

Conflict of interest

The authors declare no conflict of interest towards this chapter.

Author details

Nirubana Varatharajan[1*], Deepika Chandra Sekaran[2], Karthikeyan Murugan[1] and Vanniarajan Chockalingam[1]

1 Department of Plant Breeding and Genetics, Agricultural College and Research Institute, Tamil Nadu Agricultural University, Madurai, Tamil Nadu, India

2 Department of Genetics and Plant Breeding, Directorate of Centre for Plant Breeding and Genetics, Tamil Nadu Agricultural University, Coimbatore, Tamil Nadu, India

*Address all correspondence to: nirujayammal@gmail.com

IntechOpen

© 2021 The Author(s). Licensee IntechOpen. This chapter is distributed under the terms of the Creative Commons Attribution License (http://creativecommons.org/licenses/by/3.0), which permits unrestricted use, distribution, and reproduction in any medium, provided the original work is properly cited. CC BY

References

[1] Wakte K, Zanan R, Hinge V, Khandagale K, Nadaf A, Henry R. Thirty-three years of 2-acetyl-1-pyrroline, a principal basmati aroma compound in scented rice (*Oryza sativa* L.): A status review. Journal of the Science of Food and Agriculture. 2017;97(2):384-395. DOI: 10.1002/jsfa.7875

[2] Prodhan ZH, Faruq G, Rashid KA, Taha RM. Effects of temperature on volatile profile and aroma quality in rice. International Journal of Agriculture and Biology. 2017;19(5):1065-1072. DOI: 10.17957/IJAB/15.0385

[3] Kovach MJ, Calingacion MN, Fitzgerald MA, McCouch SR. The origin and evolution of fragrance in rice (*Oryza sativa* L.). Proceedings of the National Academy of Sciences of the United States of America. 2009;106(34):14444-14449. DOI: 10.1073/pnas.0904077106

[4] Pachauri V, Singh MK, Singh AK, Singh S, Shakeel NA, Singh VP, Singh NK. Origin and genetic diversity of aromatic rice varieties, molecular breeding and chemical and genetic basis of rice aroma. Journal of Plant Biochemistry and Biotechnology. 2010;19(2):127-143. DOI: 10.1007/BF03263333

[5] Huang N, McCouch SR, Mew T, Parco A, Guiderdoni E. A rapid technique for scent determination in rice. Rice Genetics Newsletter. 1994;11:134-137.

[6] Sood BG, Siddiq EAA rapid technique for scent determination in rice. Indian Journal of Genetics and Plant Breeding. 1978;38:268-271.

[7] Hinge VR, Patil HB, Nadaf AB. Aroma volatile analyses and 2AP characterization at various developmental stages in Basmati and non-Basmati scented rice (*Oryza sativa* L.) cultivars. Rice. 2016;9(1):38. DOI: 10.1186/s12284-016-0113-6

[8] IRGSP (International Rice Genome Sequencing Project), Sasaki T. The map-based sequence of the rice genome. Nature. 2005;436:793– 800.

[9] Lorieux M, Petrov M, Huang N, Guiderdoni E, Ghesquière A. Aroma in rice: Genetic analysis of a quantitative trait. Theoretical and Applied Genetics. 1996;93(7):1145-1151. DOI: 10.1007/BF00230138

[10] Amarawathi Y, Singh R, Singh AK, Singh VP, Mohapatra T, Sharma TR, Singh NK. Mapping of quantitative trait loci for Basmati quality traits in rice (*Oryza sativa* L.). Molecular Breeding. 2008;21(1):49-65. DOI: 10.1007/s11032-007-9108-8

[11] Verma DK, Srivastav PP. Extraction, Identification and Quantification Methods of Rice Aroma Compounds with Emphasis on 2-Acetyl-1-Pyrroline (2-AP) and Its Relationship with Rice Quality: A Comprehensive Review. Food Reviews International. 2020;1-52. DOI: 10.1080/87559129.2020.1720231

[12] Weber DJ, Rochelle R, Singh US. Chemistry and Biochemistry of Aroma in Scented Rice. In: Singh RK, Singh US, Khush GS. editors. Aromatic Rices. Oxford and IBH Publishing Co. Pvt. Ltd. New Delhi, India; 2000. p. 135-151.

[13] Singh RK, Singh US. Indigenous scented rices of India-farmers' perception and commitment. In: Proceedings of the International Conference on "Creativity, innovation, and networking at grass root level"; 11-14 January 1997; Hyderabad. p. 11-14.

[14] Verma DK, Srivastav PP. Extraction Technology for Rice Volatile Aroma Compounds. In: Meghwal M, Goyal MR,

editors. Food Engineering: Emerging Issues, Modeling, and Applications. Innovations in Agricultural and Biological Engineering. 2nd Volume. Apple Academic Press, USA; 2016. p. 246-284.

[15] Verma DK, Srivastav PP. Introduction to Rice Aroma, Flavor, and Fragrance. In: Science and technology of Aroma, Flavour and Fragrance in Rice. In: Verma DK, Srivastav PP, editors. Apple Academic Press, USA; 2018. p. 3-34.

[16] Verma DK, Mahato DK, Srivastav PP. Simultaneous Distillation Extraction (SDE): A Traditional Method for Extraction of Aroma Chemicals in Rice. In: Verma DK, Srivastav PP, editors. Science and technology of Aroma, Flavour and Fragrance in Rice. Apple Academic Press, USA; 2018a. p. 81-90.

[17] Verma DK, Mahato DK, Billoria S, Srivastav PP. Solid Phase Micro Extraction (SPME): A Modern Extraction Method for Rice Aroma Chemicals. In: Verma DK, Srivastav PP, editors. Science and technology of Aroma, Flavour and Fragrance in Rice. Apple Academic Press, USA; 2018b. p. 93-140.

[18] Buttery RG, Ling LC, Juliano BO. 2-Acetyl-1-pyrroline: an important aroma component of cooked rice. Chem Ind (Lond). 1982;12:958-9.

[19] Champagne ET. Rice aroma and flavor: A literature review. Cereal Chemistry. 2008;85:445-454. DOI: 10.1094/cchem-85-4-0445

[20] Yoshihashi TN, Huong TT, Inatomi H. Precursors of 2-acetyl-1-pyrroline, a potent flavor compound of an aromatic rice variety. Journal of the Science of Food and Agriculture. 2002;50:2001-2004. DOI: 10.1021/jf011268s

[21] Chen S H, Yang Y, Shi W W, Ji Q, He F, Zhang Z D, Cheng Z K, Liu X N, Xu M L. *Badh2*, encoding betaine aldehyde dehydrogenase, inhibits the biosynthesis of 2-acetyl-1-pyrroline, a major component in rice fragrance. Plant Cell. 2008;20(7):1850-1861. DOI: 10.1105/tpc.108.058917

[22] Bradbury L M T, Gillies S A, Brushett D J, Waters D L E, Henry R J. Inactivation of an aminoaldehyde dehydrogenase is responsible for fragrance in rice. Plant Molecular Biology. 2008;68(4-5):439-449. DOI: 10.1007/s11103-008-9381-x

[23] Huang T C, Teng C S, Chang J L, Chuang H S, Ho C T, Wu M L. Biosynthetic mechanism of 2-acetyl-1-pyrroline and its relationship with Δ^1-pyrroline-5-carboxylic acid and methylglyoxal in aromatic rice (*Oryza sativa* L.) callus. Journal of Agricultural and Food Chemistry. 2008; 56(16):7399-7404. DOI: 10.1021/jf8011739

[24] Prodhan Z H and Shu Q. Rice Aroma: A Natural Gift Comes with Price and the Way Forward. Rice Science. 2020;27(2):86-10. DOI: 10.1016/j.rsci.2020.01.001

[25] Wijerathna YMAM, Kottearachchi NS, Gimhani DR, Sirisena DN. Exploration of relationship between fragrant gene and growth performances of fragrant rice (*Oryza sativa* L.) seedlings under salinity stress. Journal of Experimental Biology and Agriculture Sciences. 2014; 2(1): 7-12.

[26] Huang ZL, Tang XR, Wang YL, Chen MJ, Zhao ZK, Duan MY, Pan SG. Effects of increasing aroma cultivation on aroma and grain yield of aromatic rice and their mechanism. Scientia Agricultura Sinica. 2012; 45(6): 1054-1065.

[27] Mo Z, Ashraf U, Tang Y, Li W, Pan S, Duan M, Tian H, Tang X. Nitrogen application at the booting

stage affects 2-acetyl-1-pyrroline, proline, and total nitrogen contents in aromatic rice. Chilean Journal of Agricultural Research. 2018; 78(2): 165-172.

[28] Yang S, Zou Y, Liang Y, Xia B, Liu S, Md I, Li D, Li Y, Chen L, Zeng Y, Liu L, Chen Y, Li P, Zhu J. Role of soil total nitrogen in aroma synthesis of traditional regional aromatic rice in China. Field Crops Research. 2012;125:151-160. DOI: 10.1016/j.fcr.2011.09.002

[29] Berner DK, Hoff BJ. Inheritance of scent in American long grain rice. Crop Science.1986;26(5):876-878. DOI: 10.2135/cropsci1986.0011183X002600050008x

[30] Ahn SN, Bollich CN, Tanksley SD. RFLP Tagging of a Gene for Aroma in Rice. Theoretical and Applied Genetics. 1992;84(7):825-828.

[31] Bollich CN, Rutger JN, Webb BD. Developments in rice research in the United States. International Rice Commission Newsletter. 1992;41:32-34.

[32] Cordeiro GM, Christopher MJ, Henry RJ, Reinke RF. Identification of microsatellite markers for fragrance in rice by analysis of the rice genome sequence. Molecular Breeding. 2002;9(4):245-250. DOI: 10.1023/A:1020350725667

[33] Jin QS, Waters D, Cordeiro GM, Henry RJ, Reinke RF. A single nucleotide polymorphism (SNP) marker linked to the fragrance gene in rice (*Oryza sativa* L.). Plant Science. 2003;165(2):359-364. DOI: 10.1016/S0168-9452(03)00195-X

[34] Kadam BS, Patankar VK. Inheritance of Aroma in Rice. Chronica Botanica. 1938;4:496-497.

[35] Dhulappanavar CV. Inheritance of scent in rice. Euphytica. 1976;25(1):659-662. DOI: 10.1007/BF00041603

[36] Geetha S, Nadu T. Inheritance of aroma in two rice crosses. International Rice Genetic analysis of a quantitative trait. Theoretical and Applied Genetics. 1994;93:1145-1151.

[37] Pinson SRM. Inheritance of Aroma in 6 Rice Cultivars. Crop Science. 1994;34(5):1151-1157. DOI: 10.2135/cropsci1994.0011183X003400050002x

[38] Vivekanandan P, Giridharan S. Inheritance of aroma and breadthwise grain expansion in Basmati and non-Basmati rices. International Rice Research Notes. 1994;19(2): 4-5.

[39] Jodon NE. The inheritance of flower fragrance and other character in rice. Journal of American Society of Agronomy. 1944;36:844-848.

[40] Kuo SM, Chou SY, Wang AZ, Tseng TH, Chueh FS, Yen He, Wang CS. The betaine aldehyde dehydrogenase (BAD2) gene is not responsible for the aroma trait of SA0420 rice mutant derived by sodium azide mutagenesis. In: 5th International Rice Genetics Symposium and 3rd International Rice Functional Genomics Symposium. International Rice Research Institute Press, Manila, Philippines. 2005.Vol.166.

[41] Ghose RLM, Butany WT. Studies on the inheritance of some characters in rice (*Oryza sativa* L.)'. Indian Journal of Genetics and Plant Breeding. 1952;12:26-30.

[42] Scod BC, Siddiq EA. Studies on component quality attributes of basmati rice, *Oryza sativa* L. Zeitschrift fuer Pflanzenzuechtung. 1980;84:299-301.

[43] Shekhar BP, Reddy GM. Genetic basis of aroma and flavour component studies in certain scented cultivars of rice. In IV International SABRO Congress. 1981.May 4.

[44] Reddy VD, Reddy GM. Genetic and biochemical basis of scent in rice (*Oryza*

sativa L.). Theoretical and Applied Genetics. 1987;73(5):699-700. DOI: 10.1007/BF00260778

[45] Song W, Chen Z, Zhang Y. Inheritance of aroma in autotetraploid and diploid rices. Report of the National Rice Research Institute. 1989. p. 277.

[46] Huang QH, Zou XY. Inheritance of aroma in two aromatic rice varieties. International Rice Research Newsletter. 1992;17:5-6.

[47] Ali SS, Jafri SJ, Khan MJ, Butt MA. Inheritance studies for aroma in two aromatic varieties of Pakistan. International Rice Research Newsletter. 1993.18:2-6.

[48] Wu AH, Gao LK, Cai XZ, Zhao Z, Chen DX. The genetic analysis of fragrance in Shangnong scented glutinous rice'. Journal of Shanghai Agricultural College. 1994;12(1):31-34.

[49] Katare NB, Jambhale ND. Inheritance of scent in rice. *Oryza*. 1995;32:193-194.

[50] Li X, Gu M, Cheng Z, Yu H. Chromosome location of a gene for aroma in rice. Chinese Journal of Rice Research. 1995;4(2): 5-6.

[51] Kato T, Itani T. Effects of the gene for scented grain in a rice cultivar BG1 on agronomic performance'. SABRO Journal (Japan). 1996;28(1):1-9.

[52] Tragoonrung S, Sheng JQ, Vanavichit A. Tagging an aromatic gene in lowland rice using bulk segregant analysis. Rice Genetics III. IRRI; 1996. p. 613-8. DOI: 10.1142/9789812814289_0073

[53] Li J, Ku D, Li L. Analysis of fragrance inheritance in scented rice variety, Shenxiangjing 4. Acta Agriculturae Shanghai. 1996;12(3):78-81.

[54] Li J, Gu D. 1997. Analysis of inheritance of scented rice variety Shenxiangjing 4 China Rice Research Newsletter. Rice Science. 1997;5(1):4-5.

[55] Sadhukhan RN, Roy K, Chattopadhyay P. Inheritance of aroma in two local aromatic rice cultivars. Environment and Ecology. 1997;15(2):315-317.

[56] Garland S, Lewin L, Blakeney A, Reinke R, Henry R. PCR-based molecular markers for the fragrance gene in rice (*Oryza sativa* L.). Theoretical and Applied Genetics. 2000;101(3):364-71.

[57] Dartey PK, Asante MD, Akromah R. Inheritance of aroma in two rice cultivars. Agricultural and Food Science Journal of Ghana. 2006;5:375-379.

[58] Lang NT, Buu BC. Development of PCR-based markers for aroma (*fgr*) gene in rice (*Oryza sativa* L.). Omonrice. 2008;16:16-23.

[59] Niu X, Tang W, Huang W, Ren G, Wang Q, Luo D, Xiao Y, Yang S, Wang F, Lu BR, Gao F. RNAi-directed downregulation of OsBADH2 results in aroma (2-acetyl-1-pyrroline) production in rice (*Oryza sativa* L.). BMC plant biology. 2008;8(1):1-0. DOI: 10.1186/1471-2229-8-100

[60] Sun SX, Gao FY, Lu XJ, Wu XJ, Wang XD, Ren GJ, Luo H. Genetic analysis and gene fine mapping of aroma in rice (*Oryza sativa* L. Cyperales, Poaceae). Genetics and Molecular Biology. 2008;31(2):532-8. DOI: 10.1590/S1415-47572008000300021

[61] Sarhadi WA, Hien NL, Zanjani M, Yosofzai W, Yoshihashi T, Hirata Y. Comparative analyses for aroma and agronomic traits of native rice cultivars from Central Asia. Journal of Crop Science and Biotechnology. 2008;11(1):17-22.

[62] Asante MD, Kovach MJ, Huang L, Harrington S, Dartey PK, Akromah R, Semon M, McCouch S. The genetic origin of fragrance in NERICA1. Molecular Breeding. 2010;26(3):419-24. DOI: 10.1007/s11032-009-9382-8

[63] Vazirzanjani M, Sarhadi WA, NWE JJ AM, Siranet R, Trung NQ. Characterization of aromatic rice cultivars from Iran and surrounding regions for aroma and agronomic traits. SABRAO Journal of Breeding and Genetics. 2011;43:15-26.

[64] Tsuzuki E, Shimokawa E. Inheritance of aroma in rice. Euphytica. 1990;46:157-159.

[65] Geetha S. Inheritance of aroma in two rice crosses. International Rice Research Notes. 1994;19(2):5.

[66] Hien N, Yoshihashi T, Sarhadi WA, Thanh V, Oikawa Y, Hirata Y. Evaluation of Aroma in Rice (*Oryza sativa* L.) using KOH Method, Molecular Markers and Measurement of 2-Acetyl-1-Pyrroline Concentration. Japanese Journal of Tropical Agriculture. 2006;50:190-198.

[67] Siddiq EA. Breeding for quality improvement in rice-Present state and strategy for future. In: Rice in West Bengal IV. Directorate of Agriculture, West Bengal. 1983. p. 73-95.

[68] Tripathi RS, Rao MJBK. Inheritance and linkage relationship of scent in rice'. Euphytica. 1979;28:319-323.

[69] Nagaraju M, Chowdhary D, Rao MJBK. A simple technique to identify scent in rice and inheritance pattern of scent. Current Science. 1975;44(16):599.

[70] Reddy PR, Sathyanarayanaiah K. Inheritance of aroma in rice. Indian Journal of Genetics and Plant Breeding. 1980;40(2):327-9.

[71] Nayak AR, Acharya US. Inheritance of scent in rice (*Oryza sativa* L.). Indian Journal of Genetics and Plant Breeding. 2004;64:59-60.

[72] Sarawagi AK, Bisne R. Inheritance of aroma in indigenous short grain aromatic rice cultivars. Journal of Rice Research. 2006;1:180-1.

[73] Chaut AT, Yutaka H, Vo CT. 2010. Genetic analysis for the Fragrance of Aromatic Rice Variete. In: 3rd International Rice Congress, November 8-12, Hanoi, Vietnam. p. 3898.

[74] Richharia RH, Mishra B, Kulkarni VA. Studies in the world genetic stock of rice. IV. Distribution of scented rice. Oryza. 1965;2:57-9.

[75] Kiani G. Marker aided selection for aroma in F2 populations of rice. African Journal of Biotechnology. 2011;10(71):15845-8.

[76] Bourgis F, Guyot R, Gherbi H, Tailliez E, Amabile I, Salse J, Lorieux M, Delseny M, Ghesquière A. Characterization of the major fragance gene from an aromatic japonica rice and analysis of its diversity in Asian cultivated rice. Theoretical and Applied Genetics. 2008;117(3):353-68.

[77] Chen S, Wu J, Yang Y, Shi W, Xu M. The fgr gene responsible for rice fragrance was restricted within 69 kb. Plant Science. 2006;171(4):505-14. DOI: 10.1016/j.plantsci.2006.05.013

[78] Shi W, Yang Y, Chen S, Xu M. Discovery of a new fragrance allele and the development of functional markers for the breeding of fragrant rice varieties. Molecular Breeding. 2008;22(2):185-92. DOI: 10.1007/s11032-008-9165-7

[79] Lanceras JC, Huang ZL, Naivikul O, Vanavichit A, Ruanjaichon V, Tragoonrung S. Mapping of genes for cooking and eating qualities in Thai jasmine rice (KDML105). DNA Research. 2000;7(2):93-101. DOI: 10.1093/dnares/7.2.93

[80] Bradbury LM, Fitzgerald TL, Henry RJ, Jin Q, Waters DL. The gene for fragrance in rice. Plant Biotechnology Journal. 2005;3(3): 363-70. DOI: 10.1111/j.1467-7652.2005.00131.x

[81] Yano M, Shimosaka E, Saito A, Nakagahra M. Linkage analysis of a gene for scent in indica rice variety, Surjamkhi, using restriction fragment length polymorphism markers. Japanese Journal of Breeding. 1992;41:338-339.

[82] Wanchana S, Kamolsukyunyong W, Ruengphayak S, Toojinda T, Tragoonrung S, Vanavichit A. A rapid construction of a physical contig across a 4.5 cM region for rice grain aroma facilitates marker enrichment for positional cloning. Science Asia. 2005;31:299-306.

[83] Li JH, Wang F, Liu WG, Jin SJ, Liu YB. Genetic analysis and mapping by SSR marker for fragrance gene in rice Yuefeng B. Molecular Plant Breeding. 2006;4:54-58.

[84] Bradbury LMT. Identification of the gene responsible for fragrance in rice and characterisation of the enzyme transcribed from this gene and its homologs. Southern Cross University ePublications@SCU Theses (2009).

[85] Yi M, Nwe KT, Vanavichit A, Chai-arree W, Toojinda T. Marker assisted backcross breeding to improve cooking quality traits in Myanmar rice cultivar Manawthukha. Field Crops Research. 2009;113:178-186.

[86] Vanavichit A, et al. in Proceedings of the 1st International Conference on Rice for the Future (Bangkok, Thailand) (2004).

[87] Vanavichit A, Yoshihashi T, Wanchana S, Areekit S, Saengsraku D, Kamolsukyunyong W, Lanceras J, Toojinda T, Tragoonrung S. Positional cloning of Os2AP, the aromatic gene controlling the biosynthetic switch of 2-acetyl-1-pyrroline and gamma aminobutyric acid (GABA) in rice. In 5th International Rice Genetics Symposium. Manila, Philippines, IRRI. 2005;44:19-23.

[88] Fitzgerald MA, Sackville Hamilton NR, Calingacion MN, Verhoeven HA, Butardo VM. Is there a second fragrance gene in rice?. Plant Biotechnology Journal. 2008;6(4): 416-423. DOI:10.1111/j.1467-7652.2008.00327.x

[89] Shao G, Tang S, Chen M, Wei X, He J, Luo J, Jiao G, Hu Y, Xie L, Hu P. Haplotype variation at *Badh2*, the gene determining fragrance in rice. Genomics. 2013;101(2):157-62. DOI: 10.1016/j.ygeno.2012.11.010

[90] He Q, Park YJ. Discovery of a novel fragrant allele and development of functional markers for fragrance in rice. Molecular breeding. 2015;35(11):1-10. DOI: 10.1007/s11032-015-0412-4

[91] He Q, Yu J, Kim TS, Cho YH, Lee YS, Park YJ. Resequencing reveals different domestication rate for BADH1 and BADH2 in rice (*Oryza sativa*). PLoS one. 2015;10(8):e0134801. DOI: 10.1371/journal.pone.0134801

[92] Pachauri V, Mishra V, Mishra P, Singh A K, Singh S, Singh R, Singh N K. 2014. Identification of candidate genes for rice grain aroma by combining QTL mapping and transcriptome profiling approaches. Cereal Research Communications. 42(3):376-388. DOI: 10.1556/crc.42.2014.3.2

[93] Talukdar PR, Rathi S, Pathak K, Chetia SK, Sarma RN. Population structure and marker-trait association in indigenous aromatic rice. Rice Science. 2017;24(3):145-54. DOI: 10.1016/j.rsci.2016.08.009

[94] Grimm CC, Champagne ET, Lloyd SW, Easson M, Condon B,

McClung A. Analysis of 2-Acetyl-l-Pyrroline in Rice by HSSE/GCIMS. Cereal Chemistry. 2011;88(3):271-277. DOI: 10.1094/CCHEM-09-10-0136.

[95] Chaintreau A. Simultaneous distillation-extraction: from birth to maturity-review. Flavour and Fragrance Journal. 2001;16(2):136-148.

[96] Schwambach SL, Peterson DG. Reduction of stale flavor development in low-heat skim milk powder via epicatechin addition. Journal of Agricultural and Food Chemistry. 2006;54:502-508.

[97] Zhang Z, Yang MJ, Pawliszyn J. Solid-phase Microextraction. Analytical Chemistry. 1994;66:844A–853A. DOI: 10.1021/ac00089a001

[98] Grimm C, Bergman CJ, Delgado JT, Bryant R. Screening for 2-acetyl-l-pyrroline in the Headspace of Rice Using SPME/GC/MS. Journal of Agricultural and Food Chemistry. 2001;49:245-249. DOI: 10.1021/jf0008902

[99] Cho S, Kays SJ. Aroma-active Compounds of Wild Rice (*Zizania palustris* L.). Food Research International. 2013;54:1463-1470. DOI: 10.1016/j.foodres.2013.09.042

[100] Yahya, FB. Extraction of aroma compound from pandan leaf and use of the compound to enhance rice flavour. PhD thesis submitted to School of Chemical Engineering, The University of Birmingham, Birmingham, UK. 2011.

[101] Duarte C, Moldao-Martins M, Gouveia AF, Costa SB, Leitao AE, Bernardo-Gil MG. Supercritical Fluid Extraction of Red Pepper (*Capsicum frutescens* L.). The Journal of Supercritical Fluids. 2004;30:155-161. DOI: 10.1016/j.supflu.2003.07.001

[102] Yahya F, Lu T, Santos RCD, Fryer PJ, Bakalis S. Supercritical carbondioxide and solvent extraction of 2-acetyl-1-pyrroline from pandan leaf: The effect of pretreatment. Journal of Supercritical Fluids. 2010;55(1):200-207.

[103] Routray W, Rayaguru K. 2-Acetyl-1-pyrroline: A key aroma component of aromatic rice and other food products. Food Reviews International. 2017;34(6):539-565. DOI: 10.1080/87559129.2017.1347672

[104] Lin, Pei-Ching. Comparison of Simultaneous Distillation and Extraction (SDE) and Headspace Solid Phase Microextraction (SPME) for Determination of Volatiles of Muscadine Grapes (*Vitis rotundifolia*). 2014.

[105] Takemitsu H, Amako M, Sako Y, Shibakusa K, Kita K, Kitamura S, Inui H. Analysis of Volatile Odor Components of Superheated Steam-cooked Rice with a Less Stale Flavor. Food Science and Technology Research. 2016;22(6):771-778. DOI:10.3136/fstr.22.771

[106] Delahunty CM, Eyres G, Dufour JP. Gas Chromatography–olfactometry. Journal of Separation Science. 2006;29(14):2107–2125. DOI: 10.1002/jssc.200500509

[107] Struve C, Christophersen C. Structural equilibrium and ring- chain tautomerism of aqueous solutions of 4-aminobutyraldehyde. Heterocycles. 2003;60(8):1907-1914.

[108] Ootsuka K, Takahashi I, Tanaka K, Itani T, Tabuchi H, Yoshihashi T, Ishikawa R. Genetic polymorphisms in Japanese fragrant landraces and novel fragrant allele domesticated in northern Japan. Breeding Science. 2014;64(2):115-124. DOI: 10.1270/jsbbs.64.115

[109] Chakraborty D, Deb D, Ray A. An analysis of variation of the aroma gene in rice (*Oryza sativa L. subsp.* indica Kato) landraces. Genetic Resources and Crop Evolution. 2016;63:953-959. DOI: 10.1007/s10722-016-0414-z

[110] Hashemi FSG, Rafii MY, Ismail MR, Mahmud TMM, Rahim HA, Asfaliza R, Latif MA. Biochemical, Genetic and Molecular Advances of Fragrance Characteristics in Rice. Critical Reviews in Plant Sciences. 2013;32(6):445-457. DOI: 10.1080/07352689.2013.807716

[111] Mo Z, Lei S, Ashraf U, Khan I, Li Y, Pan S, Tang X. Silicon fertilization modulates 2-acetyl-1-pyrroline content, yield formation and grain quality of aromatic rice. Journal of Cereal Science. 2017;75:17-24. DOI: 10.1016/j.jcs.2017.03.014

[112] Gay F, Maraval I, Roques S, Gunata Z, Boulanger R, Audebert A, Mestres C. Effect of salinity on yield and 2-acetyl-1-pyrroline content in the grains of three fragrant rice cultivars (*Oryza sativa* L.) in Camargue (France). Field Crops Research. 2010;117(1):154-160. DOI: 10.1016/j.fcr.2010.02.008

[113] Banerjee A, Ghosh P, Roychoudhury A. Salt acclimation differentially regulates the metabolites commonly involved in stress tolerance and aroma synthesis in indica rice cultivars. Plant Growth Regulation. 2019;88:87-97. DOI: 10.1007/s10725-019-00490-6

[114] Mo Z, Li W, Pan S, Fitzgerald TL, Xiao F, Tang Y. Shading during the grain filling period increases 2-acetyl-1-pyrroline content in fragrant rice. Rice. 2015;8: 9.

[115] Goufo P, Duan M, Wongpornchai S, Tang X. Some factors affecting the concentration of the aroma compound 2-acetyl-1-pyrroline in two fragrant rice cultivars grown in South China. Frontiers of Agriculture in China. 2010; 4(1): 1-9.

[116] Sriseadka T, Wongpornchai S, Kitsawatpaiboon P. Rapid method for quantitative analysis of the aroma impact compound, 2-acetyl-1- pyrroline, in fragrant rice using automated headspace gas chromatography. Journal of Agricultural and Food Chemistry. 2006;54: 8183-818.

[117] Zeng Z, Zhang H, Chen JY, Zhang T, Matsunaga R. Direct extraction of volatiles of rice during cooking using solid-phase microextraction. Cereal Chemistry. 2007;84:423-427.

[118] Xia Q, Li Y. Ultra-high pressure effects on color, volatile organic compounds and antioxidants of wholegrain brown rice (*Oryza sativa* L.) during storage: A comparative study with high-intensity ultrasound and germination pretreatments. Innovative Food Science & Emerging Technologies. 2018;45:390-400. DOI: 10.1016/j.ifset.2017.12.003

[119] Xia Q, Mei J, Yu W, Li Y. High hydrostatic pressure treatments enhance volatile components of pre-germinated brown rice revealed by aromatic fingerprinting based on HS-SPME/GC–MS and chemometric methods. Food Research International. 2017;91:103-114. DOI: 10.1016/j.foodres.2016.12.001

[120] Shi Y, Wang L, Fang Y, Wang H, Tao H, Pei F, Hu Q. A comprehensive analysis of aroma compounds and microstructure changes in brown rice during roasting process. LWT. 2018. DOI: 10.1016/j.lwt.2018.09.018

[121] Rocha-Villarreal V, Serna-Saldivar SO, García-Lara S. Effects of parboiling and other hydrothermal treatments on the physical, functional, and nutritional properties of rice and other cereals. Cereal Chemistry. 2018; 95(1):79-91. DOI: 10.1002/cche.10010

[122] Choi S, Seo HS, Lee KR, Lee SJ, Lee J. Effect of milling and long-term storage on volatiles of black rice (*Oryza sativa* L.) determined by headspace solid

phase microextraction with gas chromatography-mass spectrometry. Food chemistry. 2019;276:572-582.

[123] Mahmud MMC, Oh Y, Kim TH, Cho YH, Lee YS. Effects of milling on aromatics, lipophilic phytonutrients, and fatty acids in unprocessed white rice of scented rice cheonjihyang-1-se. food science and biotechnology. 2018;27(2):383-392.

[124] Guan B, Zhao J, Jin H. Determination of Rice Storage Time with Colorimetric Sensor Array. Food Analytical Methods. 2017;10:1054-1062. DOI: 10.1007/s12161-016-0664-6

[125] Tananuwong K, Lertsiri S. Changes in volatile aroma compounds of organic fragrant rice during storage under different conditions. Journal of the Science of Food and Agriculture, 2010;90(10):1590 – 1596.

[126] Yoshihashi T. Quantitative Analysis on 2-Acetyl-1-pyrroline of an Aromatic Rice by Stable Isotope Dilution Method and Model Studies on its Formation during Cooking. Journal of Food Science. 2002;67(2):619-622.

[127] Udomkun P, Innawong B, Niruntasuk K. The feasibility of using an electronic nose to identify adulteration of Pathumthani 1 in Khaw Dok Mali 105 rice during storage. Food Measure. 2018;12:515-2523. DOI: 10.1007/s11694-018-9868-3

[128] Shan Q, Zhang Y, Chen K, Zhang K, Gao C. Creation of fragrant rice by targeted knockout of the OsBADH 2 gene using TALEN technology. Plant biotechnology journal. 2015;13(6):791-800. DOI: 10.1111/pbi.12312

[129] Shao G, Xie L, Jiao G, Wei X, Sheng Z, Tang S, Hu P. CRISPR/CAS9-mediated editing of the fragrant gene Badh2 in rice. Chinese Journal of Rice Science. 2017;31(2):216-22. DOI: 10.16819/j.1001-7216.2017.6098

[130] Ashokkumar S, Jaganathan D, Ramanathan V, Rahman H, Palaniswamy R, Kambale R, Muthurajan R. Creation of novel alleles of fragrance gene OsBADH2 in rice through CRISPR/Cas9 mediated gene editing. PloS one. 2020;12:15(8): e0237018. DOI: 10.1371/journal.pone.0237018

[131] Usman B, Nawaz G, Zhao N, Liu Y, Li R. Generation of high yielding and fragrant rice (*Oryza sativa* L.) Lines by CRISPR/Cas9 targeted mutagenesis of three homoeologs of cytochrome P450 gene family and OsBADH2 and transcriptome and proteome profiling of revealed changes triggered by mutations. Plants. 2020;9(6):788. DOI: 10.3390/plants9060788

Chapter 3

Aromatic Rice of India: It's Types and Breeding Strategies

Aarti Sharma, Sandhya, Akanksha Srivastava, Snehanshu Singh, Subhash Mishra, Shiva Mohan, Chhavi, Akanksha Singh, Avinash Kumar Singh and Hemant Kumar Jaiswal

Abstract

The coalescence of organoleptic traits viz., pleasant aroma, cooked rice texture, and taste make aromatic rice unique and distinguished from non-aromatic rice. Aromatic rice is cultivated in every rice growing country; with each country has its own indigenous collection. International trade of rice is dominated by *Indica* (long grained), *Japonica* (short grained), aromatic rice (Basmati and Jasmine) and glutinous rice; amidst which, Basmati types from India and Pakistan; and Jasmine types from Thailand have phenomenal demand. In India all types of aromatic rice are cultivated based on Kernel length; short, medium, long and very long grained. Basmati varieties own the major market, while other types of aromatic rice besides Basmati are popular in local market only. The country inherits rich diversity of aromatic rice germplasm; with more than 300 different types, each of the rice growing states of India has its own locally popular aromatic rice varieties. India a country where two third of its population consume rice as part of their daily food; aromatic rice always remain their favorite. Basmati, by virtue of its excellent qualities it dominates both national and international market. Every year, Basmati ranks first in respect of foreign exchange earned from the export of agricultural products from India (APEDA). The phenomenal demand and export figures have augmented Basmati Breeding program. However, only few aromatic varieties are cultivated depending on their demand, and their breeding program is also limited. In India, Basmati has overshadowed other types of aromatic rice in market and in plant breeding programs too. Breeding for Basmati varieties is undertaken by prime agricultural institutions of India. The country regulates quality standards and development of Basmati varieties with the help of *Export of Basmati Rice (Quality Control and Inspection) Rules 2003*; Agricultural and Processed Food Products Export Development Authority (APEDA); and Basmati Export Development Foundation (BEDF). However, no such initiatives have been taken to promote the development of other aromatic rice varieties of India besides Basmati.

Keywords: aromatic rice, types of aromatic rice in India, basmati rice, basmati breeding program

1. Introduction

Rice is an important crop as half of the world's population depends on it. The year 1966 was declared as Year of Rice by FAO and again in 2002, United Nations General

Assembly declared 2004 as International Year of Rice. Milled rice is the third most produced crop after wheat and maize in world. Rice cultivation is source of employment to billions of people in rice cultivating countries. The global rice consumption is dominated by the countries in the Asia-Pacific region, like China, India, Indonesia, Bangladesh, and Vietnam. There are different types of rice cultivated around the world, but if we talk about global rice trade, then there are four types: long grained *Indica* (80%), short grained *Japonica* (15%), aromatic rice *viz*., Basmati and Jasmine (4%) [1] and glutinous rice (1%). Among different types of rice, aromatic rice occupies a very small group but they possess excellent quality traits of rice. However, the word aromatic literal meaning is pleasant or sweet smell, though the aromatic rice is more than aroma. They are known for soft cooked rice texture, sweet taste and tenderness which make them class apart and command premium price in comparison to non-aromatic type of rice. From Glaszmann [2] classification of rice types, aromatic rice falls under three groups; **Group I** (Jasmine, aromatic rice from South East Asia and China); **Group V** (Basmati); and **Group VI** (aromatic rice from Indonesia, Philippines, and China). Aromatic rice is cultivated in all rice growing countries; each having their own indigenous collection: like *Basmati* of India and Pakistan; *Dulhabhog* of Bangladesh; *Khao-Dawk-Mali* (*Jasmine*) from Thailand; *Azucena* and *Milfor* of Philippines; *Rodojolele* of Indonesia and *Sugandhi* of Myanmar [3] *Della* rice of USA. Major aromatic rice producing and exporting countries in the global market are India (Basmati types), Pakistan (Basmati types) and Thailand (Jasmine types). In this chapter we will learn about aromatic rice varieties of India it economic importance to India and breeding strategies used to develop these varieties.

2. Aromatic rice in India

India has nearly 300 indigenous collections of aromatic rice varieties. These varieties falls under different types based on kernel length: **Small grained (≤5.51 mm); Medium grained (5.51–6.60 mm); Long grained (6.61–7.51 mm); and Very Long grained (>7.51 mm)**. Most of the aromatic rice of India are short to long grained only few of them are Very long grained (Basmati types). In general, there are two types of aromatic rice: Basmati and non-Basmati types [4]. This differentiation is done on the basis of kernel dimensions (Kernel length, Kernel breadth, ratios of length/breadth; before and after cooking) as mentioned in **Table 1**.

2.1 Non-Basmati type of aromatic rice of India

This group mainly constitutes of small, medium and long grained types of aromatic rice. The center of diversity of non-Basmati types of aromatic rice of India is located in Himalayan foothills; Indian states of Uttar Pradesh and Bihar; and

Kernel dimensions	Basmati type	Non-Basmati type
Kernel length	6.4–7.6 mm	5.2–5.4 mm
Kernel length/breadth ratio before cooking	3.5–4.2	3.3–3.5
Kernel length/breadth ratio after cooking	4.9–5.6	2.95–3.8
Elongation ratio of kernel length	1.7–1.83	1.4–1.57
Elongation ratio of kernel breadth	1.26–1.33	1.31–1.61

Table 1.
Differentiation of Basmati and non-Basmati types.

States of India	Small grained (≤5.51 mm)	Medium grained (5.51–6.60 mm)	Long grained (6.61–7.51 mm)
Andhra Pradesh	—	Jeeragasamba	—
Assam	*Joha rice*	—	—
Bihar	Badshah Bhog, Deobhog, *Katarni*, Tulsi-Manjari, Shyam, Jeevan, Kanak Jeera, Mircha, Bramobhusi, Ranijawain, Karina, Tulsi Pasand, Dewatabhog	Gopal Bhog, Champaran Basmati (Lal), Champaran Basmati (Kali), Champaran Basmati (Bhini), Bhilahi Basmati, Amod, Abdul, Kesar, Sonachur	Baikani
Himachal Pradesh		Achhu, Begrui, Panarsa (local)	Baldhar Basmati, Madhumati, Mushkan, Seond Basmati
Kerela	*Jeeraksala, Gandhaksala*	—	—
Madhya Pradesh	Chinore, Dubrej, Kalimooch, Bishnubhog, Badshah Bhog, Tulsi-Manjari	Chatri, Modhuri, Vishnu Parag	Laloo
Maharashtra	Ambemohar, *Ajaraghansal*	Prabhavati	—
Manipur	—	*Chak Hao*	—
Uttar Pradesh	Adamchini, Badshah Pasand, Bindli, Bhartaphool, Dhania, Chhoti Chinnawar, Laungchoor, Jeerabattis, Kanak Jeeri, Yuvraj, Moongpholi, Rambhog, Ramjawain	Karmuhi, Kesar, Kesarparsom, Sonachur, Tilakchandan, Kalanamak, Vishnu Bhog	Type-3, Hansraj, Nagina-12, Safeda, Kalasukhdas, Tapovan Basmati, Type-9, Duniapat Dabraj
West Bengal	*Gobindbhog, Tulaipanji*	Kanakchur	—

Italicized and bold genotypes names indicate that they hold GI tag.

Table 2.
Indigenous aromatic varieties cultivated in different states of India.

Tarai region of Nepal [5]. Few famous and locally cultivated varieties of this group with respective kernel length are listed below (**Table 2**) along with their area of cultivation. There are a total of eight non-Basmati types of aromatic rice with GI tag in India; names of such varieties are mentioned in bold and italicized letters in the table (**Table 2**).

2.2 Basmati type of aromatic rice of India

This group includes slender and long to very long grained type of aromatic rice. It is indigenous to Himalayan foothills. The word Basmati is derived from Sanskrit word, *vas* (aroma) and *mayup* (ingrained or present from earlier). Morphologically Basmati rice is similar to *indica* type but differs from *indica* in phenol reaction and isoenzyme pattern [2]; opaque kernel appearance, intermediate amylose content and alkali spreading value. Consequently, Basmati is classified into intermediate group between *indica* and *japonica* [6].

Basmati is aromatic rice, but all aromatic rice is not Basmati. A rice variety to be called as Basmati rice has to meet all the minimum standards of Basmati rice quality traits. These minimum standards have been outlined by the recommendations of the *Central Sub Committee on Crops Standards, Notification and Release of Varieties for Agricultural Crops* constituted by the Central Seed Committee established under Section 3 of the Seeds Act, 1966 of India and *Export of Basmati Rice (Quality Control and Inspection) Rules, 2003*. These traits are mainly quality traits; and are summarized below.

Hence, any aromatic rice in India can be called as Basmati only when it meets the minimum standards given in aforementioned table (**Table 3**). Apart from, quality standards, there is another writ under *Export of Basmati Rice (Quality Control and Inspection) Rules, 2003* which defines which genotype of an aromatic rice can be called as Basmati. According to this rule; Basmati is of two types:

1. Traditional Basmati: these are pureline selection from the existed Basmati landraces which are six in number (**Table 4**).

2. Evolved Basmati: Evolved Basmati varieties are developed through hybridization or any other breeding methods in such way that at least one of the parents used to develop them was a Traditional Basmati Variety or pureline Basmati variety (**Table 4**).

Given the importance of Basmati; in year 2008 **APEDA**, an Indian government Organization has filed an application to obtain GI tag for Basmati. Basmati became a registered GI form 15th February 2016 under agricultural goods with its GI periphery confines to *seven* states of India; Jammu & Kashmir, Himachal Pradesh, Punjab, Haryana, Delhi, Uttrakhand, and Uttar Pradesh. Basmati Export Development Foundation (BEDF), is an organization founded by APEDA to promote Basmati export; regulate production of foundation and certified seeds; authorize centers for sample drawn by customs department; develop new DNA testing laboratories to monitor quality standards of newly notified Basmati rice in National Trials of India; and supervise registration of Basmati as GI product.

Traits	Value
Average pre-cooked milled rice length	>6.61 mm
Average pre-cooked milled rice breadth	<2 mm
Average length/breadth ratio of pre-cooked milled rice	>3.5
Average cooked rice length	>12 mm
Average pre-cooked rice length/pre-cooked rice length (elongation ratio)	>1.7
Average volume expansion ratio	>3.5
Amylose content range	20–25%
Alkali spreading value range	4–7
Minimum brown rice recovery	76%
Minimum milled rice recovery	65%
Minimum head rice recovery	45%
Aroma	Present
Texture of cooked grain (without surface bursting of cooked rice kernel), non-stickiness, tenderness, good taste and mouth feel	Present

Table 3.
Minimum standard of Basmati rice quality traits.

Traditional Basmati	Year of release	Pedigree
Basmati-217	1973	Pureline selection from local landraces of Punjab
Basmati-370	1976	Pureline selection from local landraces of Punjab (now in Pakistan)
Type-3	1978	Pureline selection from Dehraduni Basmati
Ranbir Basmati	1996	Pureline selection from Basmati-370
Taraori Basmati	1996	Pureline selection from Karnal local
Basmati-386	1997	Pureline selection from local landraces of Punjab
Evolved Basmati		Pedigree
Punjab Basmati-1	1984	Sona/Basmati370
Pusa Basmati-1	1989	Pusa150/Karnal Local
Kasturi	1989	Basmati370/CR 88-17-1-5
Haryana-1	1991	Sona/Basmati370
Mahi Sugandha	1995	BK-79/Basmati370
Improved Pusa Basmati-1	2007	Pusa Basmati-1
Pusa Basmati-1121	2008	Pusa 614-1-2/Pusa 614-2-4-3 (sister line of PB-1)
Vallabh Basmati-22	2009	—
Pusa Basmati-6	2010	PB-1/PB-1121
Punjab Basmati-2	2012	—
CSR-30	2012	Buraratha 4-10/Pak Basmati
HUBR10-9	2013	Taraori Basmati/Jaya
Vallabh Basmati-21	2013	—
Pusa Basmati-1509	2013	Pusa Basmati-1301/Pusa Basmati-1121
Basmati-564	2015	—
Vallabh Basmati-23	2015	—
Vallabh Basmati-24	2015	—
Pusa Basmati-1609	2015	PRR78/C101A51
Pant Basmati-1	2016	PB-1/IET-12603
Pant Basmati-2	2016	—
Punjab Basmati-3	2016	B-386/IET-17948/B-386
Pusa Basmati-1637	2016	MAS derived NIL from PB-1
Pusa Basmati-1728	2016	MAS derived NIL from PB-6
Pusa Basmati-1718	2019	PB 1121/SPS97//PB1121*IRBB59

Source: APEDA and http://drdpat.bih.nic.in.

Table 4.
Notified Basmati varieties of India.

2.3 Economic importance of aromatic rice in India

World trade of aromatic rice mostly includes Basmati and Jasmine types of varieties and India is leading exporter of Basmati in International market. In year 2019–2020, India has exported 4.45 million MT of Basmati to Iran, Saudi Arab, Iraq, UAE, Kuwait (major countries which import Basmati from India), US, UK,

Singapore and Malaysia; earning 4,330.68 million USD. Major export of Basmati from India is headed to Asian countries (Middle East) followed by Western Europe [7]. Besides, milled Basmati, parboiled Basmati called as Sella Basmati rice in India and Middle East; and Cooked Basmati in United Kingdom [8] is also exported from India. Nearly half of the exported Basmati to Gulf countries (Saudi Arabia, Kuwait, and UAE) as well as UK and USA, includes Sella types of Basmati [7]. Earlier, Traditional Basmati viz., Basmati-370 and Taraori Basmati dominated the export of Basmati from India. In early 1990s, an evolved Basmati variety PB-1 replaced them and ruled Basmati export. Currently, PB1121 is the major Basmati variety exported from India; which has an exceptional kernel length (approximately 9 mm) and elongation ratio of 2.7 [9]. Cooked kernels of PB1121 attain a maximum length of 21.0 mm to 21.5 mm which maximum known in any rice germplasm [9]. It occupies 47% of Basmati growing area, followed by PB 1509 (26%), PB-6 (9%) and PB-1 (8%); (APEDA, [10], Basmati Survey Report, *Kharif*, Volume 2).

3. Plant breeding methods used to develop aromatic varieties in India

In India, the systematic rice breeding program started with the establishment of agricultural organizations like ICAR (Indian Council of Agricultural Research, Delhi) in 1929; NRRI (National Rice Research Institute, Orissa) in 1946; Directorate of Rice Research; and Agricultural universities [3]. Aromatic rice breeding program was initiated at research stations: *Kala Shah Kaku* (Punjab state, now in Pakistan) and *Nagina* (Uttar Pradesh, India) [3] in 1920s. Further a separate program namely Basmati Variety Development Program was started at different research stations in India at Kaul, Kapurthala, Pantnagar and New Delhi, to develop new Basmati varieties by applying pureline selection in available germplasm, using dwarfening genes and hybridization techniques. These iniations, and diligence of plant breeders led to the development of few short, medium, and long grained aromatic varieties and a total of 30 notified Basmati varieties. PB-1718 is latest addition to this list; notified as Basmati variety in 2019 (APEDA [10], Basmati Crop Survey Report, *Kharif*, Volume 2).

Breeding for aromatic rice varieties is a complex task which is attributed to its quality traits. In a study Khush and Juliano [11] gave three reasons which adversely affect the aromatic rice breeding programs, 1) number of breeding objectives are more; 2) lack of equipment to measure grain quality and; 3) selection indices are not well defined. In present time the second problem has been overcome due to development of different equipments, software etc. to measure to the quality attributes of aromatic rice. Still breeding for aromatic rice is a complicated task, attributed to reasons outlined in the following paragraph; after reviewing work of famous scientists on aromatic rice:

1. Rice is staple food to half of world population, and in scenario of increasing population; increasing yield become the prime objective of any varietal development program; accordingly less emphasis is made on quality rice (aromatic rice).

2. Aromatic rice and *Indica* varieties belong to two different groups; hybridization between them is incompatible resulting into hybrid sterility [12].

3. Aromatic rice yield poor [6, 13–15]; photoperiod sensitive [14].

4. Environmental factors viz., climate, soil, temperature; and cultural practices affect the grain quality of aromatic rice [6, 15].

5. Aromatic rice grow and express quality traits best in their indigenous area only [6].

Some of the most common breeding methods practiced to develop aromatic rice varieties in India are listed below:

3.1 Pureline selection

Pureline selection is oldest breeding method used in development of new aromatic rice varieties. Breeding for Basmati rice started with pureline selection in 1920's at two research stations; Kala Shah Kaku (Punjab state, now in Pakistan) and Nagina (Uttar Pradesh, India) [3]. The very first Basmati variety Basmati-370 was developed through pureline selection in 1933, at Kala Shah Kaku Research station by Late Sardar Mohammad. Few other Basmati varieties developed at these two research station were Basmati-217, Type-3, Type-23, N-10-B, N-12, Muskan, Begumi and Hansraj [8]. Among these, Basmati-217, Type-3 and Basmati-370 are still recognized as Basmati variety in India. A list of aromatic varieties (other than Basmati) developed through pureline selection is given below (**Table 5**).

3.2 Hybridization

It is a very common breeding method utilized in development of a crop variety. Hybridization is a process in which crosses are made between two varieties of same species (inter-varietal hybridization); between two different species of same genus (inter-specific); between two different genera of same family (inter-generic). For self pollinated crop like rice, hybridization program is followed via Pedigree selection, Bulk Method, and Convergent Breeding to develop new varieties. Introduction of dwarfing gene and development of hybridization techniques in 1960s, augmented the Basmati development program Siddiq et al. [7] and other aromatic varieties too. Short, medium and long grained aromatic rice were developed at

New variety developed	From	Traits improved	State of India
3 new genotypes	Jeeraksala	Improved agronomic and yield potential	Kerala
14 new genotypes	Kalanamak	Improved agronomic and yield potential	Uttar Pradesh
C435	Jeerege Sanna	Early maturing	Karnataka
K441	Kakasali	Early maturing	Karnataka
DP33	Krishna Pasangi	Early maturing	Karnataka
Madhuri Selection A	Madhuri	Perform well in delayed planting conditions	Madhya Pradesh
N-10B	Hansraj	Better quality and high yielding	Uttar Pradesh
N-12	Safeda	Better quality and high yielding	Uttar Pradesh
Type-9	Dimnepet	Better quality and high yielding	Uttar Pradesh
Type-1	Ramjeevan	Better quality and high yielding	Uttar Pradesh
Type-23	Kalasukhdas	Better quality and high yielding	Uttar Pradesh
Sugandha	Pureline selection from local Basmati		Bihar

Source: http://drdpat.bih.nic.in.

Table 5.
Aromatic varieties developed through pureline selection.

Variety name	Kernel type	Parents
Kusuma (LS)	Long	TN-1/Basmati-370
PAU 29-295	Very long	Basmati-370/Hamsa
GR101	Very long	IR8/Pankhali 203
PNR-546	Long	PNR-125-2/PNR130-2
Narendra Sugandha Dhan NDR-6093	Long	NDR 637/Type-3
Ketkijoha	Medium	Savitri/Badsh abhog
Nua kalajeera	Short	Pureline selection for Kalajeera
Nua Dhusara	Medium	Pureline selection for Dhusara
Nua Chinikamini	Short	
CR Dhan 907	Medium	Dubraj/Pusa 44
CR Sugandh Dhan 908	Medium	Swarna/Geetanjali
CR Sugandh Dhan 909	Medium	Pankaj/Podum oni
CR Sugandh Dhan 910	Medium	Swarna/Geetanjali
Gangawati Ageti	Long	Gaurav x Kalinga III
HUBR-2-1	Long	HBR92/Pusa Basmati/Kasturi

Source: http://drdpat.bih.nic.in.

Table 6.
Aromatic varieties developed through hybridization in India.

different agricultural research Institution in India (**Table 6**). PB1 is the first, high yielding, and semi-dwarf Basmati variety, developed through convergent breeding method in 1989 [9].

Outstanding achievements of Hybrid rice breeding in China encouraged Indian plant breeders to employ hybrid breeding in aromatic rice too. In India, hybrid breeding in aromatic rice was initiated first in Basmati germpalsm. CMS lines like: **Pusa 3A** and **Pusa 4A**, were developed at IARI (Indian Agriculture Research Institute, New Delhi) from PB-1. Several high yielding varieties viz., Pusa Sugandha-2, Pusa Sugandha-3, and Pusa Sugandha-5 were developed using crosses of Pusa 3A and Haryana Basmati-1 [7]. Sugandha is a Hindi word meaning "Scented". In 2001, IARI developed first *hybrid aromatic* rice (PRH10) in world. PRH-10 was developed by crossing Pusa Sugandha with Pusa 6A (CMS line).

3.3 Mutation breeding

Mutation breeding is a useful method to produce genetic variability in crop. In mutation breeding, whole plant/plant part/seed are subjected to mutagen (physical or chemical). This method has been widely applied in developing new varieties in different crops including rice. In aromatic rice, mutation breeding is used to bring desirable change in quality traits. Many mutants line have been developed from several aromatic rice genotypes including Basmati (especially Basmati-370) but only few of them are cultivated. In India mutant lines have been developed in genotypes viz., Kalimoonch-6, Bindli, Kamal Local, Type-9, NP-49, T412, Kalanamak, Gobindbhog, Badshapasand and Basmati-370. Mutants showing certain desirable trait (dwarf stature, lodging resistance, early maturing) are conserved to be used in future breeding program. One such institute is NRRI, Cuttack, India which is maintaining more than 100 mutant lines having certain desirable traits of aromatic rice. Geetanjali and ADT 41 aromatic rice varieties were developed at NRRI; these are mutant lines of Basmati-370. A-201 aromatic variety of USA was developed by using PI457920 mutant; this mutant was developed from Basmati-370 of Pakistan.

3.4 Molecular breeding

In recent years, application of molecular breeding techniques has increased in field of plant breeding. Biotechnological tools viz., NGS, GWAS, MAS, and QTL mapping etc. have been utilized at larger scale in studies related to plant breeding. In India, only two aromatic varieties (Basmati type) have been developed by using molecular breeding method. **ImprovedPB-1** has been developed which is resistant to bacterial blight (*Xanthomonas oryzae* pv. *oryzae*). Two bacterial blight gene (Xa13 and Xa 21) has been introgressed in PB-1 through Marker Assisted Backcross Breeding from donor parent IRBB55 [16]. PB-1718 is developed through MAS from NIL of PB-1121; the variety posses bacterial blight resistance gene Xa13. However, genetic mapping [17], QTL mapping [18], whole genome wide SNP marker analysis [19] have been used to study aroma genes of aromatic rice and other quality traits too.

4. Conclusion

Aromatic rice is not only special food in India, but they are culturally auspicious too. Indian consumers prefer aromatic rice over non-aromatic rice. Basmati is premium quality of aromatic rice of India and it is mainly cultivated for export purposes. There is a huge demand of Basmati in International market, but in India, demand of aromatic rice is not limited only to Basmati; rather many non-Basmati indigenous varieties are cultivated, and are very popular among locals. A major portion of non-Basmati types of aromatic varieties had been gradually lost in course of time due to aftermath of Green Revolution in India. Unfortunately, those lost aromatic germplasm inherited enriched quality traits on par with traditional Basmati types. In context of present scenario of aromatic rice in India, there is a need to emphasize on the under-utilized non-Basmati varieties rather only aiming to develop more and more of Basmati types.

Author details

Aarti Sharma[1*], Sandhya[2], Akanksha Srivastava[3], Snehanshu Singh[4], Subhash Mishra[4], Shiva Mohan[5], Chhavi[2], Akanksha Singh[6], Avinash Kumar Singh[2] and Hemant Kumar Jaiswal[1]

1 Banaras Hindu University, Institute of Agricultural Sciences, Varanasi, Uttar Pradesh, India

2 ANDUAT, Kumarganj, Ayodhaya, Uttar Pradesh, India

3 Deputy Director Office, Agriculture Department, Varanasi, Uttar Pradesh, India

4 ICAR-IIWBR, Karnal, Haryana, India

5 Sardar Vallabhbhai Patel University of Agriculture and Technology, Meerut, Uttar Pradesh, India

6 Mangalayatan University, Aligarh, Uttar Pradesh, India

*Address all correspondence to: aartisharma4564@gmail.com

IntechOpen

© 2021 The Author(s). Licensee IntechOpen. This chapter is distributed under the terms of the Creative Commons Attribution License (http://creativecommons.org/licenses/by/3.0), which permits unrestricted use, distribution, and reproduction in any medium, provided the original work is properly cited.

References

[1] Richaria, R.H., Mishra, B., and Kulkarni, V.A. (1965). Studies in world genetic stock of rice. IV—Distribution of scented rices. Oryza 2:57-59.

[2] Glaszmann, J.C. (1987). Isozymes and classification of Asian rice varieties. Theortical and Applied Genetics 74:21-30.

[3] Ahuja, S.C., Panwar, D.V.S., Ahuja, U., and Gupta, K.R. (1995). Basmati Rice: The Scented Pearl. Hisar, Haryana, India: Directorate of Publications, CCS Haryana Agricultural University. pp. 1-61.

[4] Singh, R.K., Singh, U.S., Khush, G.S., Rohilla, R., Singh, J.P., Singh, G., and Shekhar, K.S. (2000). Small and Medium Grained Aromatic Rices of India. Oxford and IBH Publishing Co. Pvt. Ltd., New Delhi. pp. 155-178.

[5] Pachauri, V., Singh, M.K., Singh, A.K., Singh, S., Shakeel, N. A., and Singh, V.P. (2010). Origin and genetic diversity of aromatic rice varieties, molecular breeding and chemical and genetic basis of rice aroma. Journal of Plant Biochemistry and Biotechnology 19(2):127-143.

[6] Ahuja, U, Ahuja, S.C., Thakrar, R., and Rani, N.S. (2008). Scented rice of India. Asian Agri-History 12(4):267-283.

[7] Siddiq, E. A., Vemireddy, L. R., and Nagaraju, J. (2012). Basmati rices: Genetics, breeding and trade. Agricultural Research 1(1):25-36.

[8] Bhattacharjee, P., Singhal, R.S., and Kulkarni, P.R. (2002). Basmati rice: A review. International Journal of Food Science and Technology 37:1-12.

[9] Singh, A.K., Ellur, R.K., Gopala Krishna, S., Bhowmick, P.K., Nagarajan, M., Vinod K.K., Haritha, B., et al. (2018). Basmati rice variety Pusa Basmati 1718. Indian Journal of Genetics 78:151.

[10] Basmati Crop Survey Report (2019). Basmati Export Development Foundation, APEDA, New Delhi. Vol. 2. p. 11.

[11] Khush, G.S., and Juliano, B.O. (1991). Research priorities for improving rice grain quality. In: Rice Grain Marketing and Quality Issues. Manila, Philippines: IRRI. pp. 65-66.

[12] Mathure, S., Shaikh, A., Renuka, N., Wakte, K., Jawali, N., Thengane, R., and Nadaf, A. (2011). Characterization of aromatic rice (*Oryza sativa* L.) germplasm and correlation between their agronomic and quality traits. Euphytica 179:237.

[13] Bemer, D.K., and Hoff, B.J. (1986). Inheritance of scent in American long grain rice. Food Science and Technology 26:876-878.

[14] Prasad, G.S.V., Padmavathi, G., Suneetha, K., Madhav, M. S., and Muralidharan, K. (2020). Assessment of diversity of Indian aromatic rice germplasm collections for morphological, agronomical, quality traits and molecular characters to identify a core set for crop improvement. CABI Agriculture and Bioscience 1:13.

[15] Prodhan, Z.H., Faruq, G., Rashid, K. A., and Taha, R. M. (2017). Effects of temperature on volatile profile and aroma quality in rice. International Journal of Agriculture and Biology 19(5):1065-1072.

[16] Gopalakrishnan, S., Sharma, R.K., Anand Rajkumar, K., et al. (2008). Integrating marker assisted background analysis with foreground selection for identification of superior bacterial blight resistant recombinants in Basmati rice. Plant Breeding 127:131-139.

[17] Bradbury, L.M., Fitzgerald, T.L., Henry, R.J., Jin, Q., and Waters, D.L. (2005) The gene for fragrance in rice. Plant Biotechnology Journal 3:363.

[18] Prodhan, Z.H., and Qingyao, S. (2020). Rice aroma: A natural gift comes with price and the way forward. Rice Science 27(2):86-100.

[19] Kishor, D.S., Seo, J., Chin, J.H., and Koh, H.J. (2020). Evaluation of whole-genome sequence, genetic diversity, and agronomic traits of Basmati rice (*Oryza sativa* L.). Frontiers in Genetics 11:86.

Chapter 4

Abiotic Stress Tolerance in Rice: Insight in Climate Change Scenario

Manoj Kumar, Sandhya, Pawan Kumar,
Akash Gaurav Singh and Aravind Kumar Jukanti

Abstract

Rice, world's second major, staple cereal crop that feed the more than 50% of world's population. To safeguard the production and to fulfill the demand of ever the increasing population and urbanization there is need to increase the rice production. Though the rice yield has increased due to the development of modern technology and climate resilient high yielding cultivars but still it is 10–15 per cent lower than its potential yield due to various biotic and abiotic stress. Drought, extreme temperature, salinity, harmful radiation, heavy metals, gaseous pollutants are the most detrimental abiotic stresses factors that cause the morphological, physiological and biochemical changes in the rice crops and ultimate result is the reduction of rice production globally. Tolerance against these stresses through exploitation of potent biotechnological tools, molecular markers, QTL mapping omices approaches, phytohormones which could offer a more adequate and rapid solution to overcome these abiotic stresses and to enhance the ultimate grain yield of rice.

Keywords: Abiotic stress, QTL mapping, phytohormones, omices

1. Introduction

Rice (*Oryza sativa* L.) is the world's second-most-important cereal crop. Nearly half of the world's population utilizes it as a staple food. It belongs to the genus *Oryza* and family Poaceae, has 22 known species and has great economic importance [1]. These are adapted to variety of climatic conditions and can be grown both in dry and wetland habitats at high and low elevations. Green revolution substantially boosted rice productivity across Asia through a combination of modern new high yielding varieties and enhanced inputs like irrigation, fertilizer, and biocides [2]. Climate change is the only aspect that took away the breeder's concern from productivity despite plateauing yield in most of the cereals over the past decade. To feed the burgeoning population of the world, especially in Asia where the population is predicted to climb from 4.3 to 5.2 billion by 2050, rice production needs to expand and the quality of the grains must to be improved [3, 4]. Global warming and climate change has been predicted to decline irrigated rice yields by around 4% by 2020 and ~ 7% by 2050, while rainfed rice yields are expected to decrease by 6% and marginally (2.5%) by 2050, respectively [5]. Abiotic stress like drought,

IntechOpen

salinity, and heat is the dominant drivers restricting agricultural growth and output of crops around the planet. Rice plants are sensitive to various abiotic stresses. Drought stress disrupts not just morphological features in rice plants, but also physiological and biochemical processes. it has been linked to a significant drop in crop yields due to detrimental effects on plant growth, physiology, and reproduction. Research studies tend to show that abiotic stress in combination of abiotic stress factors is the most harmful [6]. In rice, drought stress is the major limiting factor for rice production in both rainfed lowland (46 Mha) and upland (10 Mha) rice ecosystems across Asia [7, 8]. Soil salinity is rising expotentially with increasing sea levels in coastal areas and in irrigated lands where soluble salts washed away underwater are brought to ground level. Almost 40 Mha of rainfed lowland under rice cultivation in South and South-East Asia including India, Bangladesh, Myanmar, and Thailand, have been ravaged by unforeseen flash floods [9, 10]. Salinity stress affects the growth and development of rice plants at three different stages; germination, vegetative, and reproductive [11]. A major problem for stable rice production is high temperature in a number of tropical and sub-tropical countries, such as India, Bangladesh, China, Pakistan, Thailand, Sudan and many African countries. For the production of rice, atmospheric temperature at anthesis is crucial. Even 38°C, which is as such not high in tropical and subtropical countries, might cause substantial yield reduction due to pollen sterility [12]. The rice has emerged as a model genomic crop in the 21st century with its smaller genome size, high-quality genomic reference sequence, large genetic and genomic resources compared to every other crop. In the last 25 years a wide range of abiotic stress tolerance loci (QTLs) have been identified in rice [13]. The development of genome-wide DNA markers, i.e., simple sequence repeats, single nucleotide polymorphisms, and the identification of QTLs and marker trait associations, have prompted to new technologies in génomic and genetic engineering tolerant to various abiotic stress. Advancements in molecular biology techniques have enabled the discovery of many genes involved in abiotic stress tolerance through spatio-temporal gene expression analysis. Transgenic approaches have further validated functionally the identified candidate genes from the genetic expression analysis. Omics approaches such as genomics, proteomics, metabolomics, transcriptomics, epigenomics have emerged as powerful biotechnological tools, used for deciphering the abiotic stress responses as well as for producing climate-resilient abiotic stress-tolerant plants [14].

At present efforts are being made to identify several stress factors in the abiotic stress tolerance and to develop rice varieties with a tolerable stress through biotechnology, molecular breeding, genomics, transcriptomics, proteomics and metabolomics [15–18]. Abiotic stresses such as drought, salinity, and heat affect the productivity of many agriculturally important crops. Therefore, to meet the food requirements of a growing world population, it is necessary to develop sustainable high-yielding varieties that can persist under abiotic stress [19]. This chapter highlights abiotic stresses such as drought, high temperatures, salinity as well as abiotic stress tolerance techniques in rice plants with emphasis on increased rice yields.

2. Major abiotic stresses affecting rice crop

Rice is the world's most significant food crop, providing calories to more than half of the world's population of 7 billion people. In most cultivable rice habitats, it has become increasingly vulnerable to losses induced by abiotic factors such as drought, floods, salt, heat, and cold. Drought or water shortages are the most damaging abiotic challenges for rice farming in rainfed habitats. Another important limitation affecting the rainfed lowlands is the submergence of rice plants for one to two weeks owing to flash floods. Long-duration cultivars are frequently affected by floods in the early phases of development, followed by drought during blooming, resulting in significant yield deficits [20]. Meanwhile, salinity, which is determined by a heavy concentration of soluble salts in the soil, is the second most common soil issue after drought, and it is a major constraint for rice production across the world. As a result of global climate change, heat stress is becoming a severe hazard to rice production. Heat stress hinders plant development, disrupts metabolic processes, and reduces output. Rice growing in temperate locations, as well as high-altitude conditions in tropical and subtropical zones, is hampered by low temperatures. Cold stress has a negative impact on rice crops throughout the germination, vegetative development, and reproductive phases, resulting in considerable production losses. Rice crops are sometimes subjected to numerous stressors (such as salt and drought, or drought followed by submergence), resulting in massive crop losses. Rice productivity would be significantly increased while water resources and soil quality were preserved if combined tolerance to several forms of abiotic stress was improved [21].

2.1 Drought stress

Drought is a severe abiotic stress that disrupts rice Morphological, Physiological, biochemical and molecular responces (**Figure 1**), resulting in considerable crop losses [22]. Drought stress has been linked to changes in plant length, biomass, and leaf area are associated with leaf senescence in a variety of crops, [23, 24] including rice (*O. sativa*) [25]. The occurrence of both drought and heat stresses in combination are more destructive (~70%) to crop production than other stresses occurring individually [26]. Drought stress causes oxidative stress by producing a buildup of reactive oxygen species in the chloroplast and mitochondria. Drought stress enhances the competitive ability of some weeds on crop plants by allowing them to use more water than crops [27]. The synergistic and antagonistic interactions between drought stress and pathogenic infection were similarly observed [28]. According to Wang et al. [29] drought before and after heading, has detrimental effects on brown and milled rice rates by influencing its quality to a great extent. It is the primary constraint to rice production in both rainfed lowland (46 Mha) and upland (10 Mha) rice ecosystems throughout Asia [7, 8]. Drought stress can alter tillering, floret initiation, and subsequent spikelet sterility, as well as grain filling, throughout vegetative growth, blooming, and the terminal period of rice cultivation [30]. Terminal drought is the most detrimental abiotic stress factor to rice grain yield [31]. Several studies on the effects of drought stress and dehydration revealed physiological acclimation of plants by altered antioxidant responses [32, 33]. Understanding the mechanisms that cause physiological responses to drought stress and dehydration conditions is critical. Drought tends to reduce the rate of cell division and expansion, leaf size, stem elongation, and root proliferation in rice crops, as well as interrupts stomatal opening and shutting periods, and plant nutrient and water absorption and utilization performance [34]. Deficiency of water and moisture in rice crops increases the rate of abscisic acid (ABA) biosynthesis which diminishes stomatal efficiency and conductance which reduces transpirational losses [35]. The complex nature of drought stress in rice and its strong interaction with the environment has slowed the breeding programs to develop drought adapted varieties.

2.2 Salt stress

By the end of 2050, the world's population will have risen to about nine billion people. On the other hand, due to the accumulation of high salt content in the soil,

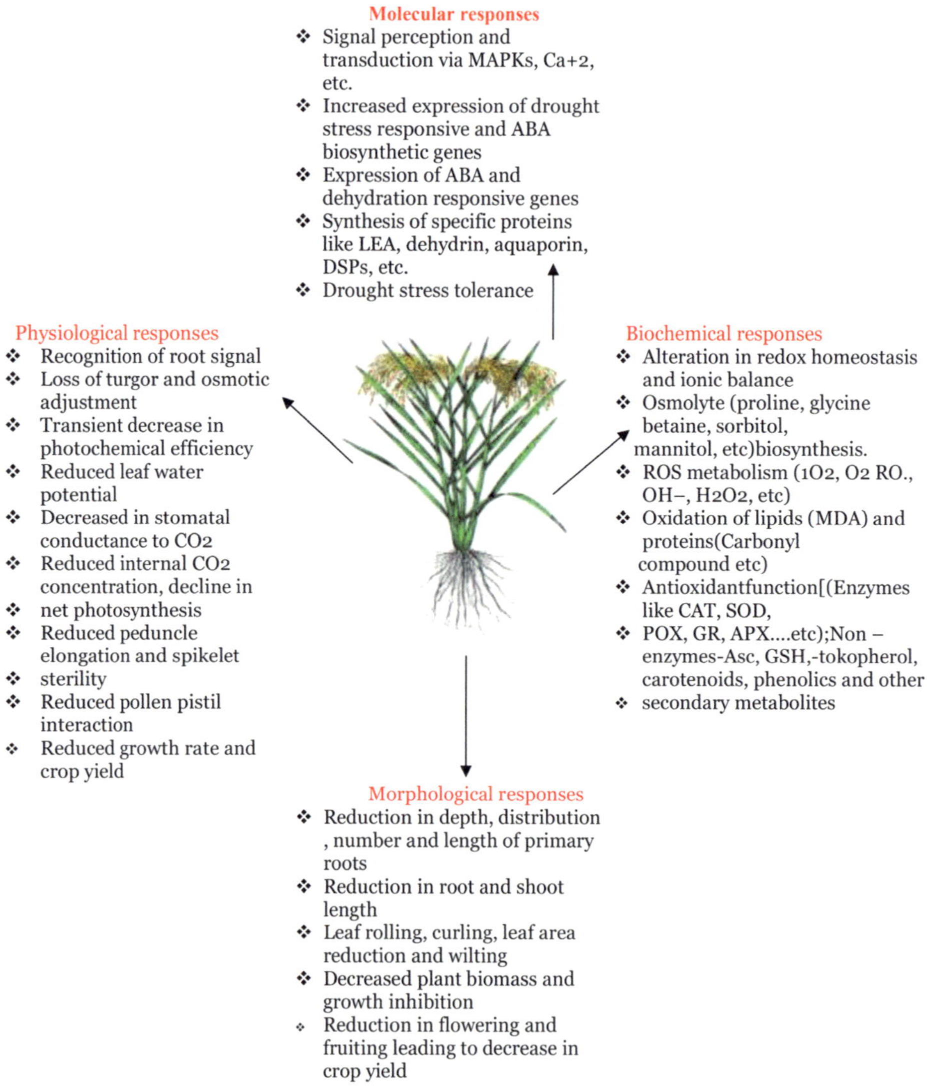

Figure 1.
Drought stress induces various morphological, physiological, biochemical, and molecular responses in rice.

worldwide agricultural production would almost probably remain static, resulting in crop growth inhibition and eventual crop mortality. Salt stress is a worrisome phenomenon because it diminishes soil agricultural productivity, leading to lower crop yields [36]. Salinity is becoming a critical threat because of diminishing irrigation water quality. (Flowers [37]. Salinity and water logging afflict 23% and 37% of worldwide cultivated land, respectively, and it is estimated that about 20% of all cultivated land and nearly half of irrigated land is salt-affected, greatly reducing yield well below the genetic potential [38]. The problem of salinity is particularly intense to agriculture in South and Southeast Asia, which produces about 90% of the world's rice [39]. By the middle of the twenty-first century, it is estimated that half of the cultivable land would be salt-affected [40]. It is suspected that the rise in soil salinity is due to poor irrigation water, its quality, and the use of brackish. Rice is a salinity-sensitive crop that performs poorly in soils with electrical conductivity (EC) as low as 3 dSm-1, however salty soil is commonly defined as EC >4 dSm-1 [41]. High-salt stress affects plants in several ways, such as ion toxicity, oxidative stress, alteration of metabolic processes, nutritional disorders, genotoxicity, membrane disorganization, reduction of cell division and expansion as well as water stress. In order to cope with the antagonistic effects of soil salinity some new strategies like remediation of salinized soils, to increase the salt tolerance of crop plants through traditional as well as molecular marker-assisted breeding techniques, and biotechnology should be adopted [37, 42, 43]. Khatun and Flowers [44] observed that salt toxicity adversely affect panicle length, spikelets per panicle, and 1000-grain weight in rice crop. Salinity also delays flowering and ripening and reduces the number of tillers, biomass, and leaf area in rice crops. According to Asch and Wopereis [45], irrespective of seasons and growth stages, salt toxicity reduces rice yield, the number of panicles, and grains and causes sterility in all rice cultivars.

2.3 Temperature stress

Heat stress, characterized by prevalence of high temperature is one of the major abiotic constraints for rice production, next only to drought and salinity [46]. Vastly increased greenhouse gases in the air are predicted to significantly affect the climate and worldwide average air temperature guesstimated to significantly raise by 1.4–5.8°C between 1990 and 2100 [47, 48]. Increasing temperature with this magnitude and severity reduced global rice production [49]. The reduction in rice yield is mainly attributable to changes in critical temperature at each specific phase of growth, namely, germination, seedling, rooting, leaf elongation, tillering, panicle initiation, primordia, panicle differentiation, anthesis and maturing [50]. One of the most sensitive phenological phases to extreme temperatures is pollination that leads to poor seed set and low grain quality [51, 52]. Rice is particularly vulnerable to heat stress during the reproductive and ripening stages, as simply a few hours of heat stress causes flowering plants to become sterile. High temperatures during ripening, on the other hand, might result in a decrease in milling quality and grain filling, resulting in reduced crop yield [53].

Increase in temperature, that leads to higher humidity, can cause spikelet sterility. Consequently, amid heat stress, the floral buds are unable to mobilize carbohydrates [53]. Low temperature stress is another similar environmental stress that can cause the plant's development and growth to be slowed. Embrane structure and function, protein synthesis, and cellular cytoskeleton structure can all be severely affected by low temperature stress. Low temperatures also impede photosynthesis in both light and dark reactions; further, electron routes are disrupted, resulting in the formation of free radical species that can be harmful to rice crops, causing membrane deterioration. Japonica genotypes are more adaptable than indicas to cold temperatures and so are prevailing in high-altitude and latitude ecologies [54]. The respiratory rate of plants increase or decrease in accordance with the temperature, short-term low temperature stress leads to high respiration rate but in the case of long-term stress the cell gets damaged and eventually dies due to the reduced respiration rate.

3. Plant responses to abiotic stress

The signals of abiotic stress are a multi-faceted phenomenon due to a wide range of environmental abuses. Plants can produce appropriate responses that cause a particular change in conjunction with a specific stress condition, whereas there is significant overlap between abiotic signals. Typically, one sort of stress happens with or is followed by other stresses. The loss of water which is due to heat stress causes drought stress and in this way, both stresses are linked to each other. Signal perception is the first phase in a signal transduction pathway which is tracked by the production of secondary signals. Secondary signals can trigger a protein phosphorylation cascade, which can then control the activation of specific transcription factors (TFs) or target genes. Additionally, these signals can modify the quantity of secondary signals; as a result, more signaling molecules are produced, providing an extra checkpoint for signals to flow in a given direction. Till date, many signaling pathways have been reported [55].

3.1 ROS signaling

Oxygen is a two-edged sword for plants, since it is a necessary element to be able to develop normal growth, but unavoidably promotes the formation of of ROS like hydrogen peroxide (H2O2), superoxide radical, hydroxyl radical, singlet oxygen, etc. as a result of aerobic metabolic activities, such as photosynthesis and respiration. During stress circumstances in a plant, organelles such as mitochondria, peroxisomes, and chloroplast generate enormous amounts of ROS, which become highly corrosive and reactionary to nucleic acids, proteins, and lipids, inevitably leading to apoptosis or cellular damage [56].

Catalase (CAT),monodehydroascorbate reductase (MDHAR), superoxide dismutase (SOD), glutathione peroxidase (GPX), glutathione reductase (GR), dehydroascorbate reductase (DHAR), peroxiredoxin, and glutathione S transferase are all ROS foraging enzymes in plants (GST) [57]. These antioxidants work in the plant cells where they are present in different locations in order to detoxify ROS. The ROS homeostasis in plants must also be accompanied by non-enzymatic antioxidants, including tocophherols, Carotenoids, GSH, flavonoids and AsA [56]. Many other mechanisms, such as photosynthetic apparatus rearrangement, leaf movement, and leaf curling, can balance a plant's energy absorption with CO2 supply, preventing ROS overproduction [57].

3.2 Phytohormone signaling

Phytohormones, a diverse set of signaling chemicals found in minute amounts inside cells, influence the responses. Plant growth, development, and nutrient allocation are all regulated by them. Responses to abiotic stress are triggered by a variety of stimuli, but phytohormones are responsible for the majority of them. Plants, which are sessile organisms, require phytohormones for survival [58]. They can carry out their function in their synthesis site or they can go to their action location [59]. Their family consist of cytokinin (CK), Indole acetic acid (IAA), salicylic acid (SA), ethylene, ABA, gibberellins (GAs), jasmonates, and brassinosteroids. The relatively newly discovered phytohormones are strigolactones. In regulating stress responses through interactions with some other hormones, ABA plays a major role. In order to regulate climate stress, it is the most exciting and indispensable hormone of the plant. In various phases of plant development, it plays a major role especially in opening and closing stomata, drought stress, seed germination and dormancy. PYR/PYL/RCAR-PP2C-SnRK2 is regarded as an ABA-generated signaling cascade, which effectively monitors seed dormancy even in occurrence of drought. ABA buildup amid drought stress restricts stomatal disclosure and modulates transpiration [60]. ABA signaling cascade consisted of 3 units, SnRK2/OST1 (Protein kinase), PP2C (protein phosphatases) and PYR/PYL/RCAR proteins have been investigated recently and their mechanism of operation was elucidated [61].

Ethylene is another major component of phytohormone. Ethylene is supposed to be the signaling path between plant growth and weather changes. Salinity, water logging, high temperature, frost, heavy metal interaction, nutritional inadequacy, and drought are examples of abiotic stressors that influence ethylene production [62]. Ethylene response factors (ERFs) in plants are members of a large family of transcription factors (TFs) that are activated in response to a variety of physiological and environmental stress.

3.3 Sos (salt overly sensitive) signaling

Salinity is the most important abiotic stress that diminishes rice crop productivity. Plants suffer from severe osmotic pressure and a scarcity of water as a result of salinity, which causes ions to accumulate in their tissues. Different rice varieties have different levels of tolerance [63]. IR29, for example, is a salt-sensitive species that accumulates 5–10 times more Na1 in the leaves than Pokkali or BK [64] . Salt tolerance in rice is achieved by limiting Na1 translocation. The gene SKC1/ HKT8 is responsible for salt tolerance and a strong K1/Na1 balance in the shoots, as well as decoding a HKT family Na1 selective transporter that regulates Na1 transport over vast distances [65]. HKT1 is a similar gene that is ubiquitously expressed in the roots and leaves and seems to be involved in the long-distance trafficking of Na1. The Na1/H1 counter transporter salt overly sensitive (SOS)1 can facilitate the outflow of Na1 across the plasma membrane [66]. Plants lacking SOS1 become particularly salt sensitive and their transfer of Na1 over long distances is hampered. The root tip epidermis and xylem parenchyma cells are the primary sites of SOS1 expression. At the root-soil interface, it extrudes too much Na1 from the root epidermal cells. SOS1 seems engaged in the Na1 translocation in the roots and shoots. It's also responsible for providing ions from the xylem to the shoots in a manageable way. The SOS2/ SOS3 kinase complex utilizes SOS1 as a substrate. Plants lacking SOS2 or SOS3 have the same salt sensitivity phenotype as SOS1 plants [64]. Ion homeostasis could be maintained as a result of this.

3.4 Calcium signaling

Calcium plays a role in a variety of abiotic stress symptoms, with varied degrees of directivity. Many of the functions in plants are governed by changes in intracellular calcium levels. Calcium is a harmful ion whose concentration in eukaryotic cells is controlled. Calcium ions travel through specific calcium ion channels, the cell membrane, or organelles into the cytosol after activation. Calcium ions have therefore been progressively increased to provide a number of final preparations for calcium-dependent proteins such as calmodulin, CDPCs and calmodulin-dependent phosphatases. Local calcium increases may also occur in specific organelles, such as chloroplasts, and can easily govern specific actions in organelles [67]. The expanded accurate determination of calcium as a result of abiotic stress, their association with downstream end progressions, and the use of calcium ion homeostasis inhibitors, however, highlight its importance.

4. Biotechnological approach for improving major abiotic stress tolerance in rice

4.1 Genetic engineering

The biotechnological approach is an appealing complement to traditional strategies for improving rice genotypes because it allows for the stacking of more genes into the genome without disrupting their genetic background [68]. Drought resistance was greatly improved by overexpressing SNAC1 (STRESS RESPONSIVE NAC 1) in rice, with 22–34 percent higher seed setting than control conditions in the field under acute drought stress during the reproductive stage, with no yield penalty or phenotypic alterations [36]. Similarly, under extreme field drought circumstances, overexpression of AP37 under the control of the OsCc1 promoter enhanced drought, salinity, and cold tolerance at the vegetative stage and also gave a 16–57 percent yield advantage over the control at the reproductive stage [69]. At the vegetative stage, overexpression of OsNAC10 with the GOS2 and RCc3 (root-specific) promoters improved drought tolerance, as well as high salinity and cold tolerance. RCc3:OsNAC10 transgenic rice cultivar showed yield advantages of 25–42 percent in the field under drought conditions [70]. OsPYL/RCAR5 (cytosolic ABA receptor) in rice plants functions as a positive inducer of abiotic stress-responsive genes [17, 18]. In contrast, rice plants exhibited a quick accumulation of soluble sugars, which act as interoperable solutes/osmoprotectants, lead to delays leaf drying and rolling [71]. Heat stress-induced gene expression and metabolite synthesis boost crop plant tolerance markedly [72]. HSFs function as molecular sensors to directly sense ROS such as H_2O_2 and control the expression of oxidative stress response genes during oxidative stress [73]. Binding of HSFs with heat shock elements (nGAAn) present in the heat responsive genes, including HSPs is critical for transcription induction of HSGs otherwise called heat shock response [74–76]. The enhanced expression of HSP70 assists in the translocation, proteolysis, translation, folding, aggregation, and refolding of denatured proteins [77]. HSP70 chaperones interact with a wide spectrum of proteins, particularly unfolded proteins generated in stressful situations [78]. Rice has 25 HSFs on 10 chromosomes other than chromosomes 11 and 12. Of these, 13 genes are class A, 8 are class B, and the remaining 4 are class C type HSFs [79]. Two HSBPs, namely OsHSBP1 and OsHSBP2, existing in rice plants and are abundantly expressed in all tissues under ordinary conditions, involved with HSR regulation, seed growth and found in considerably greater amount after heat shock recovery [80]. While considerable progress has been made in clarifying thermotolerance molecular systems, how plants perceive and translate heat stress signals is still not easy.

4.2 Marker-assisted breeding

Abiotic stress tolerance alleles were genetically eroded as a result of domestication and breeding for high yield. As a result, efforts are currently being conducted to restore allelic diversity for abiotic stress tolerance in modern high yielding varieties using locally adapted cultivars and germplasm. Stress sensitive genotypes/parents have contributed many advantageous alleles for abiotic stress tolerance, indicating the impact of a genotype's genetic background on its performance under stress [81]. A comprehensive screening and evaluation process, gene genetic background interaction, and gene environment interaction are all important factors in the utilization of QTLs in abiotic stress tolerance. The combination of whole genome expression data, QTL information, and meta-QTL analysis has proven to be a useful approach for narrowing down the search for abiotic stress tolerance candidate genes [82]. There are many success stories of introgression of QTLs for abiotic stress tolerance, and many varieties are in the advanced field trails stage [83] for tolerance to drought, salinity, and heat separately or in combination.

IRRI revealed the first important and persistent QTLs for grain yield under extreme drought stress [84]. Vikram et al [85] studied three populations: N22/IR64, N22/MTU1010, and N22/Swarna, and discovered a major consistent grain yield QTL, qDTY1.1, on chromosome 1 that can be used for marker-assisted breeding (MAB). Furthermore, in Vandana/IR64 populations, qDTY1.1 and the locus for plant height (sd1) were shown to be connected [86], suggesting that in large segregating populations, recombinant alleles with unlinked qDTY1.1 and sd1 could create drought-tolerant plants with shorter stature [87]. In Apo/Swarna, Apo/IR72, and Vandana/IR72 genetic backgrounds, another large QTL "qDTY6.1" [88] was found on chromosome 6, explaining 40–66 per cent of the genetic variation for grain yield in aerobic conditions. Swarna and IR72, both drought-prone, performed better in aerobic conditions when this QTL was present. This was also the first report of a significant QTL that increases yield and yield potential in aerobic circumstances. Nevertheless, this QTL had no effect on lowland drought stress conditions. Three grain yield QTLs under drought stress namely qDTY2.2, qDTY3.1, and qDTY12.1 were introgressed into high quality Malaysian rice cultivar MRQ74 by MAB [89]. An Indian project in collaboration with IRRI: "From QTL to variety: marker assisted breeding of abiotic stress tolerant rice varieties with major QTLs for drought, submergence and salt tolerance" has introgressed seven consistent QTLs for grain yield under drought into high yielding, submergence-tolerant elite backgrounds of Swarna-Sub1, Samba Mahsuri-Sub1, and IR64-Sub1 [83].

Saltol QTL is a key salt-tolerant QTL that has been widely exploited to create excellent rice cultivars around the world Lin et al. [90] used an F2 population resulting from a hybrid between "Nona Bokra" and "Koshihikari" to find multiple QTLs for Na_1 and K_1 absorption in shoots and roots, including a significant QTL responsible for SKC1 on chromosome 1. Ren et al. [63] cloned the SKC1 QTL, which maintains K_1 homeostasis in salt-tolerant cultivars under salt stress, and the SKC1 gene, which is a member of the HKT-type transporters and corresponds to the OsHKT8/Os01g0307500 locus. Using F_2 mapping populations, Zhou et al. [91] and Deng et al. [92] mapped QTLs qSKC-1 and qSNC-1 for SKC and SNC, respectively, between SSR markers RM283 and RM312. Deng et al. [93] used rice salt-tolerant 1 (rst1) mutant and showed that rst1 was controlled by a single recessive gene and QTL mapping between rst13Peiai 64 revealed the QTL loci on chromosome 6. Bizimana et al. [94] identified QTLs using RILs derived from IR29 (a salt-sensitive line) and Hasawi (a salt-tolerant line) and could not find Saltol or QTLs nearby this position indicating that tolerance in Hasawi is due to novel QTLs other than Saltol/SKC1. Emon et al. [95] and Kumar et al. [96] used association panel following a genome-wide association study approach to find marker-trait associations for salt stress tolerance. Kumar et al. [96] discovered 20 SNPs (loci) that were strongly related with Na1/K1 ratio at the reproductive stage, as well as the Saltol region, which is known to affect salt tolerance at the seedling stage. Many notable examples of transferring the Saltol QTL into elite rice varieties by MABC include PB1121 and PB6 [97], AS996 [98, 99], Bac Thom 7 [100, 101], Binadhan-7 [102], BRRI Dhan [103].

4.3 Omics approaches

Technological advancement in the omics area, the intrinsic genes for complicated abiotic stress in plants might be elucidated [14, 104]. Since high-strength omics approaches produced huge numbers of data, requiring both computer tools and storage resources, and data analysis, several online databases, servers and platforms were developed [105]. Proteomics and metabolomics have been shown to grow rapidly, allowing researchers to get extensive and accurate information on plant cell produced proteins and metabolites in response to environmental concerns [14, 106]. Both these emerging areas are highly expected to improve cereal crops. Similarly, profiling transcriptomics is extremely useful in ensuring a thorough understanding of regulatory molecules and their networks that are important to the communication of stress tolerance [106]. For illustrate, in order to learn more about regulatory processes and identify stress-responsive transcripts, researchers compared transcripts from tolerant and sensitive rice cultivars [107].

Despite significant improvements in high-throughput genotyping, phenotyping of complex abiotic stress responses (sometimes multigenic) remains a difficult task for molecular breeders [108]. Plants' epigenetic regulators have emerged as important regulatory mechanisms for responding to and inducing tolerance to abiotic stressors [109]. Epigenetic modulation of plant abiotic stress responses has been revealed thanks to breakthroughs in epigenomics. Short non-coding RNAs, such as miRNA, have emerged as critical epigenetic regulators of plant responses to stress [109]. However, more research is needed into how key crops, including rice, respond to abiotic stress, particularly at the epigenetic level. Overall, multiple omics techniques provide good platforms for understanding insights into plant responses and adaptation mechanisms, as well as developing abiotic stress tolerant, smart crops.

5. Perspectives and conclusion

Considering the massive losses of crop production due to severe environmental stresses, the development of crop varieties with increased tolerance or resistance to multiple stresses is presently indispensable. To date, relatively few genes reported to react to abiotic stress because agronomic characteristics of these stress tolerances have been controlled by many QTL, which show low inheritance and substantial interactions between the G/E systems. The discovery of possible genes for sustaining high pollen viability, effective gamete fertilization, and seed development in harsh conditions is not far off. With the progress of high-throughput techniques, many genes which are involved in stress regulation have been identified [110]. Plants that are subjected to several abiotic challenges at the same time must be investigated in order to comprehend the impact of various stresses. Every new combination of stresses has been suggested as a special type of stress, because it generates a totally new appropriate response. Enhanced and implementing tolerance mechanisms with the use of available low-cost sequencing and genotyping platforms, genetic and genomic resources and transgenic approaches provide huge opportunities for better rice cultivars in the near-coming future. Breeding and marker assistant selection, as well as modifying stress responses via plant hormones, are all ways that can be used to manage abiotic stress responses. Plant functional genomics perspectives which including proteomics, transcriptomic, and metabolomic analysis, as well as other high-throughput approaches and technologies, have yielded a number of drought-regulated genes, transcription factors, and cellular signaling components whose functions are crucial in rice stress tolerance.

Author details

Manoj Kumar[1*], Sandhya[1], Pawan Kumar[2], Akash Gaurav Singh[3] and Aravind Kumar Jukanti[4]

1 Agriculture University Kota, Rajasthan, India

2 ICAR-Central Institute for Arid Horticulture Bikaner, Rajasthan, India

3 Cereal Lab, IARI, New Delhi, India

4 ICAR-Indian Institute of Rice Research, Hyderabad, India

*Address all correspondence to: manoj.vishnoi108@gmail.com

IntechOpen

© 2021 The Author(s). Licensee IntechOpen. This chapter is distributed under the terms of the Creative Commons Attribution License (http://creativecommons.org/licenses/by/3.0), which permits unrestricted use, distribution, and reproduction in any medium, provided the original work is properly cited.

References

[1] Bajaj, S., Mohanty, A., 2005. Recent advances in rice biotechnology towards genetically superior transgenic rice. Plant Biotechnol. J. 3, 275_307.

[2] Hossain, M., Fischer, K.S., 1995. Rice research for food security and sustainable agricultural development in Asia: achievements and future challenges. Geo. J. 3, 286_298.

[3] Fahad, S., Hussain, S., Saud, S., Hassan, S., Chauhan, B.S., Khan, F., 2016a. Responses of rapid viscoanalyzer profile and other rice grain qualities to exogenously applied plant growth regulators under high day and high night temperatures. PLoS one 11 (7). Available from: https://doi.org/10.1371/journal.pone.0159590.

[4] Fahad, S., Hussain, S., Saud, S., Khan, F., Hassan Jr, S., et al., 2016b. Exogenously applied plant growth regulators affect heat-stressed rice pollens. J. Agron. Crop Sci. 202, 139_150

[5] Kumar, S.N., 2017. Improving crop adaptations to climate change: Contextualizing the strategy. In: Minhas, P., Rane, J., Pasala, R. (Eds.), Abiotic Stress Management for Resilient Agriculture. Springer, Singapore.

[6] Ron, M. (2016). Abiotic stress, the field environment and stress combination. Trends in Plant Science 11(1): 15-19.

[7] Pandey, S., Bhandari, H. and Hardy, B. (2007a). Economics Costs of Drought and Rice Farmers' Coping Mechanisms: A Cross-Country Comparative Analysis" 203 Pp International Rice Research Institute, Manila.

[8] Pandey, S., Bhandari, H., Ding, S., et al., 2007b. Coping with drought in rice farming in Asia: Insights from a cross-country comparative study. Agric. Econ. 37, 213_224.

[9] Sarkar, R.K., Bhattacharjee, B., 2011. Rice genotypes with SUB1 QTL differ in submergence tolerance, elongation ability during submergence and re-generation growth at re-emergence. Rice 5, 7. Available from: https://doi.org/10.1007/s12284-011-9065-z.

[10] Septiningsih, E.M., Pamplona, A.M., Sanchez, D.L., Neeraja, C.N., Vergara, G.V., Heuer, S., et al., 2009. Development of submergence tolerant rice cultivars: The Sub1 locus and beyond. Ann. Bot. 103, 151_160.

[11] Fujino, K., Sekiguchi, H., Sato, T., Kiuchi, H., Nonoue, Y., Takeuchi, Y., et al., 2004. Mapping of quantitative trait loci controlling low-temperature germinability in rice (Oryza sativa L.). Theor. Appl. Genet. 108, 794_799.Gill, S.S., Tuteja, N., 2010. Reactive oxygen species and antioxidant machinery in abiotic stress tolerance in crop plants. Plant Physiol Biochem. 48 (12), 909_930.

[12] Nachimuthu, V.V., Sabariappan, R., Muthurajan, R., Kumar, A., 2017. Breeding rice varieties for abiotic stress tolerance: Challenges and opportunities. In: Minhas, P., Rane, J., Pasala, R. (Eds.), abiotic stress Management for Resilient Agriculture. Springer, Singapore. Available from: https://doi.org/10.1007/978- 981-10-5744-1_15.

[13] Dixit, S., Yadaw, R.B., Mishra, K.K., Kumar, A., 2017. Marker-assisted breeding to develop the droughttolerant version of Sabitri, a popular variety from Nepal. Euphytica 213, 184. Available from: https://doi. Org/10.1007/s10681-017-1976-3.

[14] Jha, U.C., Bohra, A., Parida, S.K., Jha, R., 2017. Integrated "omics" approaches to sustain global productivity of major grain legumes under heat stress. Plant Breed. 136, 437_459. Available from: https://doi.org/10.1111/pbr.12489.

[15] Ansari, M.R., Shaheen, T., Bukhari, S.A., Husnain, T., 2015. Genetic improvement of rice for biotic and abiotic stress tolerance. Turk. J. Bot. 39, 911_919.

[16] De Leon, T.B., Linscombe, S., Subudhi, P.K., 2017. Identification and validation of QTLs for seedling salinity tolerance in introgression lines of a salt tolerant rice landrace _Pokkali'. PLoS ONE 12, e0175361. Available from: https://doi.org/10.1371/journal.pone.0175361.

[17] Kim, H., Lee, K., Hwang, H., Bhatnagar, N., Kim, D.Y., Yoon, I.S., et al., 2014a. Overexpression of PYL5 in rice enhances drought tolerance, inhibits growth, and modulates gene expression. J. Exp. Bot. 65 (2), 453_464.

[18] Kim, S.T., Kim, S.G., Agrawal, G.K., Kikuchi, S., Rakwa, R., 2014b. Rice proteomics: A model system for crop improvement and food security. Proteomics 14, 593_610.

[19] Leonforte, A., Sudheesh, S., Cogan, N.O.I., Salisbury, P.A., Nicolas, M.E., Materne, M., et al., 2013. SNP marker discovery, linkage map construction and identification of QTLs for enhanced salinity tolerance in field pea (Pisum sativum L.). BMC Plant biol. 13, 161.

[20] Yang, X.L., Wang, B.F., Chen, L., Li, P., Cao, C.G. 2019. The different influences of drought stress at the flowering stage on rice physiological traits, grain yield, and quality. Sci. Rep. 9, 3742.

[21] Raza, A., Razzaq, A., Mehmood, S.S., Zou, X., Zhang, X., Lv, Y., Xu, J. 2019 Impact of climate change on crops adaptation and strategies to tackle its outcome: A review. Plants, 8, 34.

[22] Todaka, D., Zhao, Y., Yoshida, T., Kudo, M., Kidokoro, S., Mizoi, J., et al., 2017. Temporal and spatial changes in gene expression, metabolite accumulation and phytohormone content in rice seedlings grown under drought stress conditions. Plant J. 90, 61_78.

[23] Upadhyaya, H., Shome, S., Tewari, S., Bhattacharya, M.K., Panda., S.K., 2016. Zinc nanoparticles induced comparative growth responses in rice (*Oryza sativa* L.) cultivars. In: Paul, S., Tewari, S. (Eds.), Frontiers of Research in Physical Sciences. pp. 71_77.

[24] Upadhyaya, H., Roy, H., Shome, S., Tewari, S., Bhattacharya, M.K., Panda, S.K., 2017. Physiological impactof Zinc nanoparticle on germination of rice (Oryza sativa L) seed. J. Plant Sci. Phytopathol. 1, 62_70.

[25] Kadam, N., Tamilselvan, A., Lawas, L.M.F., Quinones, C., Bahuguna, R., Thomson, M.J., et al., 2017. Genetic control of plasticity in root morphology and anatomy of rice in response to water-deficit. Plant Physiol. 174, 2302_2315.

[26] Prasad, P.V.V., Pisipati, S.R., Momcilovic, I., Ristic, Z., 2011. Independent and combined effects of high temperature and drought stress during grain filling on plant yield and chloroplast EF-Tu expression in spring wheat. J. Agron. Crop Sci. 197, 430_441.

[27] Valerio, M., Lovelli, S., Perniola, M., Di Tommaso, T., Ziska, L., 2013. The role of water availability on weed_crop interactions in processing tomato for southern Italy. Acta Agric. Scand. B soil Plant Sci. 63, 62_68.

[28] Sharma, M., Pande, S., 2013. Unravelling effects of temperature and soil moisture stress response on development of dry root rot [Rhizoctonia bataticola (Taub.)] Butler in chickpea. Am. J. Plant Sci. 4, 584_589.

[29] Wang, C., Wang, B., Zhang, W., 2007. Effects of drought stress at different growth stages on grain yield

and milling quality of rice. J. Rice Sci. 21, 643_649.

[30] Todaka, D., Shinozaki, K., Yamaguchi-Shinozaki, K., 2015. Recent advances in the dissection of droughtstress regulatory networks and strategies for development of drought-tolerant transgenic rice plants. Front. Plant Sci. 6, 84. Available from: https://doi.org/10.3389/fpls.2015.00084.

[31] Xangsayasane, P., Jongdee, B., Pantuwan, G., Fukai, S., Mitchell, J.H., 2014. Genotypic 10 performance under intermittent and terminal drought screening in rainfed lowland rice. Field Crops Res. 156, 281_292.

[32] Auler, P.A., do Amaral, M.N., dos Santos Rodrigues, G., Benitez, L.C., da Maia, L.C., Souza, G.M., et al., 2017. Molecular responses to recurrent drought in two contrasting rice genotypes. Planta 246, 899_914.

[33] Basu, S., Roychoudhury, A., Saha, P.P., Sengupta, D.N., 2010. Differential antioxidative responses of indica rice cultivars to drought stress. Plant growth Regul. 60 (1), 51_59.

[34] Farooq, M., Wahid, A., Kobayashi, N., Fujita, D., Basra, S.M., 2009. Plant Drought Stress: Effects, Mechanisms and Management. In: Sustainable Agriculture Springer, Dordrecht, pp. 153-188.

[35] Yamaguchi-Shinozaki, K., Shinozaki, K., 2005. Organization of cis-acting regulatory elements in osmotic- and cold-stress-responsive promoters. Trends plant Sci. 10, 88_94. Available from: https://doi. org/10.1016/j. tplants.2004.12.012.

[36] Hu, H., Dai, M., Yao, J., Xiao, B., Li, X., Zhang, Q., et al., 2006. Overexpressing a NAM, ATAF, and CUC (NAC) Transcription Factor Enhances Drought Resistance and Salt Tolerance in Rice. Proc. Natl. Acad. Sci. U.

[37] Flowers, T.J., 2004. Improving crop salt tolerance. J. Exp. Bot. 55 (396), 307_319.

[38] Munns, R., 2002. Comparative physiology of salt and water stress. Plant Cell Environ. 25 (2), 239_250.

[39] Aslam, M., Qureshi, R.H., Ahmed, N., 1993. A rapid screening technique for salt tolerance in rice (Oryza sativa L.). Plant Soil. 150, 99_107.

[40] Kamran, M., Parveen, A., Ahmar, S., Malik, Z.,Hussain, S., Chattha, M.S., Saleem, M.H., Adil, M., Heidari, P.,Chen, J.T. **2019.** An overview of hazardous impacts of soil salinity in crops, tolerance mechanisms, and amelioration through selenium supplementation. Int. J. Mol. Sci., 21, 148.

[41] Hoang, T.M.L., Tran, T.N., Nguyen, T.K.T., Williams, B., Wurm, P., Bellairs, S., et al., 2016. Improvement of salinity stress tolerance in rice: Challenges and opportunities. Agronomy 6, 54. Available from: https://doi. Org/10.3390/agronomy6040054.

[42] Borsani, O., Valpuesta, V., Botella, M.A., 2003. Developing salt tolerant plants in a new century: A molecular biology approach. Plant Cell Tissue Organ Cult. 73, 101_115.

[43] Hasegawa, P.M., Bressan, R.A., Zhu, J.K., Bohnert, H.J., 2000. Plant cellular and molecular responses to high salinity. Annu. Rev. plant Physiol. Plant Mol. Biol. 51, 463_499.

[44] Khatun, S., Flowers, T.J., 1995. Effects of salinity on seed set in rice. Plant Cell Environ. 18, 61_67.

[45] Asch, F., Wopereis, M.C.S., 2001. Responses of field grown irrigated rice cultivars to varying levels of loodwater salinity in a semi-arid environment. Field Crop Res. 70, 127_137.

[46] Zafar, S.A., Hameed, A., Nawaz, M.A., Wei, M.A., Noor, M.A., Hussain, M., Mehboob-ur, R. 2018. Mechanisms and molecular approaches for heat tolerance in rice (Oryza sativa L.) under climate change scenario. J. Integr. Agric. 17, 726-738.

[47] Gomez-Zavaglia, A., Mejuto, J.C., Simal-Gandara, J. **2020**. Mitigation of emerging implications of climate change on food production systems. Food Res. Int. 134, 109256.

[48] IPCC, 2001. Climate Change 2001 the Scientific Basis. In: Houghton, J.T., Ding, Y., Griggs, P.J., Noguer, M., Van deLinden, P.J., Dai, X., Et al., Contribution of Working Group I to the Third Assessment Report of the Inter-Governmental Panel on Climate Change. Cambridge University Press, Cambridge and New York, p. 450

[49] Welch, J.R., Vincent, J.R., Auffhammer, M., Moya, P.F., Dobermann, A., Dawe, D., 2010. Rice yields in tropical/ subtropical Asia exhibit large but opposing sensitivities to minimum and maximum temperatures. Proc. Natl. Acad. Sci. U.S.a. 107, 14562_14567. With large effects in three genetic backgrounds. Theor. Appl. Genet. 124 (2), 323_332.

[50] Satake, T., Yoshida, S., 1978. High temperature induced sterility in indica Rices at flowering. Jpn. J. Crop Sci. 47, 6_17.

[51] Shanmugavadivel, P.S., Amitha Mishra, S.V., Prakash, C., Ramkumar, M.K., Tiwari, R., Mohapatra, T., Singh, N.K. 2017. High resolution mapping of QTLs for heat tolerance in rice using a 5K SNP array. Rice 10, 28.

[52] Umesh, U., Jaiswal, H.K., Sravan, T., Waza, S.A., Bhardwaj, R. 2015. Estimation of genetic variability, heritability and genetic advance for yield and quality traits in some indigenous basmati rice (Oryza sativa L.) genotypes. Int. J. Farm. Sci., 5, 32-40.

[53] Zinn, K.E., Tunc-Ozdemir, M., Harper, J.F., 2010. Temperature stress and plant sexual reproduction: Uncovering the weakest links. J. Exp. Bot. 61, 1959_1968. Available from: https://doi.org/10.1093/jxb/erq053.

[54] Rashid, A., Sofi, N.R., Shikari, A.B., Khan, G.H., Waza, S.A., Sheikh, F.A., Parray, G.A., Bhat, M.A., Sofi, M., Hussain, A. 2019. Developing rice hybrids for temperate conditions using three-line approach. Indian J. Genet. Plant Breed., 79, 25-33.

[55] Jenks, M.A., Haseqawa, P.M., 2007. Plant Abiotic Stress. Blackwell Publishing, Oxford, UK. Available from: https://doi.org/10.1017/S0014479706224906.

[56] Gill, S.S., Tuteja, N., 2010. Reactive oxygen species and antioxidant machinery in abiotic stress tolerance in crop plants. Plant Physiol. Biochem. 48, 909_930. Available from: https://doi.org/10.1016/j.plaphy.2010.08.016.

[57] You, J., Chan, Z., 2015. ROS regulation during abiotic stress responses in crop plants. Front. Plant Sci. 6,1092. Available from:https://doi.org/10.3389/fpls.2015.01092.

[58] Fahad, S., Hussain, S., Bano, A., Saud, S., Hassan, S., Shan, D., et al., 2015. Potential role of phytohormones and plant growth-promoting rhizobacteria in abiotic stresses: Consequences for changing environment. Environ. Sci. Pollut. Res. 22, 4907_4921. Available from: https://doi.org/10.1007/s11356-014-3754-2.

[59] Peleg, Z., Blumwald, E., 2011. Hormone balance and abiotic stress tolerance in crop plants. Curr. Opin. Plant biol. 14, 290_295. Available from: https://doi.org/10.1016/j.pbi.2011.02.001.

[60] Dong, H., Bai, L., Chang, J., Song, C.P. 2018. Chloroplast protein PLGG1 is involved in abscisic acid-regulated lateral root development and stomatal movement in Arabidopsis. Biochem. Biophys. Res. Commun. 495, 280-285.

[61] Takahashi F., Kuromori, T., Sato, H., Shinozaki, K. 2018. Regulatory gene networks in drought stress responses and resistance in plants. In survival strategies in extreme cold and desiccation; springer: Berlin/ Heidelberg, Germany, pp. 189-214.

[62] Dubois M., Van den Broeck, L., Inzé, D. 2018. The pivotal role of ethylene in plant growth. Trends Plant Sci. 2018,23, 313-323.

[63] Ren, Z.H., Gao, J.P., Li, L.G., Cai, X.L., Huang, W., Chao, D.Y., et al., 2005. A rice quantitative trait locus for salt tolerance encodes a sodium transporter. Nat. Genet. 37 (10), 1141_1146.

[64] Martinez-Atienza, J., Jiang, X., Garciadeblas, B., Mendoza, I., Zhu, J.K., Pardo, J.M., et al., 2006. Conservation of the salt overly sensitive pathway in rice. Plant Physiol. 143, 1001_1012. Available from: https://doi.org/10.1104/pp.106.092635.

[65] Golldack, D., Su, H., Quigley, F., Kamasani, U.R., Mun˜oz-Garay, C., Balderas, E., et al., 2002. Characterization of a HKT-type transporter in rice as a general alkali cation transporter. Plant J. 31, 529_542. Available from: https://doi.org/10.1046/j.1365-313X.2002.01374.x.

[66] Shi, H., Quintero, F.J., Pardo, J.M., Zhu, J.K., 2002. The putative plasma membrane Na1/H1 antiporter SOS1 controls long-distance Na1 transport in plants. Plant Cell 14, 465_477. Available from: https://doi.org/ 10.1105/tpc.010371.et.

[67] Knight, H., 1999. Calcium signaling during abiotic stress in plants. Int. Rev. Cytol. 195, 269_324. Available from: https://doi.org/10.1016/S0074-7696(08)62707-2.

[68] Jain, R.K., Jain, S., 2000. Transgenic strategies for genetic improvement of basmati rice. Indian J. Exp. Biol. 38 (1), 6_17.

[69] Kim, Y.S., Kim, J.K., 2009. Rice transcription factor AP37 involved in grain yield increase under drought stress. Plant signal Behav. 4 (8), 735_736.

[70] Jeong, J.S., Kim, Y.S., Baek, K.H., Jung, H., Ha, S.H., Do Choi, Y., et al., 2010. Root-specific expression of OsNAC10 improves drought tolerance and grain yield in rice under field drought conditions. Plant Physiol. 1531, 185_197.

[71] Abdula, S.E., Lee, H.J., Kim, J., Nin˜o, M.C., Jung, Y.J., Cho, Y.C., et al., 2016. BrUGE1 transgenic rice showed improved growth performance with enhanced drought tolerance. Breed Sci. 66 (2), 226_233.

[72] Hasanuzzaman, M., Nahar, K., Alam, M.M., Roychowdhury, R., Fujita, M., 2013. Physiological, biochemical, and molecular mechanisms of heat stress tolerance in plants. Int. J. Mol. Sci. 14 (5), 9643_9684.

[73] Volkov, R.A., Panchuk, I.I., Mullineaux, P.M., Scho¨ ffl, F., 2006. Heat stress-induced H2O2 is required for effective expression of heat shock genes in Arabidopsis. Plant Mol. Biol. 61 (45), 733746.

[74] Hu, W., Hu, G., Han, B., 2009. Genome-wide survey and expression profiling of heat shock proteins and heat shock factors revealed overlapped and stress specific response under abiotic stresses in rice. Plant Sci 176, 583_590.

[75] Ouyang, Y., Chen, J., Xie, W., Wang, L., Zhang, Q., 2009. Comprehensive

sequence and expression profile analysis of Hsp20 gene family in rice. Plant Mol. Biol. 70, 341_357.

[76] Pareek, A., Singla, S.L., Grover, A., 1995. Immunological evidence for accumulation of two high-molecularweight (104 and 90 kDa) HSPs in response to different stresses in rice and in response to high temperature stress in diverse plant genera. Plant Mol. Biol. 29 (2), 293_301.

[77] Zhang, J.X., Wang, C., Yang, C.Y., Wang, J.Y., Chen, L., Bao, X.M., et al., 2010. The role of Arabidopsis AtFes1A in cytosolic Hsp70 stability and abiotic stress tolerance. Plant J. 62 (4), 539_548.

[78] Batra, G., Chauhan, V.S., Singh, A., Sarkar, N.K., Grover, A., 2007. Complexity of rice Hsp100 gene family: Lessons from rice genome sequence data. J. Biosci. 32, 611_619.

[79] Guo, J., Wu, J., Ji, Q., Wang, C., Luo, L., Yuan, Y., et al., 2008. Genome-wide analysis of heat shock transcription factor families in rice and Arabidopsis. J. Genet. Genomics 35, 105_118.

[80] Rana, R.M., Dong, S., Tang, H., Ahmad, F., Zhang, H., 2012. Functional analysis of OsHSBP1 and OsHSBP2 revealed their involvement in the heat shock response in rice (Oryza sativa L.). J. Exp. Bot. 63, 6003_6016.

[81] Tiwari, S., Krishnamurthy, S.L., Kumar, V., Singh, B., Rao, A.R., Mithra, S.V.A., et al., 2016. Mapping QTLs for salt tolerance in rice (Oryza sativa L.) by bulked segregant analysis of recombinant inbred lines using 50K SNP chip. PLoS ONE 11 (4), e0153610. Available from: https://doi.org/10.1371/journal. pone.0153610.

[82] Sandhu, M., Sureshkumar, V., Prakash, C., Dixit, R., Solanke, A.U., Sharma, T.R., et al., 2017. RiceMetaSys for salt and drought stress responsive genes in rice: A web interface for crop improvement. BMC bioinformatics 18 (1), 432. Available from: https://doi.org/10.1186/s12859-017-1846-y.

[83] Singh, R., Singh, Y., Xalaxo, S., Verulkar, S., Yadav, N., Singh, S., et al., 2016. From QTL to varietyharnessing the benefits of QTLs for drought, flood and salt tolerance in mega rice varieties of India through a multi-institutional network. Plant Sci. 242, 278_287.

[84] Bernier, J., Kumar, A., Venuprasad, R., Spaner, D., Atlin, G.N., 2007. A large-effect QTL for grain yield under reproductive-stage drought stress in upland rice. Crop Sci. 47, 507_516.

[85] Vikram, P., Swamy, B.P.M., Dixit, S., Ahmed, H.U., Teresa, M., Cruz, S., 2011. Reproductive-stage drought stress with a consistent effect in multiple elite genetic backgrounds. BMC Genet. 12, 1_15.

[86] Venuprasad, R., Bool, M.E., Quiatchon, L., Cruz, M.S., Amante, M., Atlin, G.N., 2012a. A large-effect QTL for rice grain yield under upland drought stress on chromosome 1. Mol. Breed. 30 (1), 535_547.

[87] Vikram, P., Swamy, B.M., Dixit, S., Trinidad, J., Cruz, M.T.S., Maturan, P.C., et al., 2016. Linkages and interactions analysis of major effect drought grain yield QTLs in rice. PLoS ONE 11. Available from: Https:// doi. org/10.1371/journal.pone.0151532.

[88] Venuprasad, R., Bool, M.E., Quiatchon, L., Atlin, G.N., 2012b. A QTL for Rice Grain Yield in Aerobic Environments

[89] Shamsudin, N.A.A., Swamy, B.P.M., Ratnam, W., Sta. Cruz, M.T., Sandhu, N., Raman, A.K., et al., 2016. Pyramiding of drought yield QTLs into a high quality Malaysian rice cultivar MRQ74 improves yield under reproductive stage drought. Rice 9, 21. Available from: https://doi.org/10.1186/s12284-016-0093-6.

[90] Lin, H.X., Zhu, M.Z., Yano, M., Gao, J.P., Liang, Z.W., Su, W.A., et al., 2004. QTLs for Na1 and K1 uptake of the shoots and roots controlling rice salt tolerance. Theor. Appl. Genet. 108 (2), 253_260.

[91] Zhou, J., Wang, F., Deng, P., Jing, W., Zhang, W., 2013. Characterization and mapping of a salt-sensitive mutant in rice (Oryza sativa L.). J. Integr. Plant Biol. 55 (6), 504_513.

[92] Deng, P., Shi, X., Zhou, J., Wang, F., Dong, Y., Jing, W., et al., 2015b. Identification and fine mapping of a mutation conferring salt-sensitivity in rice (Oryza sativa L.). Crop Sci. 55, 219_228.

[93] Deng, P., Jiang, D., Dong, Y., Shi, X., Jing, W., Zhang, W., 2015a. Physiological characterisation and fine mapping of a salt-tolerant mutant in rice. Funct. Plant Biol. 42 (11), 1026_1035.

[94] Bizimana, J.B., Luzi-Kihupi, A., Murori, R.W., Singh, R.K., 2017. Identification of quantitative trait loci for salinity tolerance in rice (Oryza sativa L.) using IR29/Hasawi mapping population. J. Genet. 96 (4), 571_582.

[95] Emon, R.M., Islam, M.M., Halder, J., Fan, Y., 2015. Genetic diversity and association mapping for salinity tolerance in Bangladeshi rice landraces. Crop J. 3, 440_444.

[96] Kumar, V., Singh, A., Mithra, S.V.A., Krishnamurthy, S.L., Parida, S.K., Jain, S., et al., 2015. Genome-wide association mapping of salinity tolerance in rice (Oryza sativa). DNA Res 22 (2), 133_145. Available from: https://doi.org/10.1093/dnares/dsu046.

[97] Singh, R.K., Gopalakrishnan, S., Singh, V.P., Prabhu, K.V., Mohapatra, T., Singh, N.K., et al., 2011. Marker assisted selection: A paradigm shift in basmati breeding. Indian J. Genet. 71, 120_128.

[98] Cuc, L.M., Quyen, P.T., Huyen, L.T.N., Ham, L.H., 2015. Breeding for the salinity tolerance rice variety in Vietnam. J. Agric. Technol. 11 (8), 2263_2272.

[99] Huyen, N.T.L., Cuc, M.L., Ismail, A.M., Ham, H.L., 2012. Introgression the salinity tolerance QTLs saltol into AS996, the elite rice variety of Vietnam. Am. J. Plant Sci. 3, 981_987.

[100] Linh, H.L., Linh, H.T., Xuan, D.T., Ham, H.L., Ismail, A.M., Khanh, D.T., 2012. Molecular breeding to improve salt tolerance of rice (*Oryza sativa* L.) in the red river delta of Vietnam. Int. J. Plant Genom 2012. Available from: https://doi.org/10.1155/2012/949038.

[101] Vu, H.T.T., Le, D.D., Ismail, A.M., Le, H.H., 2012. Marker-assisted backcrossing (MABC) for improved salinity tolerance in rice (Oryza sativa L.) to cope with climate change in Vietnam. Aust. J. Crop Sci. 6, 1649_1654.

[102] Mondal, U., Khanom, M.S.R., Hassan, L., Begum, S.N., 2013. Foreground selection through SSRs markers for the development of salt tolerant rice variety. J. Bangladesh Agric. Univ. 11, 67_72. Available from: https://doi.org/10.3329/jbau.v11i1.18215.

[103] Ho, V.T., Thomson, M.T., Ismail, A.M., 2016. Development of salt tolerant IR64 near isogenic lines through marker-assisted breeding. J. Crop Sci. Biotechnol. 19 (5), 373_381.

[104] Kujur, A., Bajaj, D., Upadhyaya, H.D., Das, S., Ranjan, R., Shree, T., et al., 2015. Employing genome-wide SNP discovery and genotyping strategy to extrapolate the natural allelic diversity and domestication patterns in chickpea. Front. Plant Sci. 6, 162.

[105] Deshmukh, R., Sonah, H., Patil, G., Chen, W., Prince, S., Mutava, R., et al., 2014. Integrating omic approaches

for abiotic stress tolerance in soybean. Front. Plant Sci. 5, 244.

[106] Ramalingam, A., Kudapa, H., Pazhamala, L.T., Weckwerth, W., Varshney, R.K., 2015. Proteomics and metabolomics: Two emerging areas for legume improvement. Front. Plant Sci. 6, 1116.

[107] Shankar, R., Bhattacharjee, A., Jain, M., 2016. Transcriptome analysis in different rice cultivars provides novel insights into desiccation and salinity stress responses. Sci. Rep. 6, 23719. Available from: https://doi.org/10.1038/srep23719.

[108] Kuijken, R.C.P., van Eeuwijk, F.A., Marcelis, L.F.M., Bouwmeester, H.J., 2015. Root phenotyping: From component trait in the lab to breeding. J. Exp. Bot. 66, 5389_5401.

[109] Kumar, V., Khare, T., Shriram, V., Wani, S.H., 2017. Plant small RNAs: The essential epigenetic regulators of gene expression for salt-stress responses and tolerance. Plant Cell Rep. 37, 61_75.

[110] Gujjar, R.S., Akhtar, M., Singh, M., 2014. Transcription factors in abiotic stress tolerance. Indian J. Plant Physiol. 19, 306_316. Available from: https://doi.org/10.1007/s40502-014-0121-8.

Chapter 5

Understanding the Responses, Mechanism and Development of Salinity Stress Tolerant Cultivars in Rice

Sathees Nagarajan, Nirubana Varatharajan and Renganathan Vellaichamy Gandhimeyyan

Abstract

Rice is the most important staple food crop of much of the world's population. Production and consumption of rice is higher in Asia but adverse environmental conditions critically threaten the rice production. Soil salinity has been a key abiotic constraint affecting the crop production by reducing growth, development and yield of the plant. Rice is highly sensitive to salinity specifically at the early vegetative and late reproductive stages. Therefore, studying the responses of crop at the morphological, physiological, biochemical and molecular level is an effective strategy. Understanding the mechanisms behind the salinity such as osmotic stress and osmolytes, ion exclusion, inclusion and compartmentation, antioxidant response and hormonal regulation. Different screening strategies such as phenotypic and genotypic screening for rice under salinity and select the salt tolerant lines. Using the conventional and molecular breeding approaches is a prerequisite for its effective management and to develop salt tolerant cultivars in rice.

Keywords: rice, salinity, responses, mechanisms, screening, breeding methods

1. Introduction

Rice (*Oryza Sativa)* is a diploid (2n = 12) belongs to the Poaceae family (Umed [1]). It is an important cereal food crop cultivated and consumed across the world population and most of the Asiatic peoples. Rice is a staple food crop and gives 50 to 80% calories to the more than three billion peoples (Umed [1]). The global rice production needs to be increased from 560 million tonnes to 850 million tonnes by 2025 to meet the growing demand of rice [2]. Rice cultivation is continuously threatened by several biotic and abiotic stresses [3]. Salinity is the major soil problem in rice cultivation especially saline prone coastal areas and it is a second most abiotic problem than the drought [4]. Globally, 800 million ha area affected by salt accumulation, which is about 12% of the world land and 20% of the cultivated land [5]. The 1500 million ha of dry land farming and 230 million ha of irrigated farming comes under saline condition and 33% of agricultural land affected with high salinity [6]. In earth surface, the enormous amount of water is available but all the water is not useful for to survival of plants and animals, because 72% of water available as a sea water [7].

Rice is highly is sensitive to salinity at early vegetative and late vegetative stages [8, 9]. Increasing salinity (22 dS m^{-1}) affects the seed germination mainly due to the osmatic stress [10, 11]. It reduces the photosynthetic activity, chlorophyll content, leaf area and stomata where high level of salinity occurs [12]. Yield components such as panicle length, spikelet number per panicle and grain yield were also affected; and panicle emergence and flowering affect the seed set through pollen viability [13].

The present review to show the constrains of the salinity to the rice growth and development and provide the information about the plants responses against the salinity and what are the different approaches are used to develop the salt tolerance rice cultivar.

2. Salinity -abiotic constraint devastating agricultural production

Among abiotic stresses, the salinity which adversely affects the crop growth, development and production [14]. The deposition of salts or salinization is a natural phenomenon with evaporation of saline underground water, sea water infiltration of coastal ground waters, sea water salts in wind and rain and human initiated process such as irrigation with marginal water and poor agronomic practices. Salinity affected almost the 900 M ha of land [15]. In total global cultivated land, 23% contain saline and 37% contain sodic soils and $2.5x10^8$ ha of global irrigated lands affected by salinity and water logging. The salinity mostly occurs in the south Asia and Southeast Asia which contributes the 90% of the world's rice production [16, 17]. The abundant accumulation of dissolved salts in soil and water which affect the plant growth that leads to decrease the agricultural production. In ocean water contain 480 mM of Na^+ and 560 mM of Cl^-. If saline water contains the above the optimal level of sodium and chlorine for plant growth and development, they require the techniques to maintain the quality of irrigation water [18]. In some of the plants show the wide range salinity tolerance especially against the Na and Cl salinity. Based on the salinity tolerance and sensitivity, the plants are classified into two types such as glycophytes and halophytes. Glycophytes means which plant grown in low tolerance condition. Halophytes means which plant grown in higher salinity condition. Mostly all plants coming under the glycophytes category.

Cereal crops show wide range of response to salinity, for example barley show most tolerance and rice show the most sensitive to salinity [5]. Lutts et al. [19] reported that seedlings of rice are highly sensitive to toxicity of salts. According to Shannon [8] early vegetative and later vegetative stages of rice are highly sensitive to salinity. Khatun and Flowers [20] reported that the salt toxicity affected the panicle length, spikelets per panicle and 1000-grain weight; decreased photosynthesis were observed by Munns and Tester [5] which results in unfilled spikelets in rice. Delay in flowering and ripening, reduced number of tillers and biomass and leaf area also occur due to the salinity [21].

In saline soil contains numerous soluble salts such as Ca^{2+}, Mg^{2+}, Na^+ and anions SO_4^{2-}, Cl^-, HCO_3^-, K^+, CO_3^{2-} and NO_3^- and these soil contain the EC is 4 dS/m or more [22]. Based on the electrical conductivity influence in soil salinity and crop responses are mentioned in **Table 1**.

In arid and semi-arid area where low precipitation and high evaporation leads to accumulation of more salts [24] and accumulation of salts in sea shores is a natural phenomenon of sea water flooding. Worldwide fertility status of rice growing areas were mentioned in **Table 2.**

The quality of irrigation water affected by accumulation of salts. If abundance of salt accumulation in the root zone which alter the water relations in the plant. Due

Electrical conductivity (Ds m^{-1} at 25° C)	Crop response
< 2	Salinity effect is practically zero
2 – 4	Reduction in yield of very sensitive crops
4 – 8	Reduction in yield of most crops
8 – 16	Only tolerant crops produce satisfactory yield
>16	Few highly tolerant crops produce satisfactory

Table 1.
Crop response to salinity influenced by electrical conductivity of saturated soil extract [23].

Region	Percentage of problem soils (%)	Percentage of very poor soils (%)	Percentage of poor soils (%)	Percentage of Good soils (%)	Total rice area (ha)
Western Europe	1.3	46.6	8.4	43.8	25
Southeastern Asia	8.3	43.7	18.6	29.4	49,120
Eastern Asia	1.5	37.6	9.8	51.1	33,425
Western Asia	15.0	17.1	9.4	58.5	156
Asia	5.3	29.6	18.3	46.7	143,429
Southern Europe	1.4	52.8	13.1	32.8	431
Eastern Africa	4.7	62.3	13.1	19.9	3,330
Northern Africa	16.3	5.8	50.4	27.4	558
Southern Asia	4.7	13.9	22.8	58.4	60,526
Central Africa	0.4	76.0	8.6	15.1	736
Africa	3.1	60.5	18.0	18.2	10,466
South America	5.4	61.8	12.7	20.2	5,121
Melanesia	0.9	28.8	13.4	56.9	5
Northern America	7.2	16.9	28.9	47.0	1,259
Central America	1.6	23.5	42.6	32.2	330
Caribbean	7.2	22.7	12.3	57.7	437
Americas	5.6	49.7	16.9	27.8	7,147
Central Asia	42.4	4.8	13.7	39.1	202
Eastern Europe	4.3	10.8	16.9	67.9	248
Western Africa	1.3	62.8	19.0	16.8	5,843
Europe	2.4	38.1	14.1	45.4	704
Australia and New Zealand	24.8	4.6	33.1	37.5	34
Oceania	21.9	7.5	30.7	39.9	39
World Total	5.1	32.5	18.2	44.0	161,784

Table 2.
Worldwide fertility status of Rice-growing areas [25].

to the salt irrigation water induce the some of the effects in plants such as necrosis, chlorosis and interfere the plant physiological activities with depends on the some of the environmental factors such as temperature, humidity, light intensity and soil conditions [26].

2.1 Effects of salt stress in rice

The millions hectares of lands continuously salinized and also affect the crop production. It leads to contributing to the future biological catastrophe [27]. Rice is a salt sensitive crop [28]. In rice if increase in salt stress (in terms of 5 to 7.5 dS m–1), the seedling growth and fresh weight decreased [29]. The salt stress decrease the seedling biomass production in rice [10, 11]. The increased salt stress significantly reduce the mean root length, mean root numbers per plant, and shoot length [30]. In rice early stage, leaf mortality increased with increased salt stress [31] and also observed the reduction in growth and development in later stages [32]. Salt stress cause the effects in plant cell metabolism and leaf senescence and old leaves were death which indicate the crucial for the survival of a plant [33]. Due to salt stress panicle sterility observed in pollination and fertilization stage which leads to a decline in grain setting [34]. The lack of transformation of carbohydrate to growth of spikelet which leads to decrease in grain yield due to the salt stress (Sajid [35]). The salt stress contain negative linear raltionship with important factors such as number of tillers per plant, number of spikelets per panicle, and percent of sterile florets [10, 11].

In rice, the salt stress show the physiological effects such as decrease in photosynthetically active radiation (PAR), net photosynthesis (*P*n), stomatal conductance (*G*s), transpiration rate (*T*r), degradation of pigment, and relative water content (RWC) [36]. The chlorophyll and carotenoids contents in rice leaves were decreased after the salt stress [37]. The increased salt stress affect the Na+, Ca2+, K+, and Mg2+ concentration in root and shoot in rice plant [38]. The availability of zinc were decreased and increased cadmium (Cd) toxicity observed where high level of NaCl occur [39].

3. Mechanism and their responses to the salt tolerance in rice

The potential of irrigation water and soil were affected by the accumulation of salts which interfere the uptake of water into plant from soil and presence of higher concentration of Na in soil solution will affected the uptake of some essential nutrients such as K and Ca in root zone [40]. If concentration of Na and Cl reach threshold level, some specific toxicities will alter the water relation [5]. The plants develop the some of the mechanism to grown in salinity condition. The mechanism are such as osmotic stress and osmolytes, ion exclusion, inclusion and compartmentation, antioxidant response and hormonal regulation.

3.1 Mechanism of salt tolerance: osmotic stress and osmolytes

When plants are grown under salinity condition, the plants adjust the losing of water and their potential and it will lead to decrease in osmotic potential, turgor and express the signal that trigger the adaptive responses [41]. The recovery period, osmotic potential and hydraulic conductivity of the membranes is reduced. By the accumulation of the organic and inorganic solutes and the plant turgidity is recovered after the tissue growth occur [5]. In osmotic adjustment, the cell wall elasticity was changed with decrease in RWC (Relative Water Content) and increase water content in the apoplast which decrease the salinity consequences by maintaining turgidity of the tissue. The organic osmolytes used to maintaining the osmotic potential in plants and prevent the salinity effects [42]. When salinity stress occurs, the osmolytes will synthesized and osmolyte biosynthesis and accumulation is important for the salinity tolerance. But the osmolyte biosynthesis may vary based on the plant age and rate of stress occur [43]. The most abundant and compatible osmolytes such as proline and glycine-betaine coming under the organic osmolytes [26]. Proline synthesis is enhanced under salinity [44]. Proline act as a reactive oxygen scavenger, redox buffer and molecular chaperone and stabilizes membranes and proteins under stress conditions [45]. Glycine-betaine involve in protection of enzymes and membrane structures [46].

3.1.1 Ion exclusion, inclusion and compartmentation

Ion exclusion from salt sensitive organs, inclusion in less sensitive locals such as the root and old leaves and organ compartmentation strategies used for decreasing the damaging ion-specific effects of Na and nutrient deficiencies. If the cytoplasm contain excess ions, it will transported across the tonoplast by the Na/H antiporter and compartmentalized in vacuoles which is useful for protect the plant from salinity [47]. Through the exclusion strategies the salts are removed from aerial part by saline vesicle glands in the epidermis which prevent accumulation and transported to roots which is mostly occur in halophytes and glycophytes. Salt sequestering in old leaves which will give salt-induced senescence [48]. When compared to synthesis of organic osmolytes, the ion compartmentation is low cost effective [49].

3.1.2 Antioxidant responses

The oxidative stress occur when plants under the salinity [50]. Under salinity stress, to induce increase the reactive oxygen species such as superoxide radicals and hydrogen peroxide [51]. The antioxidative system is response to the salinity exposure [50]. The chloroplasts and mitochondria are play role in the salt tolerance with the increased antioxidant defenses. The salt sensitive plants express in decreased antioxidant level [52]. Nitric oxide (NO) response to salinity tolerance with the NO donor [53].

3.1.3 Hormonal regulation

The plant hormones such as ABA (Abscisic Acid) and cytokines are increased when plant under the salt stress [54]. The negative effects of salinity such as plant growth, photosynthesis and translocation of assimilates can be reduced by the highly accumulation of hormones. The ABA also involved in compatible solutes and nutritional cations K^+ and Ca^{2+} in vacuoles of roots for salt tolerance [55].

3.2 Response of rice under salt stress

Based on the salinity responses, the plants were classified into halophytes and glycophytes. The halophytes plants tolerate to high concentration of NaCl (400 mM) when compared to glycophytes [56]. In rice, the salinity tolerance is controlled by multiple gene [57] and therefore understanding the plant responses for salinity is very much important for developing tolerant cultivars. Under salt stress condition, the rice plants exposure to different responses such as morphological, biochemical, physiological, molecular response.

3.2.1 Morphological response

Due to salinity stress, the occurrence of morphological changes such as stunted plant growth, leaf burning, chlorosis, low tillering, leaf rolling and poor root growth [58]. Decreased leaf area and changes in leaf anatomy under *invitro* condition was observed by Bahaji et al. [59]. For example, comparable levels of osmolality, the reduction in root and leaf growth were similar for both saline and osmotically-generated stress. Most of the variations in leaf anatomy features caused by the treatments could be ascribed to osmotic stress [59]. The beginning of salt stress, there is no symptoms are observed but shoot and root growth reduction occur. When plants continues exposure to salt condition, leaf senescence occur. After 3-4 days of exposure under salinity, the plants began to develop leaf symptoms such as yellowing and necrotic lesions of old leaf tips. The senescence of older leaves was observed after two weeks of stress. In Nipponbare, root and shoot growth was affected by salinity [60].

3.2.2 Biochemical response

Based on the biochemical response, the effect of salinity in plants which leads into two parts such as initial osmotic effect and later ionic stress (where accumulation salt at toxic level) [5]. The plants express some of the biochemical responses such as oxidative stress, altered metabolism, high Na^+ transport to shoot, lower K^+ uptake and low P and Zn uptake [61]. In initial osmotic effect, water potential is decreased and increased the osmotic potential due to the increased concentration of salts. The salt stressed plants contains the larger amount of proline in higher plants [62] and it act as a osmotic adjustment, shielding the enzymes, membranes and give the energy and nitrogen during salinity [63]. Soluble sugars and starch also responses to salinity as an osmoticum in plants [64]. When rice plants exposure to salinity, sugar content increased in shoot [65] and starch content increased in root, which act as a reservoir for the primary metabolism [66]. Where plants exposure to the salinity, the proteins are synthesized and accumulate as a storage food which is used as a reservoir during salt stress condition and reutilized when absence of stress [67]. The increased protein content is positively correlated to rice seedling tolerance than the sensitive one [68].

3.2.3 Physiological response

When plants under salinity, they express some of the physiological responses such as inhibition of photosynthesis, stomatal closure, decreased water content, higher amount of osmolytes and low osmotic potential. The response of rice to salinity, to study the physiological mechanism and it was associated with the plant defense mechanism activated during stress. During salinity, chloroplast and mitochondria are mostly affected compared to other organs [69]. In chloroplast, some of the potential indicators show the effects in the photosynthesis efficiency such as changes in chlorophyll fluorescence and membrane permeability [70]. Salinity affects the mesophyll tissue which leads to affect the vascular bundles. The more accumulation of sodium salts is excited by salt exclusion [71], selective ion uptake [72] and regulation of K+/Na + ratio [73]. Estimating the different plant parameters such as tiller number, leaf area, panicle length, root length, biomass, dry weight, RGR (Relative Growth Rate) and RWC (Relative Water Content) Zeng et al. [89] from different cultivars, leaf RWC is increased in paddy under salinity and suggested the role of osmo-protectants in preventing cell injury from salt stress-induced dehydration [74].

3.2.4 Molecular response

In molecular response, the main aim is to breed and to develop the salinity tolerance lines. Genetic diversity is a primary work which is used to screen the lines with the various molecular markers such as RFLP, SSLP, RAPD and SSR markers. Salinity is controlled by several genes and inheritance of salinity trait is difficult in rice. These difficulties are overcome by using the positional cloning [75] and insertional mutagenesis [76]. Many genes are identified for the salinity [77]. In rice under stress condition some of the genes were identified such as catalase and several denovo genes. The salinity tolerance controlled by major Quantitative Trait Loci (QTL) is *Saltol* which is mapped in the chromosome 1 of the FL478 Recombinant inbred Line (RIL) line. The *saltol* linked with the flanking markers RM1287 and RM6711 and these QTL region contain 15 SSR markers. The FL478 obtained from crossing between Pokkali and IR29. The *saltol* QTL responsible for the maintain the low Na+, high K+ and Na+/K+ homeostasis in shoots of rice. The *saltol* QTL can transfer into superior cultivar and these transformation confirmed by the candidate gene approaches or Marker Assisted Selection (MAS) [78]. Among the molecular marker analysis, the SSR marker is effective for salt tolerance identification in rice [79]. Several QTLs were identified for sodium uptake, potassium uptake, and sodium: potassium selectivity [80]. The molecular markers are used to identified the QTLs and it gives new platform for salinity study.

4. Screening and development of salt tolerant cultivars

The rice germplasm with diverse and significant varietal differences shows a pool for developing screening techniques. This is important technique for to select varieties for breeding purposes. This techniques are economically feasible, easy and efficient. Rice expresses highly sensitive in early seedling stage and low sensitive in reproductive or late vegetative stage [19]. The continues accumulation of NaCl in older leaves through long exposure [81] which affect the efficiency of photosynthesis and whole plant metabolism [82]. Tolerance in vegetative and reproductive stage of the rice was expressed in seedling stage [83]. So, precise screening method is necessary to identify the salt tolerance line. Seedling stage screening is acceptable and it is rapid screening but it is difficult in vegetative and reproductive stage [4]. From salinity screening, highly tolerant line was identified such as pokkali which is used as a donor for further breeding program of salt tolerance [84].

4.1 Phenotypic and genotypic screening

Salinity tolerance is made as complex in physiologically and genetically [85]. It requires more time and expensive tissue analysis. In early seedling stage of phenotypic screening done in first, second and growing leaves. The phenotypic screening in early seedling to select first, then second and growing leaves. Where salinity occur leaf elongation and new leaf formation is suppressed [86]. In phenotypic screening, for identification of potential rice lines, first step is to screen the already available germplasms of rice. Field level phenotypic screening is difficult because it contain soil heterogeneity, climatic factors and other environmental factors which affect the physiological functions. But, phenotypic screening under laboratory or green house (hydroponics) is better merits than the field screening because it was not affected by the environment and soil factors such as temperature, relative humidity, and solar radiation. In hydroponic phenotypic screening, the seedlings are grown in salinized nutrient solution and where salinity inversely proportional to photosynthesis and chlorophyll content [87]. The vegetative and reproductive stage screening is difficult because decrease in plant height, root length and biomass. In reproductive stage, sterility of florets occurs due to the effect of reduced panicle length, number of primary branches and spikelets per panicle, fertility and panicle weight thus reducing grain yield [88]. It is a rapid, perfect and easy method. Phenotypic screening for salt tolerance is not easy because of environmental effects which hinders the development of accurate and reliable screening technique. Salt tolerance screening at early stage is not correlate with further stages [89]. The germplasm are evaluated by using the high-throughput phenotyping saves time and resources compared to traditional phenotyping methods [90].

Among molecular research tools, quantitative traits loci are useful to study the genotypic of salinity tolerance. These QTL techniques useful in the marker-assisted selection [85]. Saltol is a major QTL which is accounted more than 70% of the variation in salt uptake and these QTLs are incorporated into high yielding varieties by Marker Assisted Backcrossing [91]. While transfer of salt QTL, some of the unwanted traits of linkage drag transferred which is drawback in MABB and genotypic screening [92]. In genotypic screening, the use of molecular markers such as RFLP (Restriction Fragment Length Polymorphism), RAPD (Random Amplified Polymorphic DNA), ISSR (Inter Simple Sequence Repeats), SSR (Simple Sequence Repeats) and AFLP (Amplified Fragment Length Polymorphism) for screening of germplasm is more reliable than the phenotypic screening [93]. Among the above molecular markers, microsatellite markers to be more effective [79].

4.2 Approaches for the development of salt tolerant cultivars

The salt tolerant cultivars are developed by using several ways such as, use interspecific hybridization to raise the tolerance of current crops, use the variation already present in existing crops, generate variation within existing crops by using recurrent selection, mutagenesis or tissue culture and breed for yield rather than tolerance [85]. To improve salinity in rice genotypes by the incorporation of salt tolerance genes into high yielding cultivars.

4.2.1 Conventional breeding

In conventional Plant breeding methods, different approaches are used for development of salt tolerance cultivar such as introduction, selection, hybridization, mutation and shuttle breeding approach. The beginning the salt tolerant rice cultivars (Damodar (CSR 1), Dasal (CSR 2), and Getu (CSR 3)) are developed from pure line selection from the local traditional cultivars. In segregating population, Selection pressure is gradually increased with the generation advancement simultaneously in moderate stress and high stress of sodicity and salinity [94].

In conventional plant breeding, the new and better variety was developed by the combining the different parents genes. In this breeding method, first to generate a breeding population with highly variable and identifying parents for traits of interest. Several conventional breeding programmes are involved in to developing salt tolerance lines such as in vitro selection, pooling physiological traits, interspecific hybridization, using halophytes as alternative crops [85]. Through conventional plant breeding method, some varieties such as CSR10, CSR13, CSR27, Narendra usar 2 and Narendra usar 3 were developed and released as salt tolerant for cultivation [95]. In development of salt tolerance cultivar through conventional breeding is require more time and it depends on the environmental factors. It contains very limited success, due to the complexity of the trait.

4.2.2 Genomics based approaches

In Marker Assisted Selection (MAS), some important strategies are followed such as markers are tightly linked to loci and specific marker alleles are associated with desired alleles at target loci consistently across the different breeding populations [94].

The conventional back cross breeding method drawbacks are overcome by usage of Marker Assisted Backcrossing Method (MABB). Young and Tanksley [96], demonstrated that large amount of DNA from the donor can remain around the target gene even after many generation of backcrossing. In this method, through three important approaches used to reduce the linkage drag such as foreground selection, recombinant selection and background selection.

By using the genomic approaches, the genetic map and genetic diversity in germplasm which is done by usage of the molecular markers [97]. The most efficient markers to screen the salt tolerant genotypes were SSR (Simple Sequence Repeats) markers RM8094, RM336 and RM8046 which were contain higher polymorphic information content coupled with higher marker index value [98]. Two EST markers such as CP3970 and CP6224 and Five SSR markers such as RM1287, RM8094, RM3412, RM493 and RM140 were linked to saltol QTL on chromosome 1 [99]. In salt tolerance rice, transgenic approaches are attempted. Hoshida et al. [100] have reported the transgenic salt tolerance rice by over expresses chloroplast glutamine synthetase gene. These transgenic plants shoots contain more accumulated K^+, Ca^{2+}, Mg^{2+} and less Na^+ compared with those of non-transformed controls [101, 102]. Xujun Chen and Zejian Guo [103] recorded that tobacco OPBP1 increase the salt tolerance and disease resistance in transgenic rice.

5. Conclusion

Salinity is the one of threaten to the rice cultivation, here, the development of salt tolerance cultivar is difficult because it control by many genes and associated with environmental factors. The screening of germplasm used to select the better lines

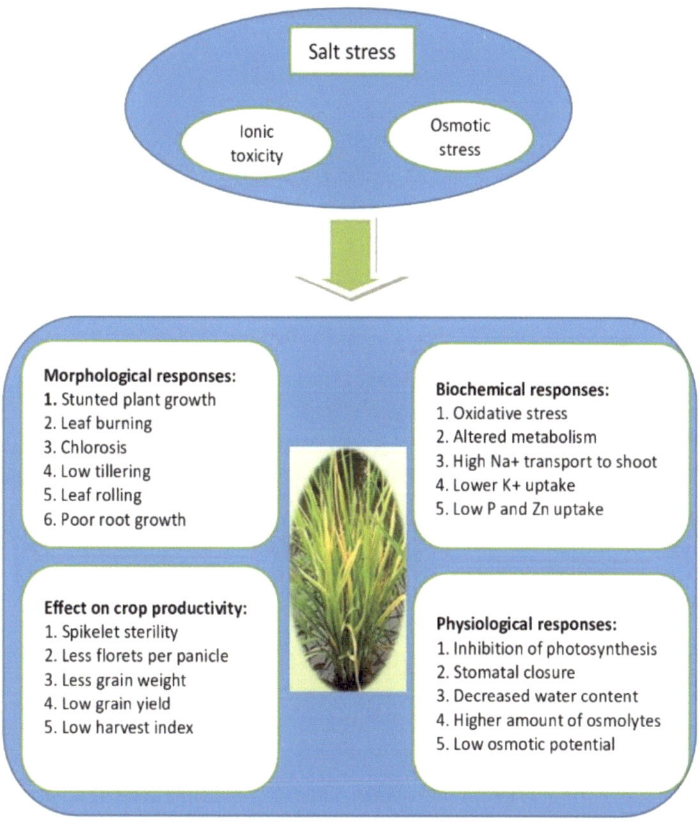

Figure 1.
Plants response to salinity.

which is used for the developing salt tolerance lines. This is done through conventional and genomic approaches. In conventional method it contain limited level of success due to the time consumption and cost. So, the genomic and molecular approaches used to develop the salt tolerant lines with cost effective and in low time period (**Figure 1**).

Author details

Sathees Nagarajan[1*], Nirubana Varatharajan[1]
and Renganathan Vellaichamy Gandhimeyyan[2]

1 Department of Plant Breeding and Genetics, Agricultural College and Research Institute, Tamil Nadu Agricultural University, Madurai, India

2 Department of Biotechnology, Agricultural College and Research Institute, Tamil Nadu Agricultural University, Madurai, India

*Address all correspondence to: sathees1063@gmail.com

IntechOpen

© 2021 The Author(s). Licensee IntechOpen. This chapter is distributed under the terms of the Creative Commons Attribution License (http://creativecommons.org/licenses/by/3.0), which permits unrestricted use, distribution, and reproduction in any medium, provided the original work is properly cited.

References

[1] Ali U, Shar T, Ahmad R, Khatoon M, Khaskheli MA, Laghari AH, Leghari AJ. Salinity Stress–A Threat to Rice Production Breeding Strategies to Develop Salinity Tolerance in Plants. Mehrgarh Journal of Sciences and Technology (MHJST). 2021; 1:13-17.

[2] Khush GS, Virk PS, Evangelista A, Romena B, Pamplona A, Lopena V, Dela Cruz N, Peng S, Cruz CV, Cohen M. Germplasm with high yield potential. 2001 annual Report, Los Baños (Philippines): International Rice Research Institute. 2001; 4-5.

[3] Singh T, Pun KB, Satapathy BS, Saikia K, Lenka S. Incremental yield and returns from rice variety Naveen in front line demonstrations-an analysis. Oryza. 2015; 52:59-64.

[4] Gregoria GB, Senadhira D, Mendoza RD. Screening rice for salinity tolerance. . IRRI, Discussion Paper Series No.22, International Rice Research Institute. Manila, Philippines 1997; 1-30.

[5] Munns R, Tester M. Mechanisms of salinity tolerance. Annual Review of Plant Biology. 2008; 59:651-681.

[6] Fao FA. Food and agriculture organisation of the United Nations. Retrieved on. 2008; 15.

[7] Harvey GW. Microlayer collection from the sea surface: A new method and initial results 1. Limnology and Oceanography. 1966; 11:608-613.

[8] Shannon MC, Rhoades JD, Draper JH, Scardaci SC, Spyres MD. Assessment of salt tolerance in rice cultivars in response to salinity problems in California. Crop Science. 1998; 38:394-398.

[9] Shannon MC. Adaptation of Plants to Salinity. Advances in Agronomy. 1998; 60: 75-120.

[10] Zeng L, Shannon MC. Effects of salinity on grain yield and yield components of rice at different seeding densities. Agronomy Journal. 2000a; 92:418-423.

[11] Zeng L, Shannon MC. Salinity effects on seedling growth and yield components of rice. Crop science. 2000b; 40:996-1003.

[12] Netondo GW, Onyango JC, Beck E. Sorghum and salinity: II. Gas exchange and chlorophyll fluorescence of sorghum under salt stress. Crop science. 2004; 44:806-811.

[13] Reddy IN, Kim SM, Kim BK, Yoon IS, Kwon TR. Identification of rice accessions associated with K+/Na+ ratio and salt tolerance based on physiological and molecular responses. Rice Science. 2017; 24:360-364.

[14] Fahad S, Hussain S, Matloob A, Khan FA, Khaliq A, Saud S, Hassan S, Shan D, Khan F, Ullah N, Faiq M. Phytohormones and plant responses to salinity stress: a review. Plant growth regulation. 2015; 75:391-404.

[15] Fageria NK, Stone LF, dos Santos AB. Breeding for salinity tolerance. InPlant breeding for abiotic stress tolerance. Springer. 2012; 103-122.

[16] Aslam M, Qureshi RH, Ahmed N. A rapid screening technique for salt tolerance in rice (Oryza sativa L.). Plant and soil. 1993; 150:99-107.

[17] Babu NN, Vinod KK, Krishnan SG, Bhowmick PK, Vanaja T, Krishnamurthy SL, Nagarajan M, Singh NK, Prabhu KV, Singh AK. Marker based haplotype diversity of Saltol QTL in relation to seedling stage salinity tolerance in selected genotypes of rice. Indian Journal of Genetics and Plant Breeding. 2014; 74:16-25.

[18] Bernstein N. Contamination of soils with microbial pathogens originating from effluent water used for agricultural irrigation. InEGU General Assembly Conference Abstracts. 2009; 8537.

[19] Lutts S, Kinet JM, Bouharmont J. Changes in plant response to NaCl during development of rice (Oryza sativa L.) varieties differing in salinity resistance. Journal of Experimental Botany. 1995; 46:1843-1852.

[20] Khatun S, Flowers TJ. Effects of salinity on seed set in rice. Plant, Cell & Environment. 1995; 18:61-67.

[21] Fahad S, Adnan M, Noor M, Arif M, Alam M, Khan IA, Ullah H, Wahid F, Mian IA, Jamal Y, Basir *A. Major* constraints for global rice production. InAdvances in rice research for abiotic stress tolerance. Woodhead Publishing. 2019; 1-22.

[22] Szabolcs I. Salt-Affected Soils. CRC Press, Boca Raton FL, Rengasamy P. Transient salinity and subsoil constraints to dry land farming in Australian sodic soils: an overview. Aust. J. Exp. Agricm. 1989; 42: 351-361.

[23] Mengel K, Kirkby EA, Kosegarten H, Appel T. Zinc. In Principles of plant nutrition. Springer. 2001; 585-597.

[24] Rengasamy P. Transient salinity and subsoil constraints to dryland farming in Australian sodic soils: an overview. Australian Journal of Experimental Agriculture. 2002; 42:351-361.

[25] Asch F, Wopereis MC. Responses of field-grown irrigated rice cultivars to varying levels of floodwater salinity in a semi-arid environment. Field Crops Research. 2001; 70:127-137.

[26] Tang X, Mu X, Shao H, Wang H, Brestic M. Global plant-responding mechanisms to salt stress: physiological and molecular levels and implications in biotechnology. Critical reviews in biotechnology. 2015; 35:425-437.

[27] Pessarakli MO. Plant and crop stress. Handbook, Marcel deckker, New York. 1994.

[28] Joseph B, Jini D, Sujatha S. Biological and physiological perspectives of specificity in abiotic salt stress response from various rice plants. Asian J. Agric. Sci. 2010; 2:99-105.

[29] Kazemi K, Eskandari H. Effects of salt stress on germination and early seedling growth of rice (Oryza sativa) cultivars in Iran. African Journal of Biotechnology. 2011; 10:17789-17792.

[30] Jiang XJ, Zhang SH, Miao LX, Tong T, Liu ZZ, Sui YY. Effect of salt stress on rice seedling characteristics, effect of salt stress on root system at seedling stage of rice. North Rice. 2010; 40:21-24.

[31] Shereen A, Mumtaz S, Raza S, Khan MA, Solangi S. Salinity effects on seedling growth and yield components of different inbred rice lines. Pak. J. Bot. 2005; 37:131-139.

[32] Munns R. Genes and salt tolerance: bringing them together. New phytologist. 2005; 167:645-663.

[33] Munns R, James RA, Läuchli A. Approaches to increasing the salt tolerance of wheat and other cereals. Journal of experimental botany. 2006; 57:1025-1043.

[34] Hasanuzzaman M, Fujita M, Islam MN, Ahamed KU, Nahar K. Performance of four irrigated rice varieties under different levels of salinity stress. International Journal of Integrative Biology. 2009; 6:85-90.

[35] Hussain S, ZHANG JH, Zhong C, ZHU LF, CAO XC, YU SM, Bohr JA, HU JJ, JIN QY. Effects of salt stress on rice growth, development characteristics,

and the regulating ways: A review. Journal of integrative agriculture. 2017; 16:2357-2374.

[36] Cattivelli L, Rizza F, Badeck FW, Mazzucotelli E, Mastrangelo AM, Francia E, Marè C, Tondelli A, Stanca AM. Drought tolerance improvement in crop plants: an integrated view from breeding to genomics. Field crops research. 2008; 105:1-4.

[37] Suriyan CU, Supaibulwattana K, Kirdmanee C. Comparative effects of salt stress and extreme pH stress combined on glycinebetaine accumulation, photosynthetic abilities and growth characters of two rice genotypes. Rice Science. 2009; 16:274-282.

[38] Razzaque MA, Talukder NM, Islam MT, Dutta RK. Salinity effect on mineral nutrient distribution along roots and shoots of rice (Oryza sativa L.) genotypes differing in salt tolerance. Archives of Agronomy and Soil Science. 2011; 57:33-45.

[39] Amanullah I, Inamullah X. Dry matter partitioning and harvest index differ in rice genotypes with variable rates of phosphorus and zinc nutrition. Rice Science. 2016; 23:78-87.

[40] Lazof DB, Bernstein N. The NaCl induced inhibition of shoot growth: the case for disturbed nutrition with special consideration of calcium. Advances in Botanical Research. 1998; 29:113-189.

[41] Hasegawa PM, Bressan RA, Zhu JK, Bohnert HJ. Plant cellular and molecular responses to high salinity. Annual review of plant biology. 2000; 51:463-499.

[42] Verslues PE, Agarwal M, Katiyar-Agarwal S, Zhu J, Zhu JK. Methods and concepts in quantifying resistance to drought, salt and freezing, abiotic stresses that affect plant water status. The Plant Journal. 2006; 45:523-539.

[43] Stepien P, Johnson GN. Contrasting responses of photosynthesis to salt stress in the glycophyte Arabidopsis and the halophyte Thellungiella: role of the plastid terminal oxidase as an alternative electron sink. Plant physiology. 2009; 149:1154-1165.

[44] Verslues PE, Sharma S. Proline metabolism and its implications for plant-environment interaction. The Arabidopsis Book/American Society of Plant Biologists. 2010; 8.

[45] Verbruggen N, Hermans C. Proline accumulation in plants: a review. Amino acids. 2008; 35:753-759.

[46] Guinn EJ, Pegram LM, Capp MW, Pollock MN, Record MT. Quantifying why urea is a protein denaturant, whereas glycine betaine is a protein stabilizer. Proceedings of the National Academy of Sciences. 2011; 108:16932-16937.

[47] Zhu JK. Regulation of ion homeostasis under salt stress. Current opinion in plant biology. 2003; 6:441-445.

[48] Reddy MP, Sanish S, Iyengar ER. Photosynthetic studies and compartmentation of ions in different tissues of Salicornia brachiata under saline conditions. Photosynthetica (CSFR). 1992; 26:173-179.

[49] Munns R. Comparative physiology of salt and water stress. Plant, cell & environment. 2002; 25:239-250.

[50] Bernstein N, Shoresh M, Xu Y, Huang B. Involvement of the plant antioxidative response in the differential growth sensitivity to salinity of leaves vs roots during cell development. Free Radical Biology and Medicine. 2010; 49:1161-1171.

[51] Acosta-Motos JR, Ortuño MF, Bernal-Vicente A, Diaz-Vivancos P, Sanchez-Blanco MJ, Hernandez JA.

Plant responses to salt stress: adaptive mechanisms. Agronomy. 2017; 7:18.

[52] Mittova V, Guy M, Tal M, Volokita M. Salinity up-regulates the antioxidative system in root mitochondria and peroxisomes of the wild salt-tolerant tomato species Lycopersicon pennellii. Journal of experimental botany. 2004; 55:1105-1113.

[53] Kaur H, Bhatla SC. Melatonin and nitric oxide modulate glutathione content and glutathione reductase activity in sunflower seedling cotyledons accompanying salt stress. Nitric Oxide. 2016; 59:42-53.

[54] Vaidyanathan R, Kuruvilla S, Thomas G. Characterization and expression pattern of an abscisic acid and osmotic stress responsive gene from rice. Plant Science. 1999; 140:21-30.

[55] Gurmani AR, Bano A, Khan SU, Din J, Zhang JL. Alleviation of salt stress by seed treatment with abscisic acid (ABA), 6-benzylaminopurine (BA) and chlormequat chloride (CCC) optimizes ion and organic matter accumulation and increases yield of rice ('*Oryza sativa*'L.). Australian Journal of Crop Science. 2011; 5:1278-85.

[56] Maas EV, Nieman RH. Physiology of plant tolerance to salinity. Crop tolerance to suboptimal land conditions. 1978; 32:277-299.

[57] Sahi C, Singh A, Kumar K, Blumwald E, Grover A. Salt stress response in rice: genetics, molecular biology, and comparative genomics. Functional & integrative genomics. 2006; 6:263-284.

[58] Haq TU, Akhtar J, Nawaz S, Ahmad R. Morpho-physiological response of rice (Oryza sativa L.) varieties to salinity stress. Pak. J. Bot. 2009; 41:2943-2956.

[59] Bahaji A, Mateu I, Sanz A, Cornejo MJ. Common and distinctive responses of rice seedlings to saline-and osmotically-generated stress. Plant Growth Regulation. 2002; 38:83-94.

[60] Chang J, Cheong BE, Natera S, Roessner U. Morphological and metabolic responses to salt stress of rice (Oryza sativa L.) cultivars which differ in salinity tolerance. Plant Physiology and Biochemistry. 2019; 144:427-435.

[61] Razzaq A, Ali A, Safdar LB, Zafar MM, Rui Y, Shakeel A, Shaukat A, Ashraf M, Gong W, Yuan Y. Salt stress induces physiochemical alterations in rice grain composition and quality. Journal of food science. 2020; 85:14-20.

[62] Lutts S, Kinet JM, Bouharmont J. Effects of salt stress on growth, mineral nutrition and proline accumulation in relation to osmotic adjustment in rice (Oryza sativa L.) cultivars differing in salinity resistance. Plant Growth Regulation. 1996; 19:207-218.

[63] Cram WJ. Negative feedback regulation of transport in cells. The maintenance of turgor, volume and nutrient supply. InTransport in plants II. Springer. 1976; 284-316.

[64] Amirjani MR. Effect of salinity stress on growth, sugar content, pigments and enzyme activity of rice. International Journal of Botany. 2011; 7:73-81.

[65] Hurry VM, Strand A, Tobiaeson M, Gardestrom P, Oquist G. Cold hardening of spring and winter wheat and rape results in differential effects on growth, carbon metabolism, and carbohydrate content. Plant Physiology. 1995; 109: 697-706.

[66] Dubey RS, Singh AK. Salinity induces accumulation of soluble sugars and alters the activity of sugar metabolising enzymes in rice plants. Biologia Plantarum. 1999; 42:233-239.

[67] Jha BN, Singh RA. Physiological responses of rice varieties to different

levels of moisture stress. Indian Journal of Plant Physiology. 1997; 2:81-84.

[68] Thorat BS, Bagkar TA, Raut SM. Responses of rice under salinity stress: A review. IJCS. 2018; 6:1441-1447.

[69] Rahman S, Matsumuro T, Miyake H, Takeoka Y. Salinity-induced ultrastructural alterations in leaf cells of rice (*Oryza sativa* L.). Plant Production Science. 2000; 3:422-9.

[70] Baker NR. Chlorophyll fluorescence: a probe of photosynthesis in vivo. Annual Review of Plant Biology. 2008; 59:89-113.

[71] Yeo AR, Flowers SA, Rao G, Welfare K, Senanayake N, Flowers TJ. Silicon reduces sodium uptake in rice (*Oryza sativa* L.) in saline conditions and this is accounted for by a reduction in the transpirational bypass flow. Plant, Cell & Environment. 1999; 22:559-65.

[72] Asch F, Dingkuhn M, Dörffling K. Physiological stresses of irrigated rice caused by soil salinity in the Sahel. Irrigated rice in the Sahel: prospects for sustainable development. 1997:247-273.

[73] Rahman SM, Matsumuro T, Miyake H, Takeoka Y. Effects of salinity stress on the seminal root tip ultrastructures of rice seedlings (Oryza sativa L.). Plant production science. 2001; 4:103-111.

[74] Ashraf MY, Wu L. Breeding for salinity tolerance in plants. Critical Reviews in Plant Sciences. 1994; 13:17-42.

[75] Ron M, Weller JI. From QTL to QTN identification in livestock–winning by points rather than knock-out: a review. Animal genetics. 2007; 38:429-439.

[76] Salvi S, Tuberosa R. To clone or not to clone plant QTLs: present and future challenges. Trends in plant science. 2005; 10:297-304.

[77] Urao T, Yakubov B, Satoh R, Yamaguchi-Shinozaki K, Seki M, Hirayama T, Shinozaki K. A trans-membrane hybrid-type histidine kinase in Arabidopsis functions as an osmo-sensor. The Plant Cell. 1999; 11: 1743-1754.

[78] Waziri A, Kumar P, Purty RS. Saltol QTL and their role in salinity tolerance in rice. Austin J. Biotechnol. Bioeng. 2016; 3:1067.

[79] Bhuiyan MA. Efficiency in evaluating salt tolerance in rice using phenotypic and marker assisted selection. M.Sc. dissertation, Department of Genetics and Plant Breeding, Bangladesh Agricultural University, Mymensingh, Bangladesh. 2005; 96.

[80] Lin HX, Zhu MZ, Yano M, Gao JP, Liang ZW, Su WA, Hu XH, Ren ZH, Chao DY. QTLs for Na+ and K+ uptake of the shoots and roots controlling rice salt tolerance. Theoretical and Applied Genetics. 2004; 108:253-260.

[81] Deinlein U, Stephan AB, Horie T, Luo W, Xu G, Schroeder JI. Plant salt-tolerance mechanisms. Trends in plant science. 2014; 19:371-379.

[82] Bhusan D, Das DK, Hossain M, Murata Y, Hoque MA. Improvement of salt tolerance in rice ('Oryza sativa'L.) by increasing antioxidant defense systems using exogenous application of proline. Australian Journal of Crop Science. 2016; 10:50-56.

[83] Hariadi YC, Nurhayati AY, Soeparjono S, Arif I. Screening six varieties of rice (Oryzasativa) for salinity tolerance. Procedia Environmental Sciences. 2015; 28:78-87.

[84] De Leon TB, Linscombe S, Subudhi PK. Molecular dissection of seedling salinity tolerance in rice (*Oryza sativa* L.) using a high-density GBS-based SNP linkage map. Rice. 2016; 9:1-22.

[85] Flowers TJ. Improving crop salt tolerance. Journal of Experimental botany. 2004; 55:307-319.

[86] Arsa IA, Aini N, Lalel HJ. Evaluation of grain yield and aroma of upland rice (Pare Wangi var.) as response to soil moisture and salinity. Current Agriculture Research Journal. 2016; 4:35.

[87] Faiyue B, AL-Azzawi MJ, Flowers TJ. A new screening technique for salinity resistance in rice (*Oryza sativa* L.) seedlings using bypass flow. Plant, cell & environment. 2012; 35:1099-108.

[88] Thitisaksakul M, Tananuwong K, Shoemaker CF, Chun A, Tanadul OU, Labavitch JM, Beckles DM. Effects of timing and severity of salinity stress on rice (Oryza sativa L.) yield, grain composition, and starch functionality. Journal of agricultural and food chemistry. 2015; 63:2296-2304.

[89] Zeng L, Shannon MC, Grieve CM. Evaluation of salt tolerance in rice genotypes by multiple agronomic parameters. Euphytica. 2002; 127:235-245.

[90] Ismail AM, Horie T. Genomics, physiology, and molecular breeding approaches for improving salt tolerance. Annual Review of Plant Biology. 2017; 68:405-434.

[91] Jing Q, Bouman BA, Hengsdijk H, Van Keulen H, Cao W. Exploring options to combine high yields with high nitrogen use efficiencies in irrigated rice in China. European Journal of Agronomy. 2007; 26:166-177.

[92] Martinez VA, Hill WG, Knott SA. On the use of double haploids for detecting QTL in outbred populations. Heredity. 2002; 88:423-431.

[93] Karp A, Seberg OL, Buiatti M. Molecular techniques in the assessment of botanical diversity. Annals of Botany. 1996; 78:143-149.

[94] Reddy MA, Francies RM, Rasool SN, Reddy VR. Breeding for tolerance to stress triggered by salinity in rice. Int. J. Appl. Biol. Pharm. Technol. 2014; 5:167-176.

[95] Sankar PD. Rice breeding for salt tolerance. Research in Biotechnology. 2011; 2.

[96] Young ND, Tanksley SD. RFLP analysis of the size of chromosomal segments retained around the Tm-2 locus of tomato during backcross breeding. Theoretical and Applied Genetics. 1989; 77:353-359.

[97] Islam MM. Mapping salinity tolerance genes in rice (Oryza sativa L.) at reproduction stage. Ph.D. Dissertation. University of the Philippines Los Baños College, Laguna, Philippines. 2004; 1-149.

[98] Ali MN, Yeasmin L, Gantait S, Goswami R, Chakraborty S. Screening of rice landraces for salinity tolerance at seedling stage through morphological and molecular markers. Physiology and Molecular biology of plants. 2014; 20:411-423.

[99] Niones JM. Five mapping of the salinity tolerance gene on chromosome 1 of rice (*Oryza sativa* L.) using near-isogenic lines.2004.

[100] Hoshida H, Tanaka Y, Hibino T, Hayashi Y, Tanaka A, Takabe T, Takabe T. Enhanced tolerance to salt stress in transgenic rice that overexpresses chloroplast glutamine synthetase. Plant molecular biology. 2000; 43:103-111.

[101] Zhao F, Zhang H. Salt and paraquat stress tolerance results from co-expression of the Suaeda salsa glutathione S-transferase and catalase in transgenic rice. Plant cell, tissue and organ culture. 2006; 86:349-358.

[102] Zhen-hua ZH, Qiang LI, Hai-xing SO, Xiang-min RO, Ismail AM.

Responses of different rice (Oryza sativa L.) genotypes to salt stress and relation to carbohydrate metabolism and chlorophyll content. African Journal of Agricultural Research. 2012; 7:19-27.

[103] Chen X, Guo Z. Tobacco OPBP1 enhances salt tolerance and disease resistance of transgenic rice. International Journal of Molecular Sciences. 2008; 9:2601-2613.

Chapter 6

Breeding Rice for Sustainable Bioenergy Production

Manasi Dash, Abinash Mishra and Mahendra Kumar Mohanty

Abstract

Bioenergy including biofuels from lignocellulosic biomass has immense potential to meet growing energy demand of the ever-growing world population. Bioenergy will help to mitigate the environmental problems arising due to burning of fossil fuels. Rice is the staple food for more than half of the world population and is grown in more than 100 countries. Rice straw is rich in lignocellulose and several technologies are available for efficient extraction and conversion of cellulose to ethanol. Thus, the surplus rice straw can be utilised to produce biofuel, so as to replace conventional fossil fuel sources. But it is reported that the present-day rice varieties showing high lignocellulosic straw biomass have low grain yield potential. Hence, it is important to re orient the breeding strategies for developing dual purpose rice varieties that are bioenergy efficient without compromising grain yield.

Keywords: Rice, Bioenergy, Cellulose, Lignin, Cell wall architecture, Genomics, QTLs

1. Introduction

After the Paris climate change agreement in 2016, its signatories are making considerable efforts towards reducing carbon emissions into the atmosphere. Production of biofuel also called as 'green energy' will be a key target to achieve this by reducing the use of petrochemicals. Now focus is to harness ethanol from the existing ample quantity of lignocellulosic feedstocks such as rice and wheat straws, which are usually burnt in the fields thereby causing air pollution and health hazards [1]. The bioenergy crops have tremendous potential to address the twin issues of climate change and energy security by eliminating the 'food verses fuel' disputes.

Of the various crops grown worldwide, rice has an immense potential to be used as a dual-purpose crop, due to it's wide geographical distribution, covering entire tropical, subtropical and Mediterranean region of the globe [2]. High amount of cellulose (32–47%) and hemicellulose (19–27%) in rice straw, which can be converted to biofuel, has made it a potential future bioenergy crop [3–6]. But the cell wall polymers (cellulose, hemicellulose and lignin) form a complex network by crosslinking with each other. Hence, various pre-treatments are employed in order to break this complex to ensure higher amount of cellulose availability for the activity of cellulosic enzymes to yield considerable sugars. These pre-treatments are costly and environment unfriendly, so various genetic approaches can be utilised to enhance cellulose availability. Lignin, comprising three main types of monolignols, serves as a promising target to alter the cell wall architecture in different ways in rice [7–9]. Cellulose synthases genes particularly, OsCesA4, OsCesA7 and OsCesA9, associating with specific phenotypes, can also be suitably engineered to enhance cellulose content without changing lignin and other polymers in cell wall [10]. So, breeding approaches that can alter plant cell wall architecture can be used to develop bioenergy efficient rice variety but, subjected to one condition that it should not affect yield contributing traits negatively.

Usually, a negative correlation has been observed between grain yield and biomass traits. Breeding for high grain yield is associated with developing cultivars with reduced plant height and short leaves and thereby, reducing the plant biomass as a whole. The plant breeding strategies have to be reoriented towards selection of higher yielding plants with moderate biomass traits including lesser ash & potassium content in vegetative biomass. Also, the role of stay green traits, fostering greater decomposition of vegetative biomass as well as rewarding higher yield, can never be underestimated in this regard. This chapter will deal with the above said issues and measures, with the prime focus on methods for developing rice genotypes for higher yield and greater biofuel production, in subsequent heads.

2. Ethanol production from rice straw

Ethanol production is primarily centered around the lignocellulosic fraction of the plant biomass. Among all the left-over waste of crop species, rice straw is the cheapest and most abundant source of lignocellulosic feed stock. Rice straw, possessing considerable amount of cellulose (32–47%), hemicellulose (19–27%), with relatively less lignin (5–24%), is considered as one of the potent bioenergy sources [3, 5]. Various enzymes have been identified in the biosynthesis of these polymers (**Figures 1–3**) which determine the type and amount of polymer production in the plant cells. The cell wall polymers form a complex network by crosslinking with each other inside the cell walls. Hence, various pre-treatments are employed in order to break these complexes, to reduce crystallinity of cellulose (crt), degree of polymerisation (DP), increase in

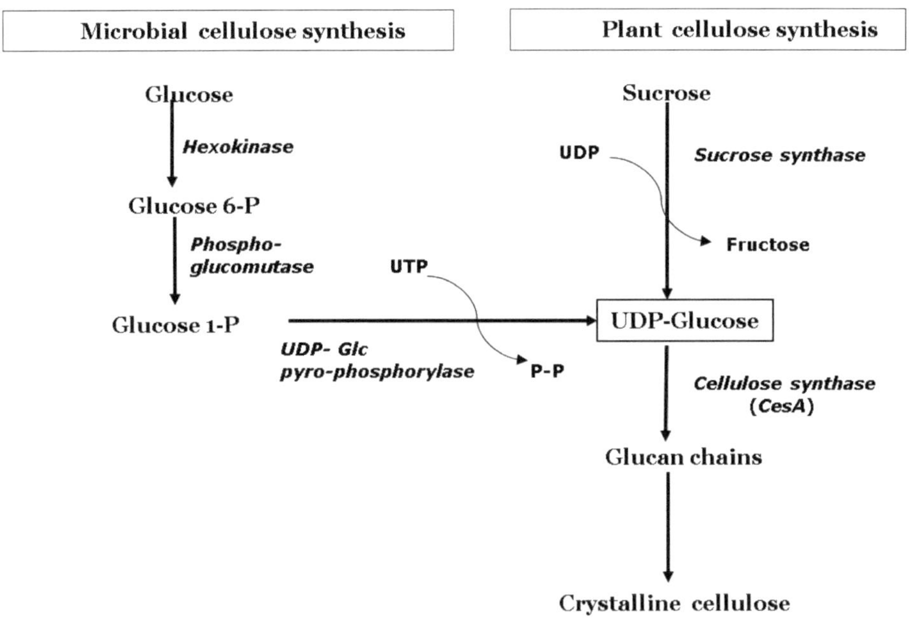

Figure 1.
Cellulose biosynthesis pathway in microbes and plants.

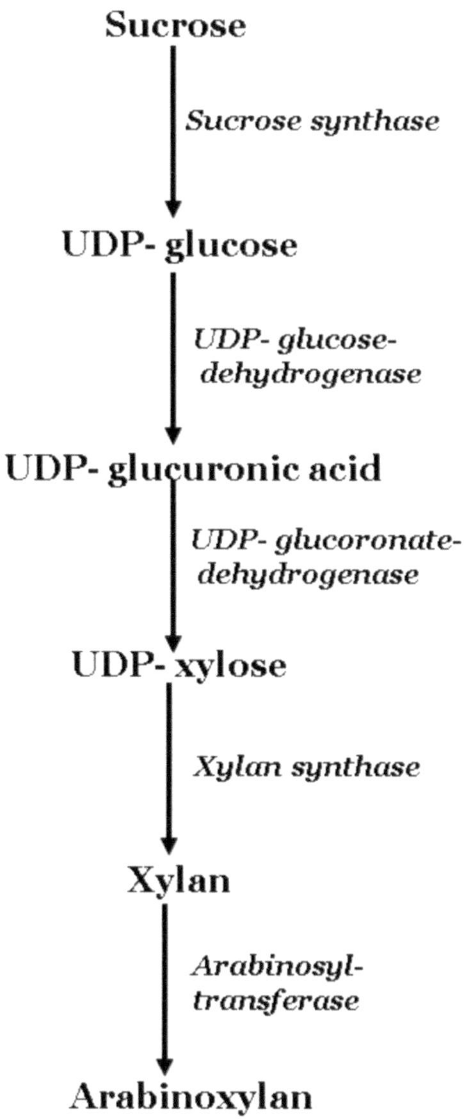

Figure 2.
Hemicellulose monomer biosynthesis pathway.

biomass surface area, and breaking the lignin seal. Chemical pre-treatment of rice straw is practiced to enable enzymatic saccharification for ethanol production [13–26].

Biological pretreatment, an eco-friendly method, overcomes the disadvantages of chemical pretreatment. White-rot fungi (*Pleurotus ostreatus*) of class Basidiomycetes are most promising microorganisms [27]. Basidiomycetes degrades lignin fraction in lignocellulosic biomass in rice straw. Patel and co-workers [28] in a study on rice straw reported that pretreatment involving *Aspergillus niger* and *Aspergillus awamori*, followed by *Saccharomyces cerevisiae* aided fermentation and recorded highest ethanol yield of 2.2 g/l. Cellulose upon hydrolysis produces glucose while hemicellulose produces hexose and pentoses [29]. Use of steam pretreatment or hydrolysis of rice straw using H_2SO_4 has also been reported [30, 31]. Pretreatment with *Aspergillus niger* increased the glucose yield from 43 to 87% [32].

Cellulose contain glucans while hemicellulose is composed of polymers of xylan, mannan, glucan, galactan and arabinan. The general process of ethanol

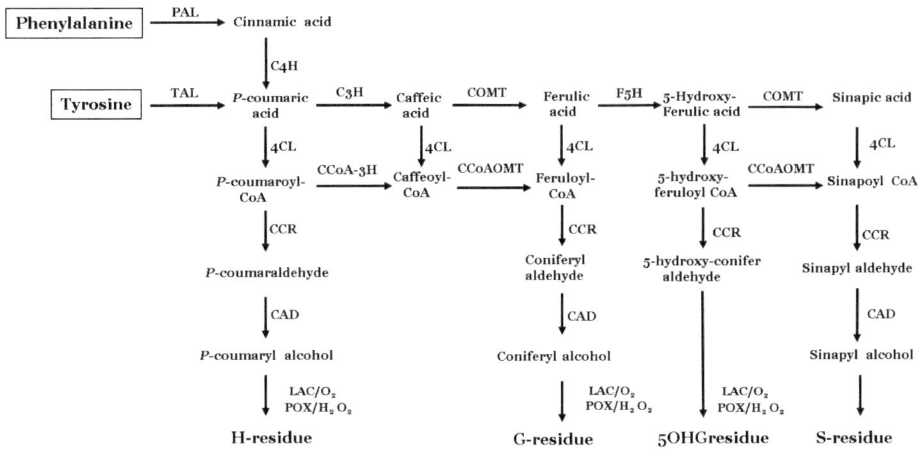

Figure 3.
Lignin biosynthesis pathways. The various enzymes are PAL [phenylalanine ammonia-lyase]; TAL [tyrosine ammonia-lyase]; C4H [cinnamate 4-hydroxylase]; C3H [4-hydroxycinnamate 3-hydroxylase]; COMT [caffeic acid 3-O- methyltransferase]; F5H [ferulate 5-hydroxylase]; 4CL [4-coumarate: CoA ligase]; CCoA-3H [coumaroyl-CoA 3-hydroxylase]; CCoAOMT [caffeoyl-CoA O-methyltransferase]; CCR [cinnamoyl-CoA reductase]; CAD [cinnamyl alcohol dehydrogenase]; LAC [laccase]; and PDX [peroxidase] (modified from Furtado et al. [11]; Vermerris and Abril [12]).

production involves conversion of cellobiose to ethanol by a series of steps of involving pre-treatment, enzymatic saccharification and fermentation as described earlier. These steps may include simultaneous saccharification and fermentation (SSF) or separate enzymatic hydrolysis and fermentation (SHF). SSF is generally used as cost incurred in the process is less [33]. In this process also, higher yield of ethanol is obtained. However, some drawbacks are observed in this process such as requirement of optimum temperature (40-50°C) for enzymatic hydrolysis, which the microorganisms cannot tolerate. This problem can be tackled by using thermophilic microorganisms such as *Kluyveromyces marxianus, Candida lusitamiae and Zymomonas mobilos* or mixed culture of *Bettanomyces clausenii* and *Saccharomyces cereviseae* [34, 35]. Shengdong and co-workers [36] employed the SSF of alkali and alkali/microwave pretreatment to generate ethanol using cellulase from *Trichoderma reesei* and *Saccharomyces cereviseae*. The ethanol concentration was 29.1 g/l and yield were 61.3% under optimum condition. Chada and co-workers [37] mentioned that SSF was superior to traditional saccharification in production of ethanol as it can improve the ethanol content by removal of end product inhibition by saccharification process. In the fermentation process alcohol is mixed with the straw to produce fermentable sugars and this is referred to as mash. This mash is fed into fractional distillation unit which differentiates alcohol from other components. The alcohol thus produced is cleaned and dehydrated to remove the water content. After cleaning and drying bioethanol is produced with a purity of 99.7% V/V.

These chemical processes for saccharification are harmful to the environment. Hence now research should be focused towards minimising or eliminating these steps by developing rice genotypes with higher saccharification efficiency (SE).

3. Role of plant breeding and biotechnology to enhance SE

As mentioned earlier the lignocellulosic biomass is primarily a complex network of various cellular constituents including cellulose, hemicellulose, lignin and interaction of a wide array of compounds like chlorophyll, waxes, oils, terpenes and phenolics, called extractives [38, 39]. It is beneficial to have knowledge on the genetics as well as correlation between biomass traits and these cellular constituents. A greater insight into the composition, structure and the synthesis of cellular constituents will help in designing suitable breeding strategies for the genetic modification of cell wall architecture and in turn development of high energy efficient rice genotypes.

3.1 Morphological and biochemical characterisation of biomass traits

In simple term, it refers to the study of various morphological, physiological, biochemical traits, associated with grain and biomass yield. Rigorous phenotyping is essential for the success of any crop improvement programme. In breeding for biofuel, although some noticeable work has been done in case of bioenergy crops like sorghum, maize and sugarcane, very limited information is available with respect to rice, which is considered as a hinderance in effective phenotyping in this regard.

It is reported that culm length, stem girth, tiller length and diameter, leaf characteristics such as leaf length, width and angle as well as leaf, stem, tiller dry weight are few key biomass traits that can be used for indirect selection [40, 41]. These traits recorded at different developmental stages will help to decipher the genetic basis of biomass partitioning and accumulation in vegetative parts.

Biochemical characterisation of rice straw cell wall polymers (cellulose, lignin and hemicellulose) is an integral part of biomass phenotyping. Many methods like use of ultrasonicator, HPLC, microarrays, Infrared absorption spectroscopy, X- ray diffraction (XRD), transmission electron microscopy (TEM) and carbon-13 nuclear magnetic resonance (C-NMR) spectroscopy can be used for quantitative estimation of cell wall polymers [42, 43].

3.2 Polymer composition of rice biomass

Cellulose is the utmost abundant organic compound available on earth. It is a linear polymer of repeating units of cellobiose molecule. Cellobiose, a β (1–4)-linked residue, is produced when two glucose molecules (one in 180 deg. rotation) are in proximity to yield a β (1–4)-linkage. These cellulose fibres impart a greater rigidity and strength to the cell wall and hence, enabling plants to exhibit a wide spectrum resistance to various biotic and abiotic factors [12]. The non-cellulosic polysaccharides further enhance the rigidity and strength of plant cell walls by cross-linking with cellulose and lignin. Various reports have suggested this cellular constituent as a mixture of various monosaccharides such as xylose, arabinose, glucose, galactose and rhamnose as well as certain acids [44, 45]. This complex nature of non-cellulosic polysaccharides as well as their involvement in cross linking with cellulose, possesses a major setback in the efficient enzymatic degradation of cellulose to produce biofuel.

Lignin, the second most abundant biopolymer after cellulose, is polymerised with three main types of monolignols namely, Syringyl alcohol (S), Coniferyl alcohol (H) and p-Coumaryl alcohol (H) [46, 47]. As a complex phenolic compound, it improves cell wall rigidity and strength, imparts resistance to a wide array of microbes [48], fosters transporting of minerals through vascular bundles [49], involves in resistance against lodging as well as abiotic anomalies [50–52].

Cellulose and hemicellulose in rice straw can be subjected to fermentation for production of biofuels. However, their efficient conversion into fermentable sugars is hindered by presence of higher amount of lignin (5–24%), ash (10–17%), silica (75% of ash) and potassium [53].

In rice, silica comprises 74.67% of the stem ash content. Both high ash and high silica ($SiO2$) silica content of ash negatively affect biochemical conversion of lignocellulosic feedstock [11, 53]. High silica content reduces the availability of cellulose to enzymatic digestion and thus, reducing saccharification efficiency. Besides this, high silica accumulation in the cell walls disrupts the cellulosic microfibrils and such aberration hinders overall sugar release and ultimately, ethanol yields in subsequent stages. Therefore, considerable efforts are required to engineer silica content along with lignin and non-carbohydrate polysaccharides content to develop rice genotypes, amenable to greater enzymatic digestibility.

Although, different enzymatic and chemical pre-treatment methods are being employed for the disruption of this complex network but these procedures are energy intensive, costly and harmful to the environment. Hence, genetically modifying the cell wall architecture by employing conventional and modern breeding methods are beneficial for sustainable biofuel production [11, 54].

3.3 Modifications in polymer composition for elevating cellulose utilisation

As discussed earlier, lignin serves as a key element in cross linking of cellulose and hemicellulosic polysaccharides. This feature is beneficial to the rice plant as it helps it to counteract biotic and abiotic stress but it is a limiting feature for biofuel production. The cross linking creates a barrier for the cellulose degrading enzymes to freely access cellulose for conversion. So, efforts are being made towards reducing the degree of lignification and cross linking through various approaches so as to enhance the efficiency of cellulose degrading enzymes.

3.3.1 Modification of lignin

Some noticeable work has been done to alter the plant cell wall architecture with the help of biotechnology in model dicot plants such as Arabidopsis and tobacco (**Table 1**). The purpose behind these experiments has been the downregulation of key genes involved in monolignol biosynthesis, as well as the essential enzymes involving in polymerisation thereof. Nearly 40% reduction in total lignin content was achieved by downregulating laccases and peroxidases in the Arabidopsis [61] and tobacco mutants [62], respectively. So, these available reports documenting successful reduction of lignin composition in model dicot plants can be judiciously used by the researchers to favourably alter the cell wall architecture of the less exploited prospective biofuel crops such as rice. The *japonica* rice 'Nipponbare' harbouring an Arabidopsis TF (SHN), was found to be deficient in total lignin content. Expression of essential genes such as *CAD* (cinnamyl alcohol dehydrogenase) and 4-CL (4- coumarate- CoA ligase) were reported to be repressed, which might have contributed in producing lower lignin content [9].

Alternatively, there is another way of altering the plant cell wall architecture, by curbing the expression of essential genes involved in lignin monomers synthesis (**Figure 3**). In rice, flexible culm (*fc1*) mutant with repressed *CAD* gene, a cinnamyl alcohol dehydrogenase gene, was reported to synthesise reduced level of H and G lignin monomers [7]. Zhang and co-workers, [60] were able to produce some transgenics with improved saccharification efficiency as compared to wildtype by targeting same *OsCAD2* gene in rice. Apart from these genes, few other genes including caffeoyl-CoA- methyl transferase (*CCoAOMT*) and caffeic acid o-methyl transferase (*COMT*) were genetically engineered in different species such as alfalfa, canola, maize, poplar, tobacco and sugarcane, to alter the lignin monomers composition [63–66, 70–74]. Several reports enumerating the modifications of some key TFs such as *OsMYB103L* are also available for improved plant architecture in rice [58, 59].

Cell wall polymer regulation	Genes	Approach	Phenotypes	Reference(s)
Cellulose synthesis	*OsCesA4*	Bc11 (G858R), NE1031, ND5658 (TOS17)	Reduce cellulose; affect growth	[10]
	OsCesA7	NC0258, ND8759 (TOS17)	Reduce cellulose; affect growth	[10]
	OsCesA9	ND2359 (TOS17)	Reduce cellulose; affect growth	[10]
	OsGH9B	Osfc4; Osfc11 (T-DNA)	Reduced Cr1(Cellulose crystallinity)	[55]
	OsGH9B 1; OsGH9B 3	pCAMBIAI11300: *OsGH9B*	Reduced Cr1; DP (Degree of Polymerisation)	[56]
	Fc17 (OsCESA4)	F_2 (fc17 × MH 63)	Reduced Cr1	[57]
Cellulose regulation	*OsMYB103L*	pUbi:: OsMYB103L (OE)	Increased secondary cell wall	[58]
	OsMYB103L: NAC29,31	pUbi:: OsMYB61; NAC29, 31 (OE)	Increased secondary cell wall	[59]
Lignin synthesis	*OsCAD2*	CRISPR/Cas9	Altered H and G residues; reduced lignin	[60]
	Fc1 (Cinnamyl-alcohol dehydrogenase)	Fc1 (T-DNA)	Reduced H and G residues; reduced lignin	[8]
	4CL; CAD	p35S::AtSHN2 (OE)	Reduced lignin	[9]
	Bc1 (COBRA like protein)	bc1 (γ-rays)	Reduced lignin	[7]
	Laccases	LAC4/LAC7 (T- DNA)	Reduced lignin; hinderance in deposition of G subunits	[61]
	Peroxidases	TP60 (RNAi)	Reduced lignin; reduced G and S residues	[62]
	COMT	pWFOsC4H::Bg4CLi (RNAi)	Reduced lignin; reduced S/G ratio	[63]
	CCoAOMT	*CCoAOMT (RNAi)*	Altered lignin subunit composition	[64–66]
Hemicellulose synthesis	*OsXAX1 (GT61)*	axa1 (T-DNA)	Reduced xylose, ferulic acid, coumaric acid	[67]
	BAHD acetyl transferase	pUbi: OsAt10	Reduction in matrix bound ferulic acid	[68]
	OsIRX10 (GT47)	*OsIRX10* (RGT6229D)	Reduced X/A; affect growth	[69]

Table 1.
Candidate genes for preferable altering the cell wall polymers (cellulose, hemicellulose and lignin) in plant system.

3.3.2 Modification of hemicellulose

A general trade-off has been discovered between saccharification efficiency and ferulic acid [75, 76]. Bartley and co-workers, [68] reported the possible role of OsAt10, a BAHD acetyltransferase gene in achieving higher sugar release by favourably modifying glucuronoarabinoxylan (GAX) in rice. Young leaf tissues of the genetically engineered plants were found to be deficient in ferulic acid (FA). The possible role of other genes such as OsXAX1 and OsIRX10 were known to reciprocate similar results in rice [67, 68].

3.4 Role of cellulose synthase genes

Cellulose synthase enzymes are pivotal for cellulose synthesis. These proteins organise to form a hexameric 'rosette' structure approx. 25–30 nm diameter [77]. The plant cellulose synthase (*Ces A*) genes were first identified during random sequencing of cotton ESTs [78] and its role in cellulose synthesis was first reported in Arabidopsis Ces *A* mutants [10, 79]. The Ces *A* gene family was also identified in rice, maize, barely and poplar [57, 80–92].

Tanaka and co-workers [10] generated four different introgressed lines, showing brittle culm phenotypes by suitably introgressing Tos17, a retroposons in the genetic background of rice wildtype. They identified three cellulose synthase genes namely, *OsCesA4*, *OsCesA7* and *OsCesA9* on three different chromosomes. The mutant *Osfc16* with a mutation on *CesA9*- conserved sequence was found with altered cellulose crystallinity (crt1), which possibly enhanced the saccharification efficiency [93]. In a similar experiment, conserved site of another potential cellulose synthase CesA4 is mutated to alter cellulose crystallinity (crt1) for enhanced cellulose synthesis in *fc17* mutants [57]. Considerable efforts have been made to alter various structural properties of cell wall constituents including cellulose crystallinity (crt1) and degree of polymerisation (DP) which usually negatively affect the saccharification potential. In this regard, some noticeable work has been done to identify and characterise few genes of glycoside hydrolase family (*OsGH9B 1, 3 and 16*), promising candidate genes for favourably modifying structural properties of cell wall polymers as well as cellulose synthesis in rice [55, 56]. Beside cellulose synthase genes, other genes including KORRIGN [94–96], COBRA-like protein [7] and KOBITO [80] need to be explored properly to develop energy efficient elite cultigens in rice.

3.5 Genomics and QTL identification for biomass traits

Correlation between biomass traits and grain yield in rice is negative. Breeding varieties for high grain yield usually involves designing the varieties for medium plant height with short erect flag leaves which in turn affect the total biomass yield. This can be addressed to some extent by crossing rice cultivars, showing high polymorphism for grain yield potential as well as biomass traits and identify the candidate genes or QTLs involved. After the successful mapping of genes or, QTLs, the linked markers can be used for marker assisted selection (MAS) as well as can be used to screen the existing wild types or landraces for dual characteristics. As we have discussed earlier, cell wall polymers i.e., cellulose, hemicellulose and lignin composition can be altered for improved saccharification traits, hence, it is essential to search for the genetic link between cell-wall polymer composition and grain yield in order to breed dual purpose rice cultigens [97–103]. Gui- Fu and co-workers [97] identified few major QTLs associated with three plant traits namely, total biomass yield, straw yield and grain yield by developing suitable doubled haploid population. A QTL co-associated with both cell wall polymer composition and heading date (HD17) has also been identified by crossing parents with considerable polymorphism for the dual characters [102]. Recently, Genome wide association survey (GWAS) involving high throughput molecular markers (SNPs) were employed to identify the genomic regions exhibiting significant association between markers and phenotypic trait and characterise the candidate genes involved [101]. Dissecting the genomic fragments involving lignin and cellulose biosynthesis is possible now with the application of GWAS technique [100].

3.6 Plant breeding strategies for improving biomass traits

Pre- existence of variability is of paramount importance in any crop improvement programme. Selection, being the core stone of plant breeding activities, is employed to harness the existing variability present in various germplasms including wild types and landraces, before creating additional variability by mutation.

A preferred high HI for good yield reduces the vegetative biomass of the rice as a whole including reduction in plant height. There is a trade-off for plant height vs. biomass yield. Hence, the role of long- culm rice cultivars in breeding high energy efficient varieties has been given due consideration [104–108]. However, an increase in culm length may increase the risk of plant lodging, which is a major factor influencing rice grain yield stability especially in direct-seeded rice. A thick culm with tolerable lignin content in cell wall will decrease the risk of plant lodging. So, there should be a balance between the cell wall constitution and morphological characters. Hence, judicious selection of genotypes for increased plant height with thick culm along with high grain yield can address the negative impact exerted by short culm height on overall biomass production. Few researches have enumerated the importance of selecting certain traits such as stem girth, plant height, leaf, sheath and stem weight for higher biomass yield in rice [40, 108, 109].

Another way for breeding dual purpose cultigens is to incorporate 'stay green' traits in cultivated type [110, 111]. Varieties possessing these traits are able to maintain higher photosynthetic activity at post-flowering stage, increasing yield thereof. At the same time, higher decomposability of these traits could serve the dual objectives as discussed above. Hence, there is a possible opportunity to exploit this stay green character in developing dual purpose rice genotypes as it has been exploited in other biofuel/bioenergy crops. Also, this character genetically enhances the photosynthetic efficiency, there may be no need to apply extra N inputs. Nevertheless, more research is required in this aspect. As of till date there are no reports of this strategy being exploited in breeding rice varieties for dual purpose.

Next it is important to identify the genetic loci (QTL) associated with these stay green trait and the markers flanking those regions [112–115]. Various biotechnological tools can then be employed for their successful integration into the plant genome or alternatively, marker aided selection (MAS) can be employed for varietal improvement. Also, heterosis breeding can be used to exploit the possible heterotic gene combinations in remodelling the plant architecture for higher biomass yield in rice, as it has been done in sorghum which possesses similar architectural traits [116–119].

4. Conclusion

There is an urgent need to address greenhouse gas emissions (GHGs) and climate changes occurring due to rampant use of fossil fuels. Rice straw, being an abundant source of lignocellulosic feedstock, has the potential to produce green energy to address the above said global concerns. Lignin and hemicellulose complexes act as hinderance to produce energy efficient cultigens and hence, various studies are being made to down regulate the gene involved in their biosynthesis without affecting the plant system and cellulose concentration. Directly engineering cellulose synthase genes also provides an alternative opportunity in designing plant cell wall architecture. Stay green traits and heterosis breeding enhance the opportunity of developing energy efficient varieties to a greater extent. Thus, the role of plant breeding can never be bypassed as careful selection of individuals for dual traits will be highly rewarding in achieving the goal of growing dual-purpose rice varieties.

Acknowledgements

The authors thank Professor, Biofuel Chair, BPCL-OUAT project, Odisha, India for his guidance and encouragement to write this chapter. The authors also acknowledge the funding provided by the BPCL-OUAT Biofuel project implemented in OUAT, Bhubaneswar, India for the research work.

Conflict of interest

The authors declare no conflict of interest.

Author details

Manasi Dash[1*], Abinash Mishra[1] and Mahendra Kumar Mohanty[2]

1 Department of Plant Breeding and Genetics, College of Agriculture, Odisha University of Agriculture and Technology, Bhubaneswar, Odisha, India

2 College of Agricultural Engineering, Odisha University of Agriculture and Technology, Bhubaneswar, Odisha, India

*Address all correspondence to: mdpbgouat@gmail.com

IntechOpen

© 2021 The Author(s). Licensee IntechOpen. This chapter is distributed under the terms of the Creative Commons Attribution License (http://creativecommons.org/licenses/by/3.0), which permits unrestricted use, distribution, and reproduction in any medium, provided the original work is properly cited.

References

[1] Mussatto SI, Roberto IC. Optimal experimental condition for hemicellulosic hydrolyzate treatment with activated charcoal for xylitol production. Biotechnology progress. 2004; 20(1): 134-139.

[2] Abraham A, Mathew AK, Sindhu R, Pandey A, Binod P. Potential of rice straw for bio-refining: an overview. Bioresource Technology. 2016 Sep 1; 215:29-36.

[3] Maiorella BI. Ethanol industrial chemicals. Biochem Fuels.

[4] Zamora R, JA SC. Production of an acid extract of rice straw. Acta cientifica venezolana. 1995 Jan 1;46(2):135-139.

[5] Garrote G, Domínguez H, Parajó JC. Autohydrolysis of corncob: study of non-isothermal operation for xylooligosaccharide production. Journal of Food Engineering. 2002 May 1;52(3):211-218.

[6] Saha BC. Hemicellulose bioconversion. Journal of industrial microbiology and biotechnology. 2003 May 1;30(5):279-291.

[7] Li Y, Qian Q, Zhou Y, Yan M, Sun L, Zhang M, Fu Z, Wang Y, Han B, Pang X, Chen M. BRITTLE CULM1, which encodes a COBRA-like protein, affects the mechanical properties of rice plants. The Plant Cell. 2003 Sep 1;15(9): 2020-2031.

[8] Li X, Yang Y, Yao J, Chen G, Li X, Zhang Q, Wu C. FLEXIBLE CULM 1 encoding a cinnamyl-alcohol dehydrogenase controls culm mechanical strength in rice. Plant molecular biology. 2009 Apr;69(6): 685-697.

[9] Ambavaram MM, Krishnan A, Trijatmiko KR, Pereira A. Coordinated activation of cellulose and repression of lignin biosynthesis pathways in rice. Plant physiology. 2011 Feb 1;155(2): 916-931.

[10] Tanaka K, Murata K, Yamazaki M, Onosato K, Miyao A, Hirochika H. Three distinct rice cellulose synthase catalytic subunit genes required for cellulose synthesis in the secondary wall. Plant physiology. 2003 Sep 1;133(1):73-83.

[11] Furtado A, Lupoi JS, Hoang NV, Healey A, Singh S, Simmons BA, Henry RJ. Modifying plants for biofuel and biomaterial production. Plant biotechnology journal. 2014 Dec;12(9):1246-1258.

[12] Vermerris W, Abril A. Enhancing cellulose utilization for fuels and chemicals by genetic modification of plant cell wall architecture. Current opinion in biotechnology. 2015 Apr 1;32:104-112.

[13] Hideno A, Inoue H, Tsukahara K, Fujimoto S, Minowa T, Inoue S, Endo T, Sawayama S. Wet disk milling pretreatment without sulfuric acid for enzymatic hydrolysis of rice straw. Bioresource Technology. 2009 May 1;100(10):2706-2711.

[14] Bak JS, Ko JK, Han YH, Lee BC, Choi IG, Kim KH. Improved enzymatic hydrolysis yield of rice straw using electron beam irradiation pretreatment. Bioresource Technology. 2009 Feb 1;100(3):1285-1290.

[15] Xiong J, Ye J, Liang WZ, Fan PM. Influence of microwave on the ultrastructure of cellulose I. Journal of South China University Technology. 2000;28(1):84-89.

[16] Azuma JI, Higashino J, Isaka M, Koshijima T. < Original> Microwave Irradiation of Lignocellulosic Materials: IV. Enhancement of Enzymatic

Susceptibility of Microwave-irradiated Softwoods. Wood research: bulletin of the Wood Research Institute Kyoto University. 1985 Feb 28;71:13-24.

[17] Wongjewboot I, Kangsadan T, Kongruang S, Burapatana V, Pripanapong P. Ethanol production from rice straw using ultrasonic pretreatment. In2010 International Conference on Chemistry and Chemical Engineering 2010 Aug 1 (pp. 16-19). IEEE.

[18] Wi SG, Choi IS, Kim KH, Kim HM, Bae HJ. Bioethanol production from rice straw by popping pretreatment. Biotechnology for biofuels. 2013 Dec;6(1):1-7.

[19] Gáspár M, Kálmán G, Réczey K. Corn fiber as a raw material for hemicellulose and ethanol production. Process Biochemistry. 2007 Jul 1;42(7):1135-1139.

[20] Zhang Q, Cai W. Enzymatic hydrolysis of alkali-pretreated rice straw by Trichoderma reesei ZM4-F3. Biomass and Bioenergy. 2008 Dec 1;32(12):1130-1135.

[21] Cha YL, Yang J, Ahn JW, Moon YH, Yoon YM, Yu GD, An GH, Choi IH. The optimized CO 2-added ammonia explosion pretreatment for bioethanol production from rice straw. Bioprocess and biosystems engineering. 2014 Sep;37(9):1907-1915.

[22] Harun S, Geok SK. Effect of sodium hydroxide pretreatment on rice straw composition. Indian J Sci Technol. 2016 Jun;9(21):1-9.

[23] Kim SB, Lee SJ, Lee JH, Jung YR, Thapa LP, Kim JS, Um Y, Park C, Kim SW. Pretreatment of rice straw with combined process using dilute sulfuric acid and aqueous ammonia. Biotechnology for biofuels. 2013 Dec;6(1):1-1.

[24] Chiranjeevi T, Mattam AJ, Vishwakarma KK, Uma A, Peddy VR, Gandham S, Ravindra Velankar H. Assisted single-step acid pretreatment process for enhanced delignification of rice straw for bioethanol production. ACS Sustainable Chemistry & Engineering. 2018 Jun 5;6(7):8762-8774.

[25] Hon DN, Shiraishi N. Wood and cellulosic chemistry, revised, and expanded. CRC press; 2000 Nov 8.

[26] Wei CJ, Cheng CY. Effect of hydrogen peroxide pretreatment on the structural features and the enzymatic hydrolysis of rice straw. Biotechnology and bioengineering. 1985 Oct;27(10):1418-1426.

[27] Taniguchi M, Suzuki H, Watanabe D, Sakai K, Hoshino K, Tanaka T. Evaluation of pretreatment with Pleurotus ostreatus for enzymatic hydrolysis of rice straw. Journal of bioscience and bioengineering. 2005 Dec 1;100(6):637-643.

[28] Patel SJ, Onkarappa R, Shobha KS. Comparative study of ethanol production from microbial pretreated agricultural residues. Journal of Applied Sciences and Environmental Management. 2007;11(4):137-141.

[29] Taherzadeh MJ, Niklasson C. Ethanol from lignocellulosic materials: pretreatment, acid and enzymatic hydrolyses, and fermentation.

[30] Gonzalez Valdes I, Lopez Planes R. Study of the hydrolysis of rice straw with sulfuric acid under moderate conditions. Rev. Cienc. Quim.;(Cuba). 1983 Jan 1;14(1).

[31] Abedinifar S, Karimi K, Khanahmadi M, Taherzadeh MJ. Ethanol production by Mucor indicus and Rhizopus oryzae from rice straw by separate hydrolysis and fermentation. Biomass and bioenergy. 2009 May 1;33(5):828-833.

[32] Aderemi BO, Abu E, Highina BK. The kinetics of glucose production from rice straw by Aspergillus niger. African journal of Biotechnology. 2008; 7(11).

[33] Wyman CE. Ethanol from lignocellulosic biomass: technology, economics, and opportunities. Bioresource Technology. 1994 Jan 1;50(1):3-15.

[34] Spindler DD, Wyman CE, Mohagheghi A, Grohmann K. Thermotolerant yeast for simultaneous saccharification and fermentation of cellulose to ethanol. Applied Biochemistry and Biotechnology. 1988 Apr;17(1):279-293.

[35] Golias H, Dumsday GJ, Stanley GA, Pamment NB. Evaluation of a recombinant Klebsiella oxytoca strain for ethanol production from cellulose by simultaneous saccharification and fermentation: comparison with native cellobiose-utilising yeast strains and performance in co-culture with thermotolerant yeast and Zymomonas mobilis. Journal of biotechnology. 2002 Jun 26;96(2):155-168.

[36] Shengdong Z, Yuanxin W, Yufeng Z, Shaoyong T, Yongping X, Ziniu Y, Xuan Z. Fed-batch simultaneous saccharification and fermentation of microwave/acid/alkali/H2O2 pretreated rice straw for production of ethanol. Chemical Engineering Communications. 2006 May 1;193(5):639-648.

[37] Chadha BS, Kanwar SS, Garcha HS. Simultaneous saccharification and fermentation of rice straw into ethanol. Acta microbiologica et immunologica hungarica. 1995 Jan 1;42(1):71-75.

[38] Browning BL. The wood-water relationship. The chemistry of wood. Interscience Publishers, Willey, New York. 1963.

[39] Carroll A, Somerville C. Cellulosic biofuels. Annual review of plant biology. 2009 Jun 2;60:165-182.

[40] Jahn CE, Mckay JK, Mauleon R, Stephens J, McNally KL, Bush DR, Leung H, Leach JE. Genetic variation in biomass traits among 20 diverse rice varieties. Plant Physiology. 2011 Jan 1;155(1):157-168.

[41] Slavov GT, Nipper R, Robson P, Farrar K, Allison GG, Bosch M, Clifton-Brown JC, Donnison IS, Jensen E. Genome-wide association studies and prediction of 17 traits related to phenology, biomass and cell wall composition in the energy grass Miscanthus sinensis. New phytologist. 2014 Mar;201(4):1227-1239.

[42] Bhattacharyya P, Bhaduri D, Adak T, Munda S, Satapathy BS, Dash PK, Padhy SR, Pattanayak A, Routray S, Chakraborti M, Baig MJ. Characterization of rice straw from major cultivars for best alternative industrial uses to cutoff the menace of straw burning. Industrial Crops and Products. 2020 Jan 1;143:111919.

[43] Harris D, Bulone V, Ding SY, DeBolt S. Tools for cellulose analysis in plant cell walls. Plant physiology. 2010 Jun 1;153(2):420-426.

[44] Esteghlalian AR, Srivastava V, Gilkes N, Gregg DJ, Saddler JN. An overview of factors influencing the enzymatic hydrolysis of lignocellulosic feedstocks.

[45] Wyman CE, Decker SR, Himmel ME, Brady JW, Skopec CE, Viikari L. Hydrolysis of cellulose and hemicellulose. Polysaccharides: Structural diversity and functional versatility. 2005;1:1023-1062.

[46] Miao YC, Liu CJ. ATP-binding cassette-like transporters are involved in the transport of lignin precursors across plasma and vacuolar membranes.

Proceedings of the National Academy of Sciences. 2010 Dec 28;107(52):22728-22733.

[47] Ralph J. Hydroxycinnamates in lignification. Phytochemistry Reviews. 2010 Mar 1;9(1):65-83.

[48] Ithal N, Recknor J, Nettleton D, Maier T, Baum TJ, Mitchum MG. Developmental transcript profiling of cyst nematode feeding cells in soybean roots. Molecular Plant-Microbe Interactions. 2007 May;20(5):510-525.

[49] Schuetz M, Benske A, Smith RA, Watanabe Y, Tobimatsu Y, Ralph J, Demura T, Ellis B, Samuels AL. Laccases direct lignification in the discrete secondary cell wall domains of protoxylem. Plant Physiology. 2014 Oct 1;166(2):798-807.

[50] Tripathi SC, Sayre KD, Kaul JN, Narang RS. Growth and morphology of spring wheat (Triticum aestivum L.) culms and their association with lodging: effects of genotypes, N levels and ethephon. Field Crops Research. 2003 Dec 1;84(3):271-290.

[51] Moura JC, Bonine CA, de Oliveira Fernandes Viana J, Dornelas MC, Mazzafera P. Abiotic and biotic stresses and changes in the lignin content and composition in plants. Journal of integrative plant biology. 2010 Apr;52(4):360-376.

[52] Liu Q, Luo L, Zheng L. Lignins: biosynthesis and biological functions in plants. International journal of molecular sciences. 2018 Feb;19(2):335.

[53] Binod P, Sindhu R, Singhania RR, Vikram S, Devi L, Nagalakshmi S, Kurien N, Sukumaran RK, Pandey A. Bioethanol production from rice straw: an overview. Bioresource technology. 2010 Jul 1;101(13):4767-4774.

[54] Wang P, Dudareva N, Morgan JA, Chapple C. Genetic manipulation of lignocellulosic biomass for bioenergy. Current opinion in chemical biology. 2015 Dec 1;29:32-39.

[55] Xie G, Yang B, Xu Z, Li F, Guo K, Zhang M, Wang L, Zou W, Wang Y, Peng L. Global identification of multiple OsGH9 family members and their involvement in cellulose crystallinity modification in rice. PloS one. 2013 Jan 4;8(1):e50171.

[56] Huang J, Xia T, Li G, Li X, Li Y, Wang Y, Wang Y, Chen Y, Xie G, Bai FW, Peng L. Overproduction of native endo-β-1, 4-glucanases leads to largely enhanced biomass saccharification and bioethanol production by specific modification of cellulose features in transgenic rice. Biotechnology for Biofuels. 2019 Dec;12(1):1-5.

[57] Li F, Liu S, Xu H, Xu Q. A novel FC17/CESA4 mutation causes increased biomass saccharification and lodging resistance by remodeling cell wall in rice. Biotechnology for biofuels. 2018 Dec;11(1):1-13.

[58] Yang C, Li D, Liu X, Ji C, Hao L, Zhao X, Li X, Chen C, Cheng Z, Zhu L. OsMYB103L, an R2R3-MYB transcription factor, influences leaf rolling and mechanical strength in rice (Oryza sativa L.). BMC plant biology. 2014 Dec;14(1):1-5.

[59] Huang D, Wang S, Zhang B, Shang-Guan K, Shi Y, Zhang D, Liu X, Wu K, Xu Z, Fu X, Zhou Y. A gibberellin-mediated DELLA-NAC signaling cascade regulates cellulose synthesis in rice. The Plant Cell. 2015 Jun 1;27(6):1681-1696.

[60] Zhang G, Wang L, Li X, Bai S, Xue Y, Li Z, Tang SW, Wang Y, Wang Y, Hu Z, Li P. Distinctively altered lignin biosynthesis by site-modification of OsCAD2 for enhanced biomass saccharification in rice. GCB Bioenergy. 2021 Feb;13(2):305-319.

[61] Berthet S, Demont-Caulet N, Pollet B, Bidzinski P, Cézard L, Le Bris P, Borrega N, Hervé J, Blondet E, Balzergue S, Lapierre C. Disruption of LACCASE4 and 17 results in tissue-specific alterations to lignification of Arabidopsis thaliana stems. The Plant Cell. 2011 Mar 1;23(3):1124-1137.

[62] Blee KA, Choi JW, O'Connell AP, Schuch W, Lewis NG, Bolwell GP. A lignin-specific peroxidase in tobacco whose antisense suppression leads to vascular tissue modification. Phytochemistry. 2003 Sep 1;64(1):163-176.

[63] Jung JH, Fouad WM, Vermerris W, Gallo M, Altpeter F. RNAi suppression of lignin biosynthesis in sugarcane reduces recalcitrance for biofuel production from lignocellulosic biomass. Plant biotechnology journal. 2012 Dec;10(9):1067-1076.

[64] Marita JM, Ralph J, Hatfield RD, Guo D, Chen F, Dixon RA. Structural and compositional modifications in lignin of transgenic alfalfa down-regulated in caffeic acid 3-O-methyltransferase and caffeoyl coenzyme A 3-O-methyltransferase. Phytochemistry. 2003 Jan 1;62(1): 53-65.

[65] Chen F, Dixon RA. Lignin modification improves fermentable sugar yields for biofuel production. Nature biotechnology. 2007 Jul;25(7): 759-761.

[66] Pinçon G, Maury S, Hoffmann L, Geoffroy P, Lapierre C, Pollet B, Legrand M. Repression of O-methyltransferase genes in transgenic tobacco affects lignin synthesis and plant growth. Phytochemistry. 2001 Aug 1;57(7):1167-1176.

[67] Chiniquy D, Sharma V, Schultink A, Baidoo EE, Rautengarten C, Cheng K, Carroll A, Ulvskov P, Harholt J, Keasling JD, Pauly M. XAX1 from glycosyltransferase family 61 mediates xylosyltransfer to rice xylan. Proceedings of the National Academy of Sciences. 2012 Oct 16;109(42):17117-17122.

[68] Bartley LE, Peck ML, Kim SR, Ebert B, Manisseri C, Chiniquy DM, Sykes R, Gao L, Rautengarten C, Vega-Sánchez ME, Benke PI. Overexpression of a BAHD acyltransferase, OsAt10, alters rice cell wall hydroxycinnamic acid content and saccharification. Plant Physiology. 2013 Apr 1;161(4):1615-1633.

[69] Chen X, Vega-Sánchez ME, Verhertbruggen Y, Chiniquy D, Canlas PE, Fagerström A, Prak L, Christensen U, Oikawa A, Chern M, Zuo S. Inactivation of OsIRX10 leads to decreased xylan content in rice culm cell walls and improved biomass saccharification. Molecular Plant. 2013 Mar 1;6(2):570-573.

[70] Meyermans H, Morreel K, Lapierre C, Pollet B, De Bruyn A, Busson R, Herdewijn P, Devreese B, Van Beeumen J, Marita JM, Ralph J. Modifications in lignin and accumulation of phenolic glucosides in poplar xylem upon down-regulation of caffeoyl-coenzyme A O-methyltransferase, an enzyme involved in lignin biosynthesis. Journal of Biological Chemistry. 2000 Nov 24;275(47):36899-36909.

[71] Zhong R, Morrison WH, Himmelsbach DS, Poole FL, Ye ZH. Essential role of caffeoyl coenzyme A O-methyltransferase in lignin biosynthesis in woody poplar plants. Plant Physiology. 2000 Oct 1;124(2):563-578.

[72] Chen F, Srinivasa Reddy MS, Temple S, Jackson L, Shadle G, Dixon RA. Multi-site genetic modulation of monolignol biosynthesis suggests new routes for formation of syringyl lignin and wall-bound ferulic

acid in alfalfa (Medicago sativa L.). The Plant Journal. 2006 Oct;48(1):113-124.

[73] Bhinu VS, Li R, Huang J, Kaminskyj S, Sharpe A, Hannoufa A. Perturbation of lignin biosynthesis pathway in Brassica napus (canola) plants using RNAi. Canadian journal of plant science. 2009 May 1;89(3):441-453.

[74] Li X, Chen W, Zhao Y, Xiang Y, Jiang H, Zhu S, Cheng B. Downregulation of caffeoyl-CoA O-methyltransferase (CCoAOMT) by RNA interference leads to reduced lignin production in maize straw. Genetics and molecular biology. 2013;36(4):540-546.

[75] Grabber JH, Ralph J, Hatfield RD. Ferulate cross-links limit the enzymatic degradation of synthetically lignified primary walls of maize. Journal of Agricultural and Food Chemistry. 1998 Jul 20;46(7):2609-2614.

[76] Lam TB, Iiyama K, Stone BA. Hot alkali-labile linkages in the walls of the forage grass Phalaris aquatica and Lolium perenne and their relation to in vitro wall digestibility. Phytochemistry. 2003 Sep 1;64(2):603-607.

[77] Taylor NG. Cellulose biosynthesis and deposition in higher plants. New Phytologist. 2008 Apr;178(2):239-252.

[78] Pear JR, Kawagoe Y, Schreckengost WE, Delmer DP, Stalker DM. Higher plants contain homologs of the bacterial celA genes encoding the catalytic subunit of cellulose synthase. Proceedings of the National Academy of Sciences. 1996 Oct 29;93(22):12637-12642.

[79] Arioli T, Peng L, Betzner AS, Burn J, Wittke W, Herth W, Camilleri C, Höfte H, Plazinski J, Birch R, Cork A. Molecular analysis of cellulose biosynthesis in Arabidopsis. Science. 1998 Jan 30;279(5351):717-720.

[80] Holland N, Holland D, Helentjaris T, Dhugga KS, Xoconostle-Cazares B, Delmer DP. A comparative analysis of the plant cellulose synthase (CesA) gene family. Plant physiology. 2000 Aug 1;123(4):1313-1324.

[81] Taylor NG, Scheible WR, Cutler S, Somerville CR, Turner SR. The irregular xylem3 locus of Arabidopsis encodes a cellulose synthase required for secondary cell wall synthesis. The plant cell. 1999 May 1;11(5):769-779.

[82] Fagard M, Desnos T, Desprez T, Goubet F, Refregier G, Mouille G, McCann M, Rayon C, Vernhettes S, Höfte H. PROCUSTE1 encodes a cellulose synthase required for normal cell elongation specifically in roots and dark-grown hypocotyls of Arabidopsis. The plant cell. 2000 Dec 1;12(12):2409-2423.

[83] Taylor NG, Laurie S, Turner SR. Multiple cellulose synthase catalytic subunits are required for cellulose synthesis in Arabidopsis. The plant cell. 2000 Dec 1;12(12):2529-2539.

[84] Kikuchi S, Satoh K, Nagata T, Kawagashira N, Doi K, Kishimoto N, Yazaki J, Ishikawa M, Yamada H, Ooka H, Hotta I. Collection, mapping, and annotation of over 28,000 cDNA clones from japonica rice. science. 2003 Jul 18;301(5631):376-379.

[85] Taylor NG, Howells RM, Huttly AK, Vickers K, Turner SR. Interactions among three distinct CesA proteins essential for cellulose synthesis. Proceedings of the National Academy of Sciences. 2003 Feb 4;100(3):1450-1455.

[86] Zhong R, Morrison WH, Freshour GD, Hahn MG, Ye ZH. Expression of a mutant form of cellulose synthase AtCesA7 causes dominant negative effect on cellulose biosynthesis. Plant physiology. 2003 Jun 1;132(2):786-795.

[87] Appenzeller L, Doblin M, Barreiro R, Wang H, Niu X, Kollipara K, Carrigan L, Tomes D, Chapman M, Dhugga KS. Cellulose synthesis in maize: isolation and expression analysis of the cellulose synthase (CesA) gene family. Cellulose. 2004 Sep;11(3): 287-299.

[88] Burton RA, Shirley NJ, King BJ, Harvey AJ, Fincher GB. The CesA gene family of barley. Quantitative analysis of transcripts reveals two groups of co-expressed genes. Plant physiology. 2004 Jan 1;134(1):224-236.

[89] Joshi CP, Bhandari S, Ranjan P, Kalluri UC, Liang X, Fujino T, Samuga A. Genomics of cellulose biosynthesis in poplars. New Phytologist. 2004 Oct;164(1):53-61.

[90] Desprez T, Juraniec M, Crowell EF, Jouy H, Pochylova Z, Parcy F, Höfte H, Gonneau M, Vernhettes S. Organization of cellulose synthase complexes involved in primary cell wall synthesis in Arabidopsis thaliana. Proceedings of the National Academy of Sciences. 2007 Sep 25;104(39):15572-15577.

[91] Persson S, Paredez A, Carroll A, Palsdottir H, Doblin M, Poindexter P, Khitrov N, Auer M, Somerville CR. Genetic evidence for three unique components in primary cell-wall cellulose synthase complexes in Arabidopsis. Proceedings of the National Academy of Sciences. 2007 Sep 25;104(39):15566-15571.

[92] Harris DM, Corbin K, Wang T, Gutierrez R, Bertolo AL, Petti C, Smilgies DM, Estevez JM, Bonetta D, Urbanowicz BR, Ehrhardt DW. Cellulose microfibril crystallinity is reduced by mutating C-terminal transmembrane region residues CESA1A903V and CESA3T942I of cellulose synthase. Proceedings of the National Academy of Sciences. 2012 Mar 13;109(11):4098-4103.

[93] Li F, Xie G, Huang J, Zhang R, Li Y, Zhang M, Wang Y, Li A, Li X, Xia T, Qu C. Os CESA 9 conserved-site mutation leads to largely enhanced plant lodging resistance and biomass enzymatic saccharification by reducing cellulose DP and crystallinity in rice. Plant biotechnology journal. 2017 Sep;15(9):1093-1104.

[94] Nicol F, His I, Jauneau A, Vernhettes S, Canut H, Höfte H. A plasma membrane-bound putative endo-1, 4-β-d-glucanase is required for normal wall assembly and cell elongation in Arabidopsis. The EMBO journal. 1998 Oct 1;17(19):5563-5576.

[95] Lane DR, Wiedemeier A, Peng L, Höfte H, Vernhettes S, Desprez T, Hocart CH, Birch RJ, Baskin TI, Burn JE, Arioli T. Temperature-sensitive alleles of RSW2 link the KORRIGAN endo-1, 4-β-glucanase to cellulose synthesis and cytokinesis in Arabidopsis. Plant physiology. 2001 May 1;126(1):278-288.

[96] Szyjanowicz PM, McKinnon I, Taylor NG, Gardiner J, Jarvis MC, Turner SR. The irregular xylem 2 mutant is an allele of korrigan that affects the secondary cell wall of Arabidopsis thaliana. The Plant Journal. 2004 Mar;37(5):730-740.

[97] Gui-Fu LI, Jian YA, Jun ZH. Mapping QTL for biomass yield and its components in rice (Oryza sativa L.). Acta Genetica Sinica. 2006 Jul 1;33(7):607-616.

[98] Liu B, Gómez LD, Hua C, Sun L, Ali I, Huang L, Yu C, Simister R, Steele-King C, Gan Y, McQueen-Mason SJ. Linkage mapping of stem saccharification digestibility in rice. PloS one. 2016 Jul 14;11(7):e0159117.

[99] Matsubara K, Yamamoto E, Kobayashi N, Ishii T, Tanaka J, Tsunematsu H, Yoshinaga S, Matsumura O, Yonemaru JI,

Mizobuchi R, Yamamoto T. Improvement of rice biomass yield through QTL-based selection. PloS one. 2016 Mar 17;11(3):e0151830.

[100] Naz AA, Reinert S, Bostanci C, Seperi B, Leon J, Böttger C, Südekum KH, Frei M. Mining the global diversity for bioenergy traits of barley straw: genomewide association study under varying plant water status. Gcb Bioenergy. 2017 Aug;9(8):1356-1369.

[101] Wood IP, Pearson BM, Garcia-Gutierrez E, Havlickova L, He Z, Harper AL, Bancroft I, Waldron KW. Carbohydrate microarrays and their use for the identification of molecular markers for plant cell wall composition. Proceedings of the National Academy of Sciences. 2017 Jun 27;114(26): 6860-6865.

[102] Xu Z, Li S, Zhang C, Zhang B, Zhu K, Zhou Y, Liu Q. Genetic connection between cell-wall composition and grain yield via parallel QTL analysis in indica and japonica subspecies. Scientific reports. 2017 Oct 2;7(1):1-3.

[103] Matsubara K, Yonemaru JI, Kobayashi N, Ishii T, Yamamoto E, Mizobuchi R, Tsunematsu H, Yamamoto T, Kato H, Yano M. A follow-up study for biomass yield QTLs in rice. PloS one. 2018 Oct 23;13(10):e0206054.

[104] Takeda T, Oka M, Agata W. Characteristics of dry matter and grain production of rice cultivars in the Warmer Part of Japan: I. Comparison of dry matter production between old and new types of rice cultivars. Japanese Journal of Crop Science. 1983 Sep 5;52(3):299-306.

[105] Kuroda E, Ookawa T, Ishihara K. Analysis on difference of dry matter production between rice cultivars with different plant height in relation to gas diffusion inside stands. Japanese Journal of Crop Science. 1989 Sep 5;58(3):374-382.

[106] Peng S, Laza RC, Visperas RM, Sanico AL, Cassman KG, Khush GS. Grain yield of rice cultivars and lines developed in the Philippines since 1966. Crop science. 2000 Mar;40(2):307-314.

[107] Yang W, Peng S, Laza RC, Visperas RM, Dionisio-Sese ML. Grain yield and yield attributes of new plant type and hybrid rice. Crop Science. 2007 Jul;47(4):1393-1400.

[108] Ookawa T, Yasuda K, Kato H, Sakai M, Seto M, Sunaga K, Motobayashi T, Tojo S, Hirasawa T. Biomass production and lodging resistance in 'Leaf Star', a new long-culm rice forage cultivar. Plant production science. 2010;13(1):58-66.

[109] Yang XC, Hwa CM. Genetic modification of plant architecture and variety improvement in rice. Heredity. 2008 Nov;101(5):396-404.

[110] Ahlawat S, Chhabra AK, Behl RK, Bisht SS. Genotypic divergence analysis for stay green characters in Wheat (Triticum aestivum L. em. Thell). The South Pacific Journal of Natural and Applied Sciences. 2008;26(1):73-81.

[111] Fu JD, Yan YF, Lee BW. Physiological characteristics of a functional stay-green rice "SNU-SG1" during grain-filling period. Journal of Crop Science and Biotechnology. 2009 Mar;12(1):47-52.

[112] Xu W, Subudhi PK, Crasta OR, Rosenow DT, Mullet JE, Nguyen HT. Molecular mapping of QTLs conferring stay-green in grain sorghum (Sorghum bicolor L. Moench). Genome. 2000 Jun 1;43(3):461-469.

[113] Cha KW, Lee YJ, Koh HJ, Lee BM, Nam YW, Paek NC. Isolation, characterization, and mapping of the

stay green mutant in rice. Theoretical and Applied Genetics. 2002 Mar 1;104(4):526-532.

[114] Jiang GH, He YQ, Xu CG, Li XH, Zhang Q. The genetic basis of stay-green in rice analyzed in a population of doubled haploid lines derived from an indica by japonica cross. Theoretical and Applied Genetics. 2004 Feb 1;108(4):688-698.

[115] Yoo SC, Cho SH, Zhang H, Paik HC, Lee CH, Li J, Yoo JH, Koh HJ, Seo HS, Paek NC. Quantitative trait loci associated with functional stay-green SNU-SG1 in rice. Molecules & Cells (Springer Science & Business Media BV). 2007 Aug 1;24(1).

[116] Meshram MP, Atale SB, Murumkar RD, Raut PB. Heterosis and heterobeltiosis studies in sweet sorghum. Annals of Plant Physiology. 2005;19(1):96.

[117] Corn RJ. *Heterosis and composition of sweet sorghum* (Doctoral dissertation, Texas A & M University).

[118] Pfeiffer TW, Bitzer MJ, Toy JJ, Pedersen JF. Heterosis in sweet sorghum and selection of a new sweet sorghum hybrid for use in syrup production in Appalachia. Crop Science. 2010 Sep;50(5):1788-1794.

[119] Packer DJ, Rooney WL. High-parent heterosis for biomass yield in photoperiod-sensitive sorghum hybrids. Field Crops Research. 2014 Oct 1;1(167): 153-158.

Section 2

Crop Production and Protection

Chapter 7

Rice Cultivation - A Way of Life for the People of North Eastern Hill Region of India

Noren Singh Konjengbam, Mayurakshee Mahanta and Andrean Allwin Lyngdoh

Abstract

Being an amazing picturesque of land, the North Eastern Hill Region of India, consisting of more than 200 ethnic groups, has only about 2.27% of the total rice area and shares only 1.96% of the total rice production in the country. Whether profitable or not, the rice cultivation is a way of life for the people of North Eastern Hill Region of India. Till today, the production and productivity of rice in this region is below the national average because of its fragile ecosystem and the varied physio-graphic conditions pertaining to this region. Neither the wider recommendation of agricultural technology such as variety nor the use of a single technology or variety can solve this problem of low yield. However, the development of location specific high yielding rice variety using the existing land races prevalent in the area can be one of the promising technique for improving the production and productivity of rice cultivation in this region.

Keywords: Rice, North Eastern Hill region, India, Rice Improvement, Fragile ecosystem

1. Introduction

Rice is the staple food crop of the North East Hill (NEH) Region inhabited by different tribes/ethnic groups. The varying agro-ecosystem and different tribe grow different land races of their preference based on quality and other religious purpose. The North-Eastern region of India has got a strategic importance and is surrounded by Bhutan and China in the north, Myanmar in the east while Bangladesh is in the south eastern side. More than 200 ethnic groups inhabit this region. This region is characterized by high rainfall, humidity, with varied topography and altitude making it a hub rich in floristic and crop diversities [1].

Rice being the way of life and culture of the people of NEH Region, irrespective of remunerative or non-remunerative, is cultivated under different rice growing situations of NEH states ranging from low lying lake areas to sloppy land of high hills in different rice growing seasons by adopting their age old indigenous rice growing methods and practices for their food and livelihood. In Arunachal Pradesh, rice is grown up to an altitude of 2000 m. Assam is characterized by both hilly and plain areas and rice is found to be cultivated in both. In Manipur, both upland and lowland local cultivars of rice are grown. In Meghalaya, soft varieties are cultivated and used as both flaked and in raw form. In Mizoram, cultivation of rice is limited only to the valley and lowland areas. Rice is also a main staple in the state of Nagaland where more than 400 accessions of rice germplasm have been collected. Rice, in Tripura, is cultivated in hills, hillocks and flat valley. In Sikkim, is a hilly stretch in the Himalayas where rice is cultivated annually [2].

1.1 Agro-climatic and physio-graphic condition of the region

The North East Hill (NEH) Region comprising the states of Arunachal Pradesh, Manipur, Meghalaya, Mizoram, Nagaland, Sikkim and Tripura, is lying between 21.57° to 29.30° North latitude and 88° to 97.30° East longitude representing a distinct agro-climatic area of the country. The region is having a total geographical area of 262185 Sq. km and it has international border with China in the north, Myanmar in the East, Bhutan in the North West and Bangladesh in the South Western part [3].

The NEH Region is divided into three geographic regions:

i. The Eastern Himalayan Region: This region includes the state of Sikkim and Arunachal Pradesh excluding Tirap District and part of Lohit District

ii. The Purvanchal Region: This region comprises Nagaland, Manipur, Tripura and Mizoram states including Tirap and Lohit Districts of Arunachal Pradesh

iii. The Meghalaya-Mikir Region: This includes Khasi, Jaintia and Garo Hills of Meghalaya.

These broad geographical regions show a wide diversity of climate due to altitudinal, physiographical and edaphic factors contributing to the diversity of agricultural crops as well as agricultural activities of the people [4]. The broad geographical regions of the North Eastern Hill States is given in **Table 1**.

The variation in the agro-climatic condition is due to its location and topography of this region. Being the hilly zone intercepted by small valley areas in between the hills and its position in the latitude and longitude pinning in the Sub-tropical areas, existence of these wide variation is the outcome. Although the agro-climatic zones show distinct differences in agro-climatic characteristics, it was difficult to draw a clear line of demarcation between two zones. However, to cater the specific needs of agriculture in these zones, the above delineation would serve the purpose and cover the maximum proportion of land, water, climatic conditions existing in the entire NEH Region.

1.2 Classification of rice in NEH region

According to rice growing seasons, rice in the region may be classified into three groups as under as given by [5]:

i. Autumn or Pre-*kharif* or *Ahu* paddy

These groups of rice are grown generally from the month of February as irrigated transplanted rice and may continue upto the month of May. Harvesting starts from May to August.

ii. Winter or *kharif* or *Sali* paddy

These groups of rice are grown during the month from July–August as rainfed direct seeded or transplanted rice and irrigated transplanted rice. The crop is harvested during the month of November and December.

DOI: http://dx.doi.org/10.5772/intechopen.99108

Zones	Altitude Range	Approx. geographic-al area (sq.km)	Annual Average Rainfall (mm)	Mean Temp. Min/Max	Areas	Remarks
1. Alpine Zone	>3500 m	47068	750	2°C / 17°C	**Arunachal Pradesh:** Gorichen Upper Tawang, Tulungla, Bumla, Sela pass areas of West Kameng District, Jidu and adjoining areas of Northern Siang. **Sikkim:** Gnathong, Chhangu, Serrathong, Thangu, Yakthan, Zema, Lachen, Heegyathang, Samsinggeling, Cholemu, Lima, Nathula range	No rice crop is generally grown in this zone.
2. Sub-alpine and Temperate zone	1500-3500 m	33564	2000	11°C / 20°C	**Arunachal Pradesh:** Tawang, Dirang, Bomdila, Shergaon, areas West Kameng District, Dibang Valley, Northern part of East Siang, Upper Subansiri District, Part of West Siang around Anini & North Eastern part of Lohit District **Meghalaya:** Upper Shillong, Mawphlang and Mairang of East Khasi Hills District **Manipur:** Mao & Maram areas of Senapati District, Ukhrul and adjoining areas of Ukhrul District **Sikkim:** Karponang, Bordong, Resi, Kangdin, Melli, Param, Lachem, Laichung, Hilley, Yoksum. **Mizoram:** Blue mountain, Halikhan, Tuipang, Nauzuarzo, Tiang **Nagaland:** Tuensang and Zunhoeboto Districts, Vangkong area of Wokha District, Higher areas of Mokokchung District.	Rice crop, mostly local varieties, are grown either as transplanted rice in the terraced fields or as direct seeded rice in *jhum* field of hill slopes.
3. Sub-tropical hill zone	1000-1500 m	29021	1600	12°C / 30°C	**Arunachal Pradesh:** Changyak, Naga and Khonsa area of Tirap District, Basar area of Siang District **Meghalaya:** Jowai Sub-division of Jaintia Hills, Part of Nongstoin Sub-division, Nokrek and Kailash area of West Garo Hills and Western part of East Garo Hills **Sikkim:** Namchi, Gayzing, Rongli, Rehnok, Mangan, Changthang, Uttre, Gangtok **Mizoram:** Whole State except lower valleys of Northern and Western part and area adjoining Cachar District and lower parts of Chhimtuipuii District **Nagaland:** Mokukchung District, Lower parts of Kohima, Wokha District and Mon District.	Rice crop is grown in *jhum* fields of hill slopes as direct seeded rice and in terraces as wetland rice mostly by transplanting.

Zones	Altitude Range	Approx. geographic-al area (sq.km)	Annual Average Rainfall (mm)	Mean Temp. Min/Max	Areas	Remarks
4. Sub-tropical plain zone (valley area)	400-1000 m	812	1375	12°C / 27°C	**Manipur:** Manipur Valley **Nagaland:** Bhaghti & Longnak valley **Meghalaya:** Umkiang area of Jaintia Hills.	Rice is grown as rainfed and irrigated wetland rice either transplanted or direct seeded with sprouted seeds.
5. Mild -tropical hill zone	200-800 m	26349	1400	12°C / 30°C	**Arunachal Pradesh:** Southern part of lower Subansiri District. **Meghalaya:** Southern part of Jowai Sub-division adjoining Karimganj, Southern part of Nongpoh Sub-division of Khasi Hills, Eastern part of East Garo Hills and West Khasi Hills. **Manipur:** Tamenglong District including Jiribam Sub-Division, Churachandpur and Thanlon of Churachandpur District. Moreh area of Chandel District. **Sikkim:** Rongpoh area of East District. **Mizoram:** Lower valley of Northern and Western parts of Chhimtuipuii District. **Tripura:** Jampui Hills. **Nagaland:** Medziphema area of Dimapur Sub-Division.	Rice is grown mostly as direct seeded upland crop in *jhum* fields of hill slopes and in terraced fields as wetland transplanted rice.
6. Mild -tropical plain zone	0-200 m	29333sq	2000	17°C / 33°C	**Arunachal Pradesh:** Pasighat area, Singphow area of Tirap District and lower parts of Lohit District. **Meghalaya:** Lower part of West Garo Hills District. **Mizoram:** Parts of adjoining Cachar District of Assam and North Tripura District. **Tripura:** Major part of Tripura excepting Jampui Hills **Nagaland:**Southern part of Dimapur Sub-Division excluding Medziphema area.	Rice is grown as rainfed and irrigated wetland rice either transplanted or direct seeded.

Table 1.
The broad geographical regions of NEH states (delineated into six distinct agro-climatic zones).

iii. Spring or Summer or *Boro* paddy

These groups of rice are grown during the month of November–December and harvested in the month of April–May mostly in water stagnated areas.

Further, rice grown in the low land areas of NEH region is further classified into two groups:

i. *Ashara* or shallow water rice

Rice is grown from March–April and harvested in November–December in shallow water rice area of 0.5 to 2.0 m water depth.

ii. *Bao* or deep water/floating rice

Rice is also grown from March–April and harvested in November–December in deep water areas as floating rice with water depth of 2 m and above.

Based on the rice agro-ecosystems prevailing in the NEH Region, rice may also be classified as under:

i. Hill or slope land rice

These groups of rice are grown in hill slopes either in the *jhum* field (**Figure 1**) as rain-fed upland direct seeded rice or in terraces (**Figure 2**) mostly as irrigated wetland transplanted rice. The hill rice may be of low-altitude (1000-1500 m MSL), mid-altitude (1500-2000 m MSL) and high-altitude (above 2000 m MSL).

ii. Valley or flat land rice

These groups of rice are grown in flat land either as rain-fed dry-land direct seeded rice or rain-fed wetland direct seeded rice with sprouted seeds (**Figure 3**) or rain-fed wetland transplanted rice or irrigated wetland

Figure 1.
Jhum *cultivation.*

Figure 2.
Terrace cultivation.

Figure 3.
Direct seeded rice.

transplanted rice (**Figure 4**). The rice in valleys are generally grown at an altitude of about 400 to 1000 m above MSL.

iii. Low lying semi-deep water and deep water rice

These groups of rice are generally adapted to semi-deep water (say from 0.5 to 2.0 m) and deep water conditions (more than 2.0 m) (**Figure 5**). This rice is generally direct seeded before monsoon rains. These areas are generally available in an altitude from 100 to 700 m above MSL.

DOI: http://dx.doi.org/10.5772/intechopen.99108

Figure 4.
Transplanted rice.

Figure 5.
Deep water rice.

2. Present status of rice production in NEH region

Rice is the most important cereal crop of the North Eastern Hill Region covering an area of about 998000 hectares producing about 2154000 tonnes of rice with an average productivity of 2.00 t/ha which is below the national average of 2.50 t/ha (average from 2014 to 2015 to 2018–2019). The rice area of NEH Region to the total area of the country is only about 2.27% with a rice production of 1.96% to the total rice production of the country. The present deficit in rice productivity of NEH Region from the national average is about 19.71% [6].

The average area, production, productivity, requirement, excess/deficit and per cent excess/deficit of rice in NEH states from the year 2014–2015 to 2018–2019 are presented in **Table 2**.

The average rice productivity in the NEH states varies from 1667 kg/ha in Mizoram to 2941 kg/ha in Tripura with a total rice excess of about 1.39 lakh metric tonnes. With the availability of land for rice cultivation becoming a limiting factor, increasing the level of productivity wherever possible by adopting the best available rice production technology to meet the ever increasing demand of rice has now become a great concern for NEH Region.

Name of state	Area ('00000 ha)	Production ('00000 tonne)	Productivity (kg/ha)	Requirement ('00000 tonne)	Percentage excess (+) / Deficit (−)	Requirement of quality seed at 25% SRR ('000 tonne)
Arunachal Pradesh	1.30	2.36	1814.80	2.02	(+) 14.40	1.63
Manipur	2.35	4.23	1791.80	4.16	(+) 1.65	2.94
Meghalaya	1.11	2.62	2361.80	4.33	(−) 65.27	1.39
Mizoram	0.37	0.61	1667.00	1.60	(−) 162.30	0.46
Nagaland	2.06	3.63	1771.60	2.88	(+) 13.22	2.58
Sikkim	0.10	0.18	1701.40	0.89	(−) 394.44	0.13
Tripura	2.69	7.92	2941.00	5.36	(+) 32.32	3.36
Total (NEH)	9.98	21.54	2007.05	21.24	**(+) 1.39**	12.48
All India	439.066	1097.648	2499.80			

Source: *[6].*
Per capita consumption of cereals: *146 kg/person/year (based on RDA of NIN, Hyderabad).*
Population: *As per 2011 census.*

Table 2.
Average rice area, production, productivity (2014–2015 to 2018–2019) and quality seed requirement of rice crop at 25% seed replacement rate in NEH states of India.

The total gross annual domestic income from rice agriculture for the NEH states of India, by taking average yield of rice from 2014-15 to 2018-19, is presented in **Table 3**.

As per **Table 2**, it is observed that rice production is in short from the requirement of cereals in the Mizoram, Meghalaya and Sikkim states of this region. However, when we look NEH states as a whole, the region is self-sufficient in rice considering the shortage is made up from the surplus states. The rice economy of this region is about 1.96% of the whole country and a lesser productivity from the national average productivity of this crop. The requirement here is also calculated on the basis of total cereals required per capita as per [7]. The only cereals which are staple food that consumed twice a daily is rice in this region.

3. Problems and opportunities through SWOT analysis on rice production in NEH region

As the rice crop is grown in the NEH Region under widely diverse agro-climatic conditions resulted from varied altitudinal, physiographic, edaphic, rainfall, etc. conditions, there is immense variability among the rice cultivation practices as well as rice cultivars from one place to another. Hence, the SWOT analysis on rice cultivation in the NEH Region may clarify the problems and opportunities of rice production in the region as under as given in [8, 9]:

Strength:

i. Rice is the staple food of NEH region and consumed in different forms *viz.*, cooked rice, popped rice, rice flakes, rice cake, rice flour, etc.

ii. A large number of land races/local cultivars including wild relatives *viz.*, *Zizania* spp., *O. rufipogon*, *Oryza nivara*, *O. sativa f. spontenea,* etc. are available in the region.

Name of state	Area ('00000 ha)	Production ('00000 tonne)	Productivity (kg/ha)	Requirement ('00000 tonne)	Gross income per hectare (INR)	Gross state income from rice (INR in crores)	Remarks
Arunachal Pradesh	1.30	2.36	1814.80	2.02	31,759.00	413.00	NEH States of India share only 1.96% of country's gross income from rice
Manipur	2.35	4.23	1791.80	4.16	31,356.50	740.25	
Meghalaya	1.11	2.62	2361.80	4.33	41,317.50	458.50	
Mizoram	0.37	0.61	1667.00	1.60	29,172.50	106.75	
Nagaland	2.06	3.63	1771.60	2.88	31,003.00	588.00	
Sikkim	0.10	0.18	1701.40	0.89	29,774.50	31.50	
Tripura	2.69	7.92	2941.00	5.36	51,467.50	1386.00	
Total (NEH)	9.98	21.54	2007.05	21.24	**35,123.37**	**3769.50**	
All India	439.066	1097.648	2499.8		**43,746.50**	**192088.40**	

Procurement price during 2018–2019 = Rs. 1750 per quintal.

Table 3.
Rice agriculture economy of NEH states by taking average yield of 2014–2015 to 2018–2019.

iii. Rice grows from deepwater conditions (e.g. deep water rice) to high hills (e.g. *Jhum* rice) in the region.

iv. Monsoon rain spreads from the month of May–June to October–November in the region.

v. Rice can grow in all the soils available in the region.

vi. Two Central institutes viz., Indian Council of Agricultural Research-Research Centre (ICAR-RC) for NEH Region and Central Agricultural University (CAU). with its constituent colleges in all the states of NEH Region, are located in the region.

Weakness:

i. Most of the available land races with consumer's preference in regard to eating quality are of low productivity. Availability of semi-glutinous or glutinous high yielding rice varieties suitable in NEH Region are minimum.

ii. Most of the rice production areas are rainfed and dependent on monsoon rains. Many a time, there is shortage of moisture in upland and even in lowland due to erratic monsoon rains.

iii. Most of the rice soil in the region is acidic in nature and farmers lack awareness of improved soil fertility management for sustainable production.

iv. Rice variety bred in mainland is not widely adapted in NEH region due to varied altitudinal and physiographical conditions of the region. Hence, availability of high yielding rice varieties for specific rice agro-ecosystem of the region is lacking.

v. Unavailability of quality seeds for different agro-ecosystem specific H.Y.V. rice varieties and lack of properly functioning Seed Certification Agency in the NEH Region.

vi. Poor communication system of the region to reach rice production and protection inputs in time to the rice farmers.

vii. Lack of technological intervention and effective extension services to the desired level for rice crop production, crop protection, farm mechanization, processing, value addition, etc.

viii. Rice monoculture is practiced on the same field year after.

ix. Existence of community land ownership system.

Opportunities:

i. Availability of land races adapted to different rice growing systems for use in breeding programme to develop rice variety suitable for different rice agro-ecosystems of the region as well as eating quality preferred by NEH people.

ii. CAU with its different constituent colleges at different NEH states with a mandate of teaching (HRD), Research and extension and ICAR Research Centre for NEH Region with its research centres in each NEH states is the opportunities for this region to address rice related problems and technological interventions.

iii. ICAR Seed Project can be converted into opportunity for this region to multiply and distribute good quality seeds of released rice varieties for different rice agro-ecosystems of the region.

iv. Availability of medium to high rainfall under monsoon is the opportunity for rice production, although the region has poor irrigation system.

v. Possibilities for adequate rain water harvesting in the region is an opportunity for rice based multiple cropping systems.

Threats:

i. Most of the rice growing areas in this region are rainfed with acidic soil.

ii. Requirement of agro-ecosystem specific rice variety is too many, demanding small quantity of seeds.

iii. Private Sector Seed Companies are less interested as the requirement of seed is low.

iv. Due to climate change, distribution of rainfall is rather erratic and unpredictable leading to unprecedented flood and drought.

v. Due to poor road connectivity, production inputs cannot reach farmers timely and in adequate quantities.

vi. High diseases and pests incidence on rice due to conducive environment prevailing in the region.

vii. Widespread prevalence of shifting cultivation which causes soil and nutrient loss.

viii. Farming community are small and marginal farmers.

Problems are many at the same time equal opportunities are also available for increasing rice production and productivity in the North Eastern Hill States.

4. Conclusion

To feed the ever increasing population in the NEH Region as well as to maintain the rice self-sufficiency in the region to-day, the present productivity of rice has to be increased at least to the tune of national average. The less productivity of rice in this region may be due to its varied physiography including hilly terrain forming different rice ecosystem. Any single improved variety cannot be grown suitably in all the ecosystem demanding lot many number of varieties with less quantity of seeds. The region is inhabited by more than 200 different ethnic groups who have different preference of crop variety and quality, taste preference along with spiritual and religious difference. Because of this feature of land and people, no seed industry comes forward to establish their business for profit. When the yield and cost involved per hectare is looked in to, the rice cultivation in this region usually landed with less profit, sometimes loss due to harsh environment and low productivity. But, whether it is profitable or not, people of this region have to continue to grow rice which have been continuing since time immemorial because of its subsistence nature of agriculture, simple life and a sense of food security with rice.

Being rice the staple food of the NEH region, its cultivation under varying agro-ecosystem ranging from rain-fed lowland to deep water and fragmented land holding system of NEH farmers, a strong breeding program and quality seed production looking into the demand of rice farmers has to be initiated. Proper linkage for better input supply system with the concerned State Govt. and Research institutes should be strengthen leaving each other ego behind for the interest of improving livelihood of rural poor farmers of this region.

Acknowledgements

The authors are thankful to Dr. M. Rohinikumar Singh, former Director of Research, Central Agricultural University, Imphal-795004, Manipur, India for his guidance, support and his valuable inputs for the preparation of this manuscript.

Conflict of interest

The authors declare no conflict of interest.

Author details

Noren Singh Konjengbam*, Mayurakshee Mahanta and Andrean Allwin Lyngdoh
College of Post Graduate Studies in Agricultural Sciences, Central Agricultural University (Imphal), Meghalaya, India

*Address all correspondence to: norensingh27@gmail.com

IntechOpen

© 2021 The Author(s). Licensee IntechOpen. This chapter is distributed under the terms of the Creative Commons Attribution License (http://creativecommons.org/licenses/by/3.0), which permits unrestricted use, distribution, and reproduction in any medium, provided the original work is properly cited. (cc) BY

References

[1] Hore D. Rice Diversity Collection, Conservation and Management in Northeastern India. Genetic Resources and Crop Evolution. 2005;52: 1129-1140. DOI: 10.1007/s10722-004-6084-2.

[2] Roy A, Das S, Tripathi A, Singh NU, Barman HK. Biodiversity in North East India and Their Conservation. Progressive Agriculture- An International Journal. 2015;15:182-189. DOI: 10.5958/0976-4615.2015.00005.8

[3] NEDFi Databank. North East India. 2008. Available from: http://databank.nedfi.com/content/general-information [Accessed: 2021-06-20]

[4] Panjab S. Agro-climatic Zonal Planning Including AgricultureDevelopment in North-Eastern India. 2006. Available from: https://niti.gov.in/planningcommission.gov.in/docs/aboutus/committee/wrkgrp11/wg11_agrclim.pdf [Accessed on 2021-06-20]

[5] Directorate of Rice Development. Status Paper on Rice. 2016. Available from: http://drdpat.bih.nic.in/Downloads/Status-Paper-on-Rice.pdf [Accessed: 2021-06-20]

[6] Directorate of Rice Development. Statistical Hand-Book. 2016. Available from: http://www.drdpat.bih.nic.in/Downloads/Statewise-APY-of-Rice-2014-15-to-2018-19.pdf [Accessed: 2020-12-17]

[7] National Institute of Nutrition. 2011. Dietary Guidelines for Indians – A Manual. 2nd Edition. NIN, Hyderabad.

[8] Rohinikumar SM. Rice self sufficiency in NEH region – Present status, problems, opportunities and Strategies. In: Rohinikumar SM, Noren SK, editors. Location specific advances in rice production technology for N.E.H. region. Published by Directorate of Research, CAU, Imphal (Manipur); 2013. ISBN 978-81-928554-0-0.

[9] Noren SK, Ranjit SP, Pramesh K. Quality seed production in rice and their strategies for various rice agro-ecosystems of N.E.H. region. In: Rohinikumar SM, Noren SK, editors. Location specific advances in rice production technology for N.E.H. region. Published by Directorate of Research, CAU, Imphal (Manipur); 2013. ISBN 978-81-928554-0-0.

Chapter 8

Rainfed Rice Farming Production Constrains and Prospects, the Kenyan Situation

Al-Imran Dianga, Ruth N. Musila and Kamau W. Joseph

Abstract

Kenya experiences huge production-consumption deficit in relation to rice. This is due to changing eating habits that has adopted more rice in the menu and rapidly rising population. Rice production has remained low being unable to meet consumption. Rice ecosystems in Kenya include irrigated, rainfed lowland and rainfed upland. Irrigated ecosystem has over the years been given more emphasis despite rainfed rice farming having double the potential over irrigation system. Ecologically rice grows well in abundant water supply, warm to high temperatures and in Clay sandy to loamy soils with slightly acidic to neutral pH. Rice varieties grown in Kenya are mainly traditional, introduced improved, hybrids and land-races. Rainfed rice farming faces constraint's key among them being; drought and erratic rainfall, weeds, pest and diseases, cheap imports, land ownership and poor infrastructure. Mitigating against drought and erratic rainfall, improving farm inputs and equipment, increasing germplasm production and distribution, credit support and marketing to farmers, improving farmers skills through technological transfers and infrastructural development are prospects that if adopted could increase rainfed rice productivity. More attention towards improvement of rainfed rice farming could greatly contribute to bridging the production-consumption deficit that is bridged through imports. It is with this, that this review updates our understanding of rain fed rice farming in Kenya in terms of ecological conditions, ecological systems, varieties, constraints and prospects.

Keywords: Rainfed, rice, rainfed, ecology, constraints

1. Introduction

Cultivated rice is grouped in to the genus *Oryza* taxonomically. The genus *Oryza* (*O.*) contains 23 species of which two are cultivated while the rest are wild [1]. The two cultivated species, are *Oryza glaberrima*, commonly referred to as African rice, and *Oryza sativa* L., commonly referred to as Asian rice [2]. *O. glaberrima* is indigenous to sub-Saharan Africa and its domestication is believed to have been from *Oryza barthii* (which used to be referred to as *Oryza brevilugata*). On the other hand, *O. sativa* was domesticated independently probably in China [3]. Cultivated rice is not limited to the two *Oryza* species, other inter-specifics for example the New Rice for Africa marketed as NERICA'S that arose due to crossing of *O. sativa,* and *O. glaberrima* are also cultivated [2].

Rice is an essential food crop that provides for most of world's population. It is also the second most consumed among cereal crops that include maize, wheat, barley, sorghum and millet. Rice is a major cereal crop with high economic and nutritional importance [4, 5]. Worldwide the leading producers of rice are Indonesia, India and China who together account for 50% world production [6]. Africa accounts for only 3% of the world's total production, with biggest producing countries being found in West Africa and they include Cote d'Ivoire, Nigeria and Mali. Mozambique and Malawi are the leading producers in Southern Africa. Madagascar and Egypt are also other substantial producers. In East Africa Tanzania ranks top in production followed by Kenya then Uganda [6].

In Kenya rice is mainly consumed as food with byproducts having other roles, for example rice hull is used as animal feeds, rice straw is also used as animal feed and substrate for growing mushrooms, while rice husks are used as cooking fuel [7]. In Kenya rice consumption has increased tremendously at an annual rate of 12% in comparison to wheat and maize that have increased at about 4% and 1% respectively. This is credited to changes in eating habits mostly among people living in urban centers [8]. Therefore, demand for rice is expected to increase further. In 2019 the annual rice consumption in Kenya was approximated to be 800,000 metric tons compared 130,000 metric tons produced the same year (**Figure 1**), the deficit was met through imports [6]. Current rice imports are estimated to be about $87.5 million consequently stretching other parts of the economy [6].

Irrigated rice production land potential is about 540,000 ha while the production land potential for rain-fed ecology is 1.0 million ha [7, 9] currently area under production is estimated to be 30,000 ha. This indicates that if rain fed ecology potential is fully explored it will contribute to towards bridging production and consumption gap. Rice yield for irrigated rice is approximated at 4–6 t ha^{-1} while for rain fed is 1 t ha^{-1} which are below optimal production capability of about 10 t ha $^{-1}$ and 7 t ha $^{-1}$ respectively [7].

In Kenya like in many other sub-Sahara Africa counties rainfed rice farming has not been given priority [10]. With a potential of over 1 million ha about only 250, 000 ha are under rainfed rice crop [11]. Increasing Rain fed rice production is likely to increase the national rice production thus decreasing the rice import bill. This

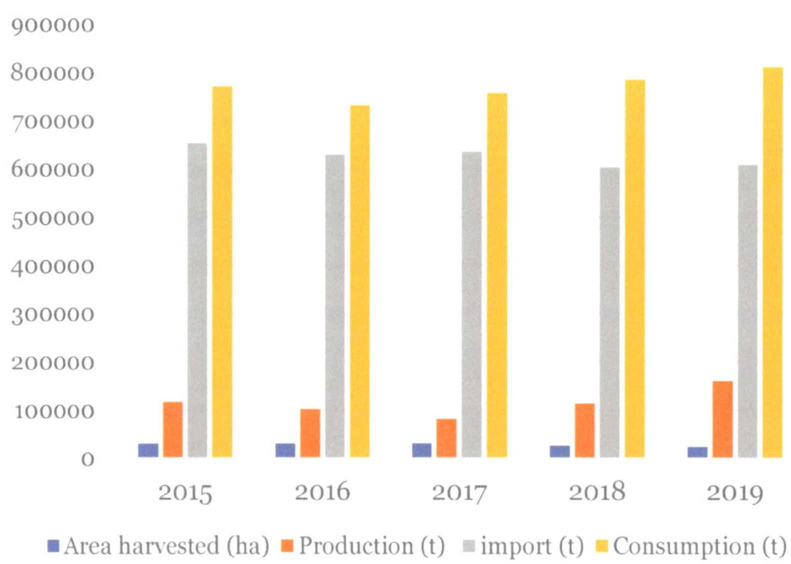

Figure 1.
Rice production, imports and consumption in Kenya between 2015 and 2019.

chapter aims at reviewing rainfed rice farming in Kenya by highlighting ecological conditions, ecological systems, the constraints faced by rainfed rice farming and discussing their potential solutions which if adopted can increase rice production in Kenya.

2. Rice ecological conditions

Rice plant water requirements is based on ecosystems under which it is cultivated. When it is grown as an upland crop under rain-fed conditions it needs 100 mm monthly rainfall and when grown as lowland crop it requires 200 mm rainfall per month. Rice can also be grown as lowland crop with standing water. Rice crop will need 125 mm monthly rainfall during vegetative stage while during ripening stage no standing water is needed. It is thus best adapted to grow with abundant water supply [12]. Rice grows in different soil conditions ranging from black clay that is heavy to sandy loam with a pH range of 4.5–7.0 and can tolerate water logged soils. Hot and humid condition with temperature ranging between 22° centigrade to 40° centigrade is the ambient climatic condition of rice. It grows well in altitudes of between 0 and 1700 meters above sea level [13].

3. Rice ecosystems

Based on IRRI, rice farming is categorized into four ecosystems depending on source and water supply. This are irrigated, flood prone, rainfed lowland and rainfed upland and [14].

Irrigated ecosystems are the most widely utilized rice farming ecosystems accounting for over 75% percent of total yield. The ecosystem includes lager parts of Europe, Australia, America, Asia and Africa. Irrigation ecosystem is again grouped into irrigated wet season and irrigated dry season. Irrigated wet season involves cultivation of rice during the wet season and irrigation water supplements rainfall. Irrigated dry season involves planting rice when rainfall is usually low and water is majorly supplied by irrigation in places that usually experience high solar radiation and evapotranspiration. In irrigated ecosystem the fields are bunded and leveled, water level maintained at between 2.5 cm to 1.5 cm determined by availability of water. Rice seeding is by either transplanting or direct seeding. In Kenya irrigation farming is done in irrigation schemes under the management of the National Irrigation Board (NIB). Major irrigation schemes include Ahero, Bunyala, West Kano, Perkera, Hola, Bura and Mwea. Small holder irrigation farming is practiced along river valleys namely; Kore, Alungo Nyachoda, Wanjare, Anyiko and Gem Rae in Western Kenya and Kipini, Malindi, Shimoni and Vanga at the coastal region [11]. Dry irrigation entails continuous flooding and it is practiced in Mwea, Ahero, Bunyala and Western Kano irrigation farming. Dry irrigation must have continuous water supply and soils must have high water retention capacity. During drought water is rationed hence reducing productivity though currently, System Rice Intensification (SRI) has been introduced.

Flood prone ecosystems involves paddy-fields being subjected to unbounded flooding for a duration that is about 5 months and water depth might range up to a maximum of about 5.0 M during plant growth. In this deep-water condition rice plants, mostly floating rice varieties outstretch their stems to get to the water surface. Flood prone ecosystem is largely practiced in Africa and Asia and accounts for 7% of the world land under rice cultivation. The cultivation is mostly located in river deltas for instance the Ganges in India, Brahmaptura in Bangladeshi, the Mekong in Vietnam and Cambodia, Niger delta in Niger and Chao Phraya of Thailand. Deep water rice system is also extensively practiced in coastal areas of India, West Africa, Vietnam and Bangladeshi based on daily tidal inundation. Key constraint in this environment is soil and water salinity and flash floods. This ecosystem extremely variable due to unpredictable flooding and drought. Farmers in this ecosystem records about 1.5 t ha^{-1} average yield with the main stress being environmental making most applicable farming inputs ineffective [14].

Rainfed lowland ecosystem involves slightly bunded and leveled field where water supply is mainly by rainfall and the water depth and the duration depends on the rain season. The water level cannot be controlled and rice plants are severely exposed to drought, deep floods, and alterations between anaerobic and aerobic environments [14]. In Kenya Rain-fed lowland rice cultivation is practiced in Kwale, Kilifi and Tana River counties at the Coast region.

Upland ecosystems involve rice fields in straighten valley bottoms to hilly mountainous lands with slopes ranging from 40% to about 0% descend. In Upland ecosystem rice cultivation done by preparing fields that are seeded when dry. These ecosystems form about 13% of harvested rice areas worldwide but accounts for only 4% of the world's total production. Upland rice is largely for a subsistence crop with yields approximated at 1 t ha^{-1} in areas with little inputs to 3–4 t ha^{-1} in situations where fertilizer application and supplementary irrigation is practiced. An estimated population of 100 million people are believed to depend on upland rice as their staple food. Upland rice is mostly grown in Asia (Bangladeshi and India), Africa and Latin America. These ecosystems have many constraints, mostly attributed to insufficient soil fertility, weed invasion and disease infection. Worldwide, rain fed ecosystem accounts for approximately 54 million ha of rice, mostly found in Africa and Asia [14]. In Kenya rain fed upland is grown in Kisumu, Busia counties in western Kenya and Kilifi, Kwale and Tana river counties of Coastal Kenya [15].

4. Rice varieties under rainfed ecology in Kenya

Rice varieties under rainfed conditions are categorized as traditional, introduced improved and hybrid rice. Traditional varieties are characterized by late maturity, low yields, lodging. However, they are adapted and are able to tolerate stresses such as pests and diseases, drought, weeds, salinity and even bird's attack. Some traditional lines possess farmers preferred traits like aroma and good gelatinization temperature.

4.1 Rainfed lowland ecology varieties

Under rainfed lowland ecology traditional lines include Madevu, Kitumbo, Kichana chawa, macho ya wanda, kijego, Matako Nyeusi, Moshi and Mtumbatu. Introduced improved lines show improved yield, earliness and less lodging. Rainfed lowland introduced improved lines include Komboka and MWIR 2. Komboka was introduced by Kenya Agricultural and Livestock Research Organization (KALRO) in co-operation with International Rice Research Institute (IRRI) in 2013. It is high yielding, good grain quality, semi aromatic and has high tillering ability. Supaa, a local landrace that is aromatic and late maturing is also grown particularly at the Kenyan coast. Highbred low land ecosystem rice lines are Arize Tej Gold and Arize 6444 Gold from Bayer East Africa that were evaluated and found promising by National Irrigation Board (NIB) however, there adoption remain low.

4.2 Rainfed upland ecology varieties

Rainfed upland introduced improved rice lines include MWUR 4, *Dourado precoce,* NERICA 4, NERICA 1, NERICA 10, NERICA 11 and NERICA 2. *Dourado precoce* was introduced by Kenya Agricultural and Livestock Research Organization, and its attributes include good grain characteristics and early maturity. The NERICAS have been the most successful lines that have been introduced in Kenya. The NERICAS were developed from interspecific fixed lines *O. sativa* and *O. glaberrima.* The NERICAS combine the hardiness of the African rice *O. glaberrima* in terms of pest, diseases, and weeds resistance with the high yielding trait of the Asian rice *O. sativa*. The NERICAS are; high yielding with small amount of fertilizer application, they are suitable for African soil and matures faster shortening growth cycle by 30–35 days enabling double cropping and minimizing drought effect [16, 17]. Improved rice lines offer a feasible option to traditional rainfed varieties. Yield of up to 4.4 t ha^{-1} was realized from on farm field test in Kerio Valley for the NERICA lines, this showed promising results. Though NERICA 4 emerged as the best line for most parts of Kenya, others like NERICA 1, NERICA 10 and NERICA 11 performed quite well [8].

5. Rain-fed Rice production constraints in Kenya

A wide range of constraints affect rice production in Kenya mostly a biotic, biotic, socio-economic and management [18]. Abiotic constraints include drought and erratic rainfall. Biotic constraints comprise of pests, diseases and weeds while socio-economic includes land ownership, unfavorable trans-border trade, high cost of machineries and inputs, poor infrastructure, unskilled farmers, slow technological advance transfer, poor access to credit and uncoordinated marketing.

5.1 Drought and erratic rainfall

In Kenya drought and erratic rainfall is a major constraint that has limited production and led to low yields for rainfed rice farming [19, 20]. In reports done by [19, 21] at the coastal and central regions of Kenya, they both conclude that drought is a constraint of great importance in rain fed rice production in the country. During drought years in Kenya rice yield in the paddy system potential drops to 1.4 t ha^{-1} from a potential of between 2.7 t ha^{-1} to 5.4 t ha^{-1} in a good year. Rice is very sensitive to drought especially during the reproductive stage where if there is drought then it leads to significant yield losses. Drought stress reduces peduncle rate of elongation and length at the booting stage. Reduction in peduncle elongation majorly predisposes reduction in panicle exertion rate [22, 23]. This results in either incomplete or failure of the panicles to exsert from the boot. Moreover, there is spikelet sterility from the damaged and abnormal development of the reproductive organs [23].

5.2 Weeds

Weeds compete for vital nutrients with rice plants. Weeds serve as alternative hosts for diseases, pests and rodents. Weeds have a cumulative effect of suppressing rice plants growth thus reducing yield. Common weeds in Kenya include; *Echinochloe colona* and *Echinochloe crus-galli* of family Poacea, *Ishaem rogusum*, *Leptochlea chinesis*, *Cyperus deformis*, *Fimbristylis miliacea*, false finger millet and striga. Weeds also reduces harvests quality and makes farmers to utilize more resources in terms of time and money to control weeds which reduces returns [24].

5.3 Diseases and pests

Diseases are also a major constraint to a Kenyan rainfed farmer. Common diseases include; Blast caused by *Pyricularia oryza*, Rice Yellow Mottle Virus (RYMV), Brown Spot (*Helminthosporium oryza*) Sheath rot caused by *Sarocladium sp*., Sheath blight caused by *Thanetophorus cucumeris*, Bacterial blight caused by *Xanthomonas oryzae* pv. oryzae, Glume discoloration caused by *Sarocladium sp*. and *Curvularia sp*., Leaf scald caused by *Rhyncosporium oryzae.* Pests also have a major effect on rice yield, major pests that attacks rice include Stem Borers, Leaf Miners and Root Cutting Insects. Others are Rice hispa, termites, Rice root aphid, seed corn maggot, cut worm, Rice water weevil, Rice leaf beetle, and rice green caterpillars. Birds e.g., quelea and rodents such as mice and rats also cause a substantial field loses [25].

5.4 Land ownership

Land ownership system has led to land fragmentation in potential areas as population increases. This has made it difficult to utilize mechanization in rice farming processes leading to reliance on manual labor which rises production costs narrowing profits margins for farmers. Furthermore, women who are key players in rice production are traditionally not allowed to own land though Kenyan laws provide for women land ownership [26].

5.5 Unfavorable trans-border trade practices and cheap imports

There is a lot of informal trade with Tanzania and Uganda. There is uncertified rice seeds movement which presents challenges to the rice sub-sector development. With no harmonized tariffs on germplasm trade between the East African community neighbors controlling this type of trade has been a challenge. There has also been illegal importation of cheaper milled rice from other countries which leads to low prices for the Kenyan farmer hence hurting profits.

5.6 High cost of machineries and inputs

The cost of acquiring machineries such as tractors and farm inputs such as fertilizers and pesticides is so high. This has been a disincentive to farmers on use of machineries and farm inputs. This has driven production cost high reducing farmers profits margins.

5.7 Poor infrastructure

In Rainfed rice systems poor infrastructure has been a major constraint to farmers. Rice mills are unevenly distributed forcing farmers to rely on traditional milling methods which are labor intensive, and lead to low quality and low percentage of milled rice recovery from paddy rice. Poorly developed roads, drainage, communication and viable public-private sector partnerships contribute to low rice productivity. More improvement in rice milling value chain could improve rice production. A study done in Rwanda by [27] showed that the system of processing rice in small hullers did not to contribute to increasing domestic supply. This was attributed to hulled rice of poor quality that was demonstrated by 30% decrease in prices of domestic rice compared to imported rice. Other aspects in which millers affects rice production is in relation to their location. Rice mills located far from farms implies high cost of transportation and this drives up production costs. The high milling costs implies farmers being unable to recover their production costs since cost of transporting paddy rice which is bulky than rice that has been milled by about 40% is expensive [28].

5.8 Unskilled farmers and slow technology transfers

Most farmers lack modern rice farming skills instead relying on traditional farming methods that have been overtaken by time. As new advancements in technology in rice are made the rate at which the technology is transferred to farmers is slow. Extension staff services are inadequate and at times the staff themselves have limited capacity on educating the farmers.

5.9 Poor access to credit and uncoordinated marketing

Most rainfed rice farmers do not have access to credit as most are small scale and subsistence farmers. Marketing is done individual farmers unlike in irrigation ecosystem where it is coordinated. Individual uncoordinated marketing makes the farmer lose ability to bargain for better prices exposing them to brokers who exploit them.

6. Rain fed rice production prospects

6.1 Mitigating against drought and erratic rainfall

Breeding of drought tolerant lines can effectively address frequent droughts problem in rainfed lowland and upland rice ecosystems [29]. The technology is cheap; costs less to grow drought tolerant lines than to grow a susceptible one. Yield performance under both drought stressed and non-drought stressed environments are realized, with the drought tolerant line having the ability to be cultivated in all seasons with no yield penalties in the good years [20]. Farmers should be encouraged to adopt early maturing lines like the NERICAS. Breeding for early maturing lines can be used to come up with genotypes that mature faster thus evading drought stress especially in rain fed production [30]. Using improved water storage, harvesting and underwater could supplement rain in rainfed system avoiding total crop failure when rains fail or are inadequate. This can also increase irrigation potential to 1.3 ha [15].

6.2 Farm inputs and equipment

Enhancing access to farm inputs and equipment's could increase yields. The government needs to subsidize fertilizers and pesticides. Farmers must also have access to appropriate germplasm and variety maintenance. The government should ensure sufficient production, supply and marketing of high-quality equipment's. The County governments need to have facilities that allow for hiring of expensive equipment's and machinery e.g., tractors to farmers.

6.3 Germplasm production and distribution

To discourage farmers from using low yielding long durational lines, lines development should be specified based on agro-ecological zones though seed multiplication should be in areas that experience low abiotic stresses. Researchers should develop breeder and foundation seed that is maintained by research institutions. Scientist certified seed should be reproduced by seed merchants who should in turn sell it to seed stockiest in rice growing areas as per projected requirements to ensure they are easily accessed by farmers. This should be followed by massive sensitization of extension officers and farmers on the new lines in the market [25].

6.4 Credit support and marketing

Organize farmers into cooperative societies and common interest groups. This makes it easier to market their rice and access credit facilities. State funded credit agencies e.g., Agricultural Finance Cooperation (AFC) should be encouraged to lend to farmers.

6.5 Infrastructural development

Construction of modern mills will promote rainfed rice farming. Improving roads, construction of health facilities to provide health services to curb water-borne diseases are other infrastructural improvement that could promote rice farming. Furthermore, both national and County governments must provide incentives and formulate policies that encourages private sector partnerships. Temperature regulated bulk seed storage facilities should also be built. Fully equipped soil analysis laboratories as well as rice harvesting machines be made available to farmers.

Improving rice mills also contributes to improved income by offering employment. These mills support food security, and increase competition that will bring down milling costs to farmers [31]. There is need to put in efforts to modernize and improve rice milling subsector. Efforts should be put in place to promote setting up of mult-pass rice mills with recovery rate of about 70% of un-husked rice compared to single pass mills with recovery rate of about 57%. In addition to that, mult-pass mills have a lower split rice percentage of about 14% compared to 27% in single pass mills. To support farmers in this situations, possible approaches to be employed include; supplying multi-stage rice mills to farmers co-operatives societies, using rural social entrepreneur to supply rural mills, assisting millers and farmers to set up out grower agreements and developing models to upgrade central and decentralized local milling technologies [31, 32].

The government should commission studies on inventories on post-harvest facilities for rice, losses assessment and information gathering that supports government planning and other stake holder's intervention to the sub-sector. Better storage facilities need to be developed and promoted, to support the milling section further the government and county governments should promote technological knowhow on agronomic applications and post-harvest technologies that entails agricultural processing to reduce losses.

Furthermore, a need also arises to utilize other energy and drying technologies like solar drying systems and hybrid's systems that use both rice straws and solar, collapsible dryers, portable thermal dryers and other renewable energy technologies. This will greatly reduce post-harvest losses. Another way of pushing profits margin up for farmers is utilizing rice by-products mostly husks that make up to about 20% of paddy in animal feeds production, bio-fertilizers and briquettes [31].

6.6 Improving farmers skills and technological know-how

Researchers' farmers and extension officers should be trained on modern rice production techniques and utilization. Setting up new training institutions and revitalizing existing ones to undertake capacity building in rice specific courses. Extension officers be posted to rice growing areas to improve quality inspection and its enforcement. Fully functional research and extension infrastructure should be set up to promote development, packaging, and timely disseminating of appropriate technology to extension officers, farmers organizations and other stakeholders. Farmers-extension-research linkages should also be improved and strengthened.

7. Conclusion

To increase Kenyan rice production further emphasis, need to be on small scale rainfed farmers. By addressing the constraints like drought and erratic rainfall, weeds, pest and diseases, cheap imports, land ownership and poor infrastructure through; mitigation against drought and erratic rainfall, improving farm inputs and equipment, increasing germplasm production and distribution, credit support and marketing to farmers, improving farmers skills through technological transfers and infrastructural development. Rainfed rice farming production potentials could be unlocked resulting in improved rice production.

Conflict of interest

The authors declare no conflict of interest.

Acronyms and abbreviations

KALRO	Kenya Agricultural and livestock Research organization
IRRI	International Rice Research Institute
SRI	System Rice Intensification
AFC	Agricultural Finance Cooperation
NIB	National Irrigation Board
$t\ ha^{-1}$	Tons per hectare

Author details

Al-Imran Dianga[1*], Ruth N. Musila[2] and Kamau W. Joseph[1]

1 Department of Plant Sciences, Kenyatta University, Nairobi, Kenya

2 Rice and Maize Breeder, Kenya Agricultural and Livestock Research Organization, Kerugoya, Kenya

*Address all correspondence to: alimrandianga@gmail.com

IntechOpen

© 2021 The Author(s). Licensee IntechOpen. This chapter is distributed under the terms of the Creative Commons Attribution License (http://creativecommons.org/licenses/by/3.0), which permits unrestricted use, distribution, and reproduction in any medium, provided the original work is properly cited.

References

[1] Khush, G. S. 1997. Origin, dispersal, cultivation and variation of rice. Plant Mol. Bioi. 35

[2] Teeken B, Nuijten E, Temudo M, Okry F, Mokuwa A, Struik P, and Richards P. (2012) Mantaining or abandoning African Rice: Lessons for understanding processes of seed innovation. Human ecology, 40(6): pp. 879-892. Retrieved from http://www.jstor.org/stable/23353260

[3] Bakare O. (2016). Production systems: Biophysical and economic environment constraints. In Takeshima H, Gyimah-Brempong K, & Johnson M. (Eds), *The Nigerian Rice Economy: Policy Options for Transforming Production, Marketing, and Trade* pp. 51-84. Philadelphia: University of Pennsylvania Press. Retrieved from Http://www.jstor.org/stable/j.ctt1dr36s2.9

[4] Mottaleb K.A, and Mishra A. (2016). Rice consumption and grain type preferences by household: A Bangladeshi case. *Journal of agricultural and applied economics,* 48(3): 298-319, htttps://doi.org/10.1017/aae.2016.18

[5] Nalley J, Tack A, Durand G, Thoma F, Tsiboe A, Shew C, and Barkely A. (2017) The production consumption and environmental impact of Rice published in *Agronomy Journal*. 109: Pp 193-203

[6] FAOSTAT (2021). (food and agriculture Organization of the United Nations) database. Available online *http://www.faostat.fao.org/site/339/default.aspx*. Accessed on 27 march 2021

[7] MoA (2009) Ministry of Agriculture *National Rice Development Strategy* (NRDS 2008-2018). p. 35. *Modern cultivars.* Plant Breeding. 108: 1-11.

[8] Onyango A. O. (2014) Exploring options for improving Rice production to reduce hunger and poverty in Kenya. World Environment 4: pp. 172-179.

[9] Rosemary A.E, Bibiana M.W, Njuguna N, Dominic M.K, Daniel A. (2010) Rice value chain study report of Kenya. *Ministry of Agriculture (MoA) and Kenya Agricultural Research Institute (KARI)*

[10] Tanaka, A., Johnson, J., Senthilkumar, K., Akakpo, C., Segda, Z., Yameogo, L., Bassoro, I., Lamare, D., Allarangaye, M.D., Gbakatchetche, H., Bayuh, B.A., Jaiteh, F., Bam, R.K., Dogbe, W., Sékou, K., Kamissoko, N., Mossi, I.M., Bakare, O.S., Mabone, F.L., Gasore, E.R., Cisse, M., Baggie, I., Kajiru, G.J., Mghase, J., Ablede, K.A., Nanfumba, D., Saito, K., (2017). On-farm rice yield and its association with biophysical factors in sub-Saharan Africa. European Journal of Agronomy. 85,

[11] MoW (2017) Ministry of Water and Irrigation. Irrigation Sub-Sector MTP111 Zero Draft (2); Ministry of Water and Irrigation: Nairobi, Kenya, 2017.

[12] Aryal Suman (2013) Rainfall and Water Requirement of Rice During Growing Period *Journal of Agriculture and Environment* 13: DOI: 10.3126/aej.v13i0.7576.

[13] Ghadirnezhad S., Seyede R., Fallah, A., (2020) Temperature effect on yield and yield components of different Rice cultivars in flowering stage international journal of agronomy 2020:https://doi.org/10.1155/2020/1798358 DOI-10.1155/2020/1798358

[14] Aldo F. and Antonio T (2007) Rice cultivation in the E.U ecological conditions and agronomic practices. *https//www.researchgate.net/publication/279426078*

[15] MoA (2010) Ministry of Agriculture *Economic Review of Agriculture*. Central Planning and project monitoring Unit. Nairobi.

[16] Africa Rice Centre (2013) New generation rice varieties unveiled for Africa. Available at: *http://www.cgiar.org/224023/new-generation-ricevarieties-u*

[17] Magoti R., Mayumi K, Kikutu Chimeningwa G, Kinama J, Kimani J, Hirouki S, Gicheru P, Daigo M. (2019) Growth of rice varieties in different Kenyan soil types under water deficit conditions. Journal of agricultural science 11 (6): Pp 1916 192

[18] Ngala G.N. (2013). Constraints to increased quality rice production for food, market and development with special reference to tropical Africa. *African Crop Science Conference Proceedings*, 11: pp 617 – 620

[19] Kimani J.M, Tongoona P, Derera J, Nyende A.B. (2011) Upland rice variety development through participatory breeding. ARPN *Journal of agricultural and biological science* 6(9): 39-49

[20] Musila R.N, Sibiya J, Derera J. (2015) Genetic analysis of drought tolerance and yield stability in interspecific and *Oryza sativa* L. rice germplasm. *https:// researchspace.ukzn.ac.za/handle/10413/13454*

[21] Kega, V. M. and A. R. Maingu (2008) Evaluation of New Rice for Africa (NERICA) Cultivars in Coastal Lowlands Kenya. In: A. O. Esilaba Et al., Editors, Proceedings of the 10th KARI Biennial Scientific Conference. Responding to Demands and Opportunities through Innovative Agricultural Technologies, Knowledge and Approaches, KARI Headquarters. 12-17 November. Kenya Agricultural Research Institute (KARI), Loresho, Nairobi, Kenya. p. 1-5.

[22] Rang, Z. W., S. V. K. Jagadish, Q. M. Zhou, P. Q. Craufurd and S. Heuer. (2011) Effect of high temperature and water stress on pollen germination and spikelet fertility in rice. Environmental and Experimental Botany 70: pp. 58-65.

[23] He, H. and Serraj R. (2012) Involvement of Peduncle Elongation, Anther Dehiscence and Spikelet Sterility in Upland Rice Response to Reproductive-Stage Drought Stress. *Environmental*

[24] Bruce TJA (2010). Tackling the threat to food security caused by crop pests in the new millennium. Food Sec. 2: pp. 133-141.

[25] Atera E. A., Onyancha F. N., Majiwa E. B. O. (2017) Production and marketing of rice in Kenya: Challenges and opportunities. *Journal of Development and Agricultural Economics*. 10(3): Pp 64-70, DOI: 10.5897/JDAE2017.0881 article number: 1820CDE55981 ISSN 2006-9774

[26] Cheseriek G.J, Kipkorir E.C, Webi P.O.W, Daudi I, kiptoo K.K.G, Mugalaraic E.M, Kiplagat L.K, Songok C.K (2017) Assesment of farmers challenges with the rice productivity in selected irrigation schemes in Western Kenya. International journal of current research. 4 (08): pp. 025-033

[27] Stryker J. D (2010). Developing competitive rice value chains. Presented in the second Africa Rice congress, Bamako, Mali, 22-26 march 2010: Innovation and partnerships to realize Africa's Rice potential. Associates for International Resources and Development (AIRD), 60 Bartlett avenue, Arlington, MA 02476, USA

[28] Kilimo Trust (2014) Expanding Rice Markets in the East African Community: An Opportunity for Actors in the Rice Value Chain. *Competitive Africa*

[29] Verulkar S., N. Mandal, J. Dwivedi, B. Singh, P. Sinha, R. Mahato, P. Dongre, O. Singh, L. Bose and P. Swain. (2010) Breeding resilient and productive genotypes adapted to drought-prone rainfed ecosystem of India. Field Crops Research 117: pp. 197-208.

[30] Upadhyay M.N, and Jaiswal H.K, (2015) Combining ability analysis for yield and earliness in hybrid Rice (*Oryza sativa* L.). *Asian Journal of Crop Science, 7: pp 81-86.* Using a double haploid population. Breeding Science 52:309-317.

[31] Ndungu N.S and Wilson A.O (2019) Analysis of rice millers value *chain Journal of agriculture and veterinary medicine* 10 (I) pp 38-47 and *Experimental Botany* 75: pp120-127.

[32] Dalberg (2014). Technology Study and Product Incubation: Agriculture Sector Phase 1: BoP Technology Selection.

Chapter 9

Rice Blast Disease in India: Present Status and Future Challenges

Deepak Chikkaballi Annegowda,
Mothukapalli Krishnareddy Prasannakumar,
Hirehally Basavarajegowda Mahesh,
Chethana Bangera Siddabasappa,
Pramesh Devanna, Sahana Nagaraj Banakar,
Haniyambadi Basavegowda Manojkumar
and Siddegowda Rajendra Prasad

Abstract

Rice (*Oryza sativa* L.) is the staple food of the majority of Indians, and India is both the major producer and consumer of rice. Rice cultivation in India is confronted with diverse agro-climatic conditions, varying soil types, and several biotic and abiotic constraints. Among major fungal diseases of Rice in India, the blast caused by Magnaporthe oryzae is the most devastating disease, with the neck blast being the most destructive form. Most of the blast epidemic areas in India have been identified with a mixture of races blast fungus resulting in the resistance breakdown in a short period. At present, a more significant number of the rice varieties cultivated in India were bred by conventional breeding methods with blast resistance conferred by a single resistance gene. Therefore, the blast disease in India is predominantly addressed by the use of ecologically toxic fungicides. In line with the rest of the world, the Indian scientific community has proven its role by identifying several blast resistance genes and successfully pyramiding multiple blast resistance genes. Despite the wealth of information on resistance genes and the availability of biotechnology tools, not a great number of rice varieties in India harbor multiple resistance genes. In the recent past, a shift in the management of blast disease in India has been witnessed with a greater focus on basic research and modern breeding tools such as marker-assisted selection, marker-assisted backcross breeding, and gene pyramiding.

Keywords: *Magnaporthe oryzae*, blast, resistance breeding, Marker Assisted Selection (MAS), Pyramiding, disease management, Marker-assisted backcross breeding (MABB)

1. Introduction

As the theme "Rice is life" reflects, Rice *(Oryza sativa* L.) is the single most important staple food crop for more than one-third of the world's population and more than half of the population India. Rice is grown in a wide range of

IntechOpen

agro-ecological conditions in India. Rice provides 21% of global human per capita energy and 15% per capita protein [1]. Amongst the important rice-producing nations in the world, India ranks second in terms of area and production. Out of 782 million tons (mt) of global rice production from 167.1 million hectares (m ha), India produced 116.42 m t in 44.5 m ha (rainy season: 102.13 m t from 39.27 m ha) [2]. For food insecurity to recede, agricultural production on currently cultivated land will increase by 70% globally and 100% in the developing countries by 2050 [3]. Of the various biotic factors limiting rice production and productivity, diseases continue to be an enigmatic problem in several rice-growing ecosystems of the world's tropical and temperate regions. The annual losses due to rice diseases are estimated to be 10–15% on an average basis worldwide. Rice blast fungus infects host plants at various crop growth stages, including leaf, stem, neck, collar, node, and root. The biggest challenge for rice breeders is the breakdown of resistance in existing rice varieties over the years. Therefore, breeding durable and broad-spectrum resistant cultivars is again a challenging task. The broad host range, continuous genetic variation, evolution, and host shifts are the main reasons behind the emergence of virulent pathotypes of *Magnaporthe*, which make blast management a daunting task. Hence, the Rice-*Magnaporthe* interaction pathosystem emerged as a model system to study host-pathogen interaction for several reasons, including the economic importance of blast disease in rice production and human diet.

2. Blast disease of rice and its economic importance

The blast disease affects almost all parts of the rice plant and occurs in different crop growth stages, starting from nursery to harvesting. The symptoms at different stages are called by different names, *viz*., nursery blast, leaf blast, node blast, neck blast, and panicle blast (**Figure 1**, panels a-e). The disease was first reported as "rice fever" in China by Soong Ying-shin in 1637, and later, it was reported from Japan by Imochi-byo during 1704. It is presently found in approximately 85 countries in the world and India. It was first recorded in 1913, and the first devastating epidemic was reported in 1919 in the Tanjore delta of erstwhile Madras state [5]. Later, the disease

Figure 1.
(a) Blast disease symptoms at nursery stage (b) typical leaf blast symptoms under field condition (c) typical node blast symptoms (Photo Curtsey: http://www.knowledgebank.irri.org) (d) Neck blast infection leading to the choppiness and breakage of panicle (e) Panicle blast where symptoms appears on grains (f) life cycle of rice blast fungi [4].

has been reported to occur in different regions of India [6, 7]. Blast disease occurs in all rice ecosystems. However, it is more damaging in upland rice than in irrigated ecosystem of rice cultivation. It is the major contributor to the yield gap. It causes more losses, especially in the humid rice-growing areas of India, including the cool season crop in Karnataka, Andhra Pradesh, Tamil Nadu, and Kerala. With the introduction and spread of semi-dwarf high-yielding varieties in the 1960s, its incidence became almost insignificant, especially in plains of North India during the *Kharif* season. The relative losses from this disease vary in different production zones depending on the physical environment, crop management, and pathogen population dynamics. The upland rice, grown in about 6.3 M ha in Eastern India and hill rice, is more prone to blast disease, and in many cases, the disease is left uncontrolled due to non-remunerative management options. Severe epidemics of the blast have occurred between 1980 and 1987 in Himachal Pradesh, Andhra Pradesh, Tamil Nadu, and Haryana, resulting in huge financial losses. As per estimation, the extent of annual yield reduction caused by rice blast disease is sufficient to provide food to around 60 million people [8]. Among the different stages of the disease, drastic yield reductions are reported in neck and panicle blast, reducing the grain weight, the percentage of ripe spikelets, and the percentage of fully mature grains [9]. The infection of the panicle base (Neck blast) by the blast pathogen until 20 days after heading was found to cause more than 50 percent yield loss [10]. In India, yield losses due to blast could be as high as 50% when the disease attains an epidemic proportion [11]. During natural epidemics of blast disease in the wet season, disease incidence ranged from 14 to 27% (above the economic threshold), resulting in yield loss of about 27–35 percent [12].

3. Rice blast fungus: *Magnoporthe oryzae*

The fungus causes rice blast disease *Pyricularia oryzae* Cavara [synonym *P. grisea* Sacc, teleomorph *Magnaporthe oryzae* [(Hebert) Barr]. The genus *Magnaporthe*, which consists of five species (*M. grisea, M. oryzae, M. salvinii, M. poae,* and *M. rhizophila*), has shared morphological traits such as three-septate fusiform ascospores and black non-stromatic perithecia (ascocarp) with long hairy necks [13]. However, due to the limited host range of the individual isolates, all *Magnaporthe* sps were regarded as the *M. grisea* species complex (Mg complex) [14, 15]. Blast disease of rice and other gramineous species is caused by the members of the Mg complex [15]. Recently, based on phylogenetic analyses and mating tests, isolates from crabgrass were separated from the Mg complex and named *M. grisea,* and other isolates from grasses, including rice, were named as *M. oryzae* [16–19]. One hundred thirty-seven family members, including rice, are affected by *M. oryzae* where it causes blast disease [20, 21].

3.1 Disease cycle and epidemiology

The pathogen perpetuates as mycelium and conidia on diseased straw, seed, rice ratoons, volunteer rice plants, and weed hosts. The initiation of the primary infection process begins with the attachment of the conidium of *M. oryzae* to the leaf cuticle. Later stages of the infection process include the formation of appressorium, which assists in pathogen penetration, generation of turgor, formation of penetration peg, and finally penetration into host tissue (**Figure 1**, panel f) [4, 22, 23]. After penetration, hyphae grow through the plant tissue, resulting in the disease lesions and typical rice blast symptoms. Under congenial weather conditions (high relative humidity and low night temperature), the fungus produces an enormous number of conidia which brings the secondary spread and infection to other healthy plants nearby and spreads rapidly to adjacent fields by wind leading to field epidemic [4, 20, 24]. New blast lesions appear within 4–5 days after landing spores at the optimum temperature on the leaf surface. New conidia are produced under warm and wet weather conditions on the disease lesion within few hours of lesion appearance. The sporulation continues for several days and provides the inoculum (secondary conidia) for secondary infection.

Although blast disease is distributed across all the parts of India, some parts of the country have been identified as hotspots of blast disease. Sub-Himalayan regions of Jammu and Kashmir, Himachal Pradesh, hill districts of Uttaranchal, and West Bengal are often associated with the northern part of India repeated epidemics of blast disease. In the eastern part of India, the blast is in its destructive form in upland rice-growing areas of Arunachal Pradesh, Manipur, Mizoram, Meghalaya, Assam, Chotanagpur belt, and Jaypore tract of Orissa. While blast is of much importance in the Konkan region of Maharashtra and Gujarat in the west, the disease is frequently reported from Andhra Pradesh, Telangana, Tamilnadu, and Coorg region of Karnataka in peninsular India. From several blast disease incidence reports and surveys, blast disease occurs in different agroclimatic conditions in the country. In North and North-Eastern India, the blast disease occurs in June to September in high rainfall areas with 20–24°C. In medium rainfall areas (1000 mm per annum) and temperatures ranging between 24 and 34°C in Western and Central India, blast occurrence is reported from August to October. However, the disease is associated with Andhra Pradesh, Telangana, Karnataka, Tamilnadu, and Kerala states in dry periods with cooler nights (18–22°C).

3.2 Pathogen variability

One of the main strengths of a blast pathogen in its interaction with the host and overpowering of the host defense system is the existence of several races. The Indian subcontinent is a center of origin and diversity for the *Magnaporthe* species complex. The pathogen is highly variable and evolves into new pathotypes within a short period. There is a nationally coordinated system (All India Coordinated Rice Improvement Programme) for regular monitoring of virulence pattern of blast disease using twenty-five rice cultivars that include international blast differentials, recombinant inbred lines, donors, and commercial cultivars. Cluster analysis of the *M. oryzae* reaction on these cultivars in different rice growing ecosystems revealed that the pathogen population could be clustered into four separate groups.

Further, there was a considerable variation within the groups, also suggesting the significant variability in the virulence of the *M. oryzae* population of India [25]. Efforts were made during the 1970s, where race profiling of Indian isolates of *M. oryzae* was carried out, and a new race group IJ was identified [6]. During the 1970s, race IC3 and ID 1 were predominantly distributed in India [6]. In another report, five pathogenic race groups, ID-1, ID-2, IB-4, IC-17, and IC-25, were identified from India and group ID-17 to be predominantly distributed in the Indian paddy ecosystem [26]. A total of 72 isolates of *M. oryzae* from rice in different districts of Karnataka were examined for identifying sexual mating alleles MAT1, MAT2, and understanding the genetic diversity based on the DNA fingerprint of pot2 an inverted repeat transposon. Among 72 isolates, 44 isolates belonged to MAT1 type (male fertile), and 28 isolates were of MAT2 (female fertile), and there were no hermaphrodite isolates [27]. Multi-marker systems including Simple Sequence Repeats (SSRs), repetitive DNA-based markers (Pot2), pathogenicity genes were used to study genetic variability of *Magnaporthe* species in rice and finger millet ecosystems from southern India. Data from multiple markers revealed high genetic diversity and clustering based on geographical location and host species [28]. Interestingly, major cluster I is dominated by Indian isolates whereas cluster II is dominated by isolates from different rice growing region of the world. Similarly, the blast isolates from the same geographical location did not belong to the same sub-cluster while genetically similar isolates from different geographical location were grouped together. Same authors grouped most of Indian isolates in one group whereas blast isolates from other parts of the world in other group might be due to presence of distinct strain in India than rest of the world [29]. Despite few studies, the race distribution of the rice blast fungus is poorly understood in India. It demands enormous attention in the context of deploying suitable resistance genes to confront the pathogen.

4. Pathogenomics

4.1 Sequencing of rice blast fungus

The whole genome of *M. oryzae* strain 70–15 was the first to be sequenced among plant pathogenic fungi using the Sanger sequencing method [30]. Subsequently, several field isolates of the blast have been sequenced using next-generation sequencing (NGS). While Field isolates from Japan (Ina168 and P131) and China (Y34) [31, 32] were sequenced using the 454 sequencing platform, more recently, two field isolates, FJ81278 and HN19311 from China, have been sequenced using Illumina technology [33]. A highly diverse *Magnaporthe* species complex and multiple field isolates of *Magnaporthe* infecting different hosts such as rice (leaf and neck), finger millet (leaf and neck), foxtail millet (leaf), and buffelgrass (leaf) have recently been sequenced from India using Illumina sequencing technology [34, 35]. The majority of these isolates included virulent field isolates from southern India and a commonly used virulent reference strain B157 isolated from rice [36]. The genomes were extensively analyzed to compare the variability in gene content, repeat element distribution, candidate effectors, genes involved in carbohydrate metabolism, and single nucleotide polymorphisms (SNPs). This study has shed light on genomic factors contributing to genome variation, pathogenic strain evolution, and host-specificity. It was the first to compare blast fungal isolates from different hosts and different host tissues in India at the genome level.

Interestingly, whole-genome sequencing of multiple isolates has revealed large chunks of novel genomic regions and multiple novel genes. In another report, the whole-genome assembly of *M. oryzae* RMg-Dl yielded 34.82 Mb genome sequence by PacBio single-molecule and Illumina HiSeq2500 sequencing, which aids in better understanding the genetic determinants of host range, host jump, survival, pathogenicity, and virulence factors of *M. oryzae* [37]. The genomic variation was attributed to race evolution over a period of time by geographical separation, chromosomal variation, and variability in repetitive elements [30, 31, 33]. The availability of pathogen genomes will undoubtedly be helpful to breeders and researchers to understand *Magnaporthe* virulence spectrum and improve blast resistance in rice and other important food crops affected by blast disease.

4.2 Rice genome sequencing

Pathogen and host are the two faces of a coin in the context of host-pathogen interactions and disease management. Hence, characterizing the rice genotypes for novel resistant genes (R) should be done parallel with that of the pathogen as host and pathogen evolve simultaneously for their survival in nature. The discovery of novel 'R' genes and understanding their mutation in evolving novel alleles/genes is an important step in resistance breeding. Allele discovery/mining could be made using high throughput technologies like whole genome sequencing using next-generation sequencing (NGS) technologies. Rice is a model cereal crop, and several rice cultivars have been sequenced at the genome level, with Nipponbare as the first rice cultivar to be sequenced and published in 2002 [38]. Further, the indica cultivar 93-11 was also sequenced and published in the same year [39]. These initial efforts laid the foundation for the genomic era in rice. Subsequently, several whole-genome sequencing efforts of rice cultivars like IR-64 [40], Kasalath [41], and HR-12 [42] also added quantum of genomic information to the existing genomic resources. HR-12 genome was assembled using a combination of Illumina short reads and PacBio long reads. This was the first report in the world to sequence rice genome using third-generation sequencing technology. The power of long-read technologies helped in repeat resolution compared to second-generation technologies. Large-scale discovery of novel alleles by resequencing of 3000 rice germplasm accessions belonging to 89 countries contributed significantly to the rice genomic resources [43]. Exploiting natural variation existing among rice landraces is an ideal method to map R genes. Mapping of R genes based on avirulent (Avr) genes pattern in the rice-growing areas is the best strategy to mitigate *Magnaporthe* via exploiting host plant resistance. The product of the avirulence (*Avr*) gene of *Magnaporthe* can be detected by the corresponding resistance (*R*) gene of rice and activates immunity to rice mediated by the *R* gene. The high degree of variability of *M. oryzae* isolates in pathogenicity makes the control of rice blast difficult. That resistance of the *R* gene in rice has been lost ascribed to the instability of the *Avr* gene in *M. oryzae*. Further study on the variation of the *Avr* genes in *M. oryze* field isolates may yield valuable information on the durable and effective deployment of *R* genes in rice production areas. *AvrPiz-t* and *Piz-t* are a pair of valuable genes in the Rice-*Magnaporthe* pathosystem. *AvrPiz-t* is detectable by *Piz-t* and determines the effectiveness of *Piz-t* [44]. Rice SNP-seek database developed based on 3000 rice genomes, possessing a large-scale single base level variation across three geographical rice ecotypes (*japonica*, *indica*, and *javanica*) been made available to the public [45]. These variants could be harnessed to study the genetic diversity and development of subspecies-specific rice cultivars. Also, rice breeders can focus on allele mining for corresponding R genes and pyramiding these genes in commonly grown cultivars in a given location to help develop resistant rice varieties.

5. Resistance genes and QTLs for rice blast disease

The resistance for blast disease is two types: i) qualitative or complete resistance governed by a major R gene, and ii) quantitative or partial resistance governed by many quantitative trait loci [46]. While qualitative resistance confers resistance against a specific race of blast pathogen, the quantitative resistance is non-race specific. To date, 109 major blast resistance genes have been identified in rice. Out of these, 25 R genes have been successfully cloned and characterized, with Pi9 being the first cloned R gene (**Table 1**). Japan and China have lead the race in identification of major R genes by identifying 34 and 27 blast resistance genes, respectively. Followed by these, a significant contribution has also been made by USA, France, Philippines and India. To date eight R genes have been mapped in India that include Pi10, Pi157f, Pi38, Pi42(t), Pikh (Pi54)[129], Pitp [117], Pi54rh [127], and Pi54of [128]. The details of the genes mapped in India and the rice varieties in which they were identified are provided in **Table 1**. Majority of the genes identified in India and rest of the world encode proteins with NBS-LRR (Nucleotide Binding Site and Leucine Rich Repeats) and Zinc finger domains that confer disease resistance. Among all the genes that were mapped in India, pi54 is of great importance as it a major blast resistance gene and provides durable resistance against Indian races of blast fungus. These qualitative and major R genes have been extensively used in blast resistance breeding programs worldwide (**Table 2**). For instance,

DOI: http://dx.doi.org/10.5772/intechopen.98847

Sl. No	Gene	Chr. No.	Position Location (bp)	Source	Country	References
1	*Mpiz*	11	4073024-16730739	Zenith (J)	Japan	[47]
2	*Pb1*	11	21711437-21361768	Modan (I)	Japan	[48, 49]
3	*PBR*	11	–	St-No 1 (J)	Japan	[50]
4	*Pi(t)*	4	–	P167 (I)	–	[51]
5	*Pi1*	11	26498854-28374448	LAC23 (J)	Philippines	[52]
6	*Pi10*	5	14521809-18854305	Tongil (I)	India	[53]
7	*Pi11*	8	–	Zhai-Ya-Quing8 (I)	China	[51]
8	*Pi12*	12	6988220–15120464	K80-R-Hang (J) Moroberekan (J)	Japan	[54]
9	*Pi13(t)*	6	12456009-16303608	*O. minuta* (W), Kasalath (I), Maowangu	Philippines	[55–57]
10	*Pi14(t)*	2	1-6725831	Maowangu	Japan	[58]
11	*Pi15*	9	9641358-9685993	GA25 (J)	China	[59]
12	*Pi157*	12	8826555-18050447	Moroberekan (J)	India	[53]
13	*Pi16(t)*	2	1-6725831	Aus373 (I)	Japan	[60]
14	*Pi17*	7	22250443-24995083	DJ123 (I)	Philippines	[61, 62]
15	*Pi18(t)*	11	26796917-28376959	Suweon365 (J)	Korea	[63]
16	*Pi19(t)*	12	8826555-13417087	Aichi Asahi (J)	Japan	[64]
17	*Pi20*	12	6988220-10603823	IR24 (I)	Philippines	[65]
18	*pi21**	4	5242654-5556378	Owarihatamochi (J)	Japan	[66]
19	*Pi22(t)*	6	4897048-6023472	Suweon365 (J)	Korea	[67]
20	*Pi23*	5	10755867-19175845	Suweon365 (J)	Korea	[67]
21	*Pi24(t)*	1	5242654-5556378	Azuenca (J)	France	[68]
22	*Pi25**	6	18080056-19257588	Gumei 2 (I)	China	[69]
23	*Pi25(t)*	2	34360810-37725160	IR64 (I)	France	[68]

Sl. No	Gene	Chr. No.	Position Location (bp)	Source	Country	References
24	*Pi26*	6	8751256-11676579	Gumei 2 (I)	China	[70]
25	*Pi26(t)*	5	2069318-2760202	Azucena (J)	France	[68]
26	*Pi27*	1	5556378-744329	Q14 (I)	France	[68]
27	*Pi27(t)*	6	6230045-6976491	IR64 (I)	France	[68]
28	*Pi28(t)*	10	19565132-22667948	IR64 (I)	France	[68]
29	*Pi29(t)*	8	9664057-16241105	IR64 (I)	France	[68]
30	*Pi3(t)*	6	–	Pai-kan-tao (J)	Philippines	[71]
31	*Pi30(t)*	11	441392-6578785	IR64 (I)	France	[68]
32	*Pi31(t)*	12	7731471-11915469	IR64 (I)	France	[68]
33	*Pi32(t)*	12	13103039-18867450	IR64 (I)	France	[68]
34	*Pi33*	8	5915858-6152906	IR64 (I)	France	[68]
35	*Pi34*	11	19423000-19490000	Chubu32 (J)	Japan	[72]
36	*Pi35(t)**	1	–	Hokkai 188 (J)	Japan	[73]
37	*Pi36**	8	2870061-2884353	Q61 (I)	China	[74]
38	*Pi37**	1	33110281-33489931	St-No 1 (J)	China	[75, 76]
39	*Pi38*	11	19137900-21979485	Tadukan (I)	India	[77]
40	*Pi39(t)*	4, 12	-	Chubu 111 (J) Q15(I)	China	[78]
41	*Pi40(t)*	6	16274830-17531111	*O. australiensis* (W)	Philippines	[79]
42	*Pi41*	12	33110281-34005652	93-11 (I)	China	[80]
43	*Pi42(t)*	12	19565132-22667948	DHR9 (I)	India	[81]
44	*Pi44*	11	20549800-26004823	Moroberekan (J)	USA	[82]
45	*Pi47*	11	–	Xiangzi 3150 (I)	China	[83]
46	*Pi48*	12	–	Xiangzi 3150 (I)	China	[83]

Sl. No	Gene	Chr. No.	Position Location (bp)	Source	Country	References
47	*Pi5(t)*	9	–	Moroberekan (J)	Philippines	[84]
48	*Pi6(t)*	12	4053339-18867450	Apura (I)	USA	[85]
49	*Pi62(t)*	12	2426648-18050026	Yashiro-mochi (J)	Japan	[68]
50	*Pi67*	–	–	Tsuyuake	Philippines	[68]
51	*Pi8*	6	6230045-8751256	Kasalath (I)	Japan	[57]
52	*Pi9**	6	10386510-10389466	*O. minuta* (W)	China	[86]
53	*Pia**	11	4073024-8078510	Aichi Asahi (J)	Japan	[87, 88]
54	*Pib**	2	35107768-35112900	Tohoku IL9 (J)	Japan	[89, 90]
55	*Pib2*	11	26796917-28376959	Lemont (J)	Philippines	[91]
56	*PiCO39(t)*	11	6304007-6888870	CO39 (I)	USA	[92]
57	*Pid(t)1*	2	20143072-22595831	Digu (I)	China	[93]
58	*Pid2**	6	17159337-17163868	Digu (I)	China	[94]
59	*Pif*	11	24695583-28462103	Chugoku 31-1 (J)	Japan	[95]
60	*Pig(t)*	2	34346727-35135783	Guangchangzhan (I)	China	[96]
61	*PiGD1*	8	–	Sanhuangzhan 2 (I)	China	[97]
62	*PiGD-2*	10	–	Sanhuangzhan 2 (I)	China	[98]
63	*PiGD3*	12	–	Sanhuangzhan 2 (I)	China	[98]
64	*Pigm(t)**	6	10367751-10421545	Gumei4 (I)	China	[99]
65	*Pii*	9	2291804-28431560	Ishikari Shiroke (J) Fujisaka 5 (J)	Japan	[95, 100]
66	*Pii1*	6	2291804-28431560	Fujisaka 5 (J)	Japan	[56, 57]
67	*Pii2*	9	1022662-7222779	Ishikari Shiroke (J)	Japan	[101]
68	*Piis1*	11	2840211-19029573	Imochi Shiraz (J)	Japan	[102]
69	*Piis2*	–	–	Imochi Shiraz (J)	Japan	[102]

Sl. No	Gene	Chr. No.	Position Location (bp)	Source	Country	References
70	*Piis3*	–	–	Imochi Shiraz (J)	–	[102]
71	*Pik**	11	27314916-27532928	Kusabue (I)	China	[103, 104] [-48]
72	*Pikg*	11	27314916-27532928	GA20 (J)	Japan	[56]
73	*Pikh (Pi54)**	11	24761902-24762922	Tetep (I)	India	[105]
74	*Pikm**	11	27314916-27532928	Tsuyuake (J)	China	[106, 107]
75	*Pikp**	11	27314916-27532928	HR22 (I)	China	[108]
76	*Piks*	11	27314916-27532928	Shin 2 (J)	Japan	[109]
77	*Pikur1*	4	24611955-33558479	Kuroka (J)	Japan	[102]
78	*Pikur2*	11	2840211-18372685	Kuroka (J)	Japan	[110]
79	*Pilm2*	11	13635033-28377565	Lemont (J)	USA	[111]
80	*Pir2-3(t)*	2	–	IR64 (I)	Indonesia	[112]
81	*Pirf2-1(t)*	2	–	*O. rufipogan* (W)	Indonesia	[112]
82	*Pise*	11	5740642-16730739	Sensho (J)	Japan	[102]
83	*Pise2*	–	–	Sensho (J)	Japan	[102]
84	*Pise3*	–	–	Sensho (J)	Japan	[102]
85	*Pish**	1	33381385-35283446	Shin 2 (J)	Japan	[65]
86	*Pi-sh*	11	33381385-35283446	Nipponbare (J)	Japan	[65]
87	*Pit**	1	2270216-3043185	Tjahaja (I), K59 (I)	Japan	[108, 113]
88	*Pita**	12	10603772-10609330	Tadukan (I),	USA	[114]
89	*Pita2*	12	10078620-13211331	Shimokita (J)	Japan	[115, 116]
90	*Pitp(t)*	1	25135400-28667306	Tetep (I)	India	[117]
91	*Pitq1*	6	28599181-30327854	Tequing (I)	USA	[111]
92	*Pitq2*	2	–	Teqing (I)	USA	[91]

Sl. No	Gene	Chr. No.	Position Location (bp)	Source	Country	References
93	*Pitq3*	3	–	Teqing (I)	USA	[91]
94	*Pitq4*	4	–	Teqing (I)	USA	[91]
95	*Pi-tq5*	2	34614264-35662091	Tequing (I)	USA	[111]
96	*Pitq6*	12	5758663-7731471	Tequing (I)	USA	[111]
97	*Piy1(t)*	2	–	Yanxian No 1 (I)	China	[118]
98	*Piy2(t)*	2	–	Yanxian No1 (I)	China	[118]
99	*Piz*	6	10155975-10517612	Zenith (J), Tadukan (I), Toride 1 (J), Fukunishiki (J)	Japan	[119]
100	*Pizh*	8	4372113-21012219	Zhai-Ya-Quing8 (I)	China	[51]
101	*Pi2**	6	10076481-10204423	Jefferson	China	[120]
102	*Pid3**	6	13055253-13058027	93-11(I),Nipponbare (J)	China	[121]
103	*Pi4*	12	-	Pai-kan-tao	Philippines	[122]
104	*Pizt**	6	-	Wild rice	Japan	[123]
105	*Pi5**	9	S04G03-C1454 (Map position)	RIL260	Korea	[124]
106	*Pb1**	11	*Os11g0597700*	Kanto 209, Koshihikari Aichi	Japan	[125]
107	*NLS1**	11	AC134922 (Accession number)	Wild rice	China	[126]
108	*Pi54rh**	11	-	*Oryza rhizomatis* (Wild rice)	India	[127]
109	*Pi54of**	11	HE589448 (Accession number)	*Oryza officinalis*	India	[128]

Note: *Chr No. = Chromosome number; I = Indica; J = Japonica; – = Not Known.*Cloned and characterized blast resistance genes in rice.*

Table 1.
Blast resistance genes identified so far in different rice cultivars.

Sl. No	Gene/ QTL	Marker type	Technique	Trait	Application	Reference
1.	*Pi1, Piz-5, Pita*	RFLP	MAS	Blast resistance	Pyramiding of three near isogenic lines (C101LAC, C101A51 and C101PKT) for blast resistance into a single cultivar CO39, each carrying the major genes *Pi1, Piz-5* and *Pita*, respectively	[130]
2.	*Pi1*	SSR, ISSR	MABB	Blast resistance	Applied for backcross breeding of variety	[131]
3.	*Xa21, Piz*	SSR	MAS	Bacterial blight and blast resistance	Pyramiding of target traits	[132]
4.	*Pid1, Pib, Pita, Pi2*	SSR	MAS	Blast resistance	*Pid1, Pib* and *Pita* genes were introduced into G46B, while *Pi2* was introduced into Zhenshan 97B	[93]
5.	*Piz*	SSR	MAS	Blast resistance	Successfully used for selection of blast resistance in a wide array of rice germplasm	[133]
6.	*Xa13, Xa21, Pi54, qSBR11*	SSR	MAS	Blast, Bacterial blight and Sheath blight resistance	Transfer genes conferring the resistances toward three different diseases in rice	[134]
7.	*Pita*	Gene specific	MAS	Blast resistance	Existence of *Pita* gene in 141 rice germplasms was determined, but the results were more articulated when *Pita* gene was introduced through advanced breeding lines	[135]
8.	*Pi1, Pi2, Pi33*	SSR	MABB	Blast resistance	Introgressed into Jin23B	[136]
9.	*Pi1, Pi2, Xa23*	SSR	MABB	Blast resistance and bacterial blight	Successfully applied for breeding variety Rongfeng B	[137]

Sl. No	Gene/ QTL	Marker type	Technique	Trait	Application	Reference
10.	*Piz-5*, *Pi54*	SSR	MABB	Blast resistance	Transfer blast resistance genes from donor lines (C101A51 and Tetep) into PRR78 to develop Pusa 1602 (PRR78 + *Piz-5*) and Pusa 1603(PRR78 + *Pi54*), respectively	[138]
11.	*Pi9*	Gene specific	MABB	Blast resistance	Applied to introgress the cultivar Luhui 17	[139]
12.	*Pi1*, *Piz*	SSR	MABB	Blast resistance	Pyramiding of *Pi1* and *Piz-5* genes into introduced PRR78	[140]
13.	*Pi39*	InDel	MABB	Blast resistance	Introgressed into Chinese cultivar Q15	[141]
14.	*Pi2*, *Xa21*, *Xa33*	SSR	MABB	Blast and Bacterial blight resistance	Introgressed into RPHR-1005	[142]
15.	*Pi40*	SSR	MABB	Blast resistance	Introgressed into elite cultivars Turkish, Osmancik-97 and Halilbey	[143]
16.	*Pi1*, *Pi2*	SSR	MABB	Blast resistance	Introgressed into Intan variety and BPT5204	[144]
17.	*Pi46*, *Pita*	SSR	MABB	Blast resistance	Introgressed into Hang hui 179 (HH179)	[145]
18.	*Pi2*, *Pi9*	SNP	MABB	Blast resistance	Introgressed into R179	[146]
19.	*Pizt*, *Pi2*, *Pigm*, *Pi40*, *Pi9*, *Piz*	SSR	MAS	Blast resistance	Introgressed into Yangdao 6	[147]
20.	*Pi1*, *Pi2*, *Pi33*	SSR	MAS	Blast resistance	Introgressed into Russian rice varieties	[148]
21.	*Pi9*, *Pizt*, *Pi54*	SNP	MABB	Blast resistance	Introgressed into japonica rice 07GY31	[145]
22.	*Pi-b*, *Pik-h*	SSR	MAS	Blast resistance	Introgressed into MR219	[149]
23.	*Pi-ar*	RAPD	MAS	Blast resistance	Double haploid technique was used for the introgression of *Pi-ar* gene	[150]

Sl. No	Gene/ QTL	Marker type	Technique	Trait	Application	Reference
24.	*Pi2, Pi54, xa13* and *Xa21*	SSR	MABB	Blast and Bacterial blight resistance	Improvement of Basmati rice varieties	[151]
25.	*Piz-5* and *Pi54*	SSR	MABB	Blast resistance	Incorporation of blast resistance into "PRR78", an elite Basmati rice restorer line	[152]
26.	*Pi1, Pi2* and *Pi33*	SSR	MABB	Blast resistance	Improve blast resistance in Indian rice (*Oryza sativa*) variety ADT43	[153]
27.	*Pi-2* and *Pi-54*	SSR	MABB	Blast resistance	Introgression of blast resistance genes into the genetic background of elite, bacterial blight resistant indica rice variety, Improved Samba Mahsuri	[154]
28.	*Xa21, xa13* and *Pi54*	Gene specific	MABB	Blast and Bacterial blight resistance	Marker-Assisted Pyramiding of Genes Conferring Resistance Against Bacterial Blight and Blast Diseases into Indian Rice Variety MTU1010	[155]
29.	Xa 5 and 4 blast QTLs	SSR	MAS	Blast and Bacterial blight resistance	Pyramiding of blast and bacterial leaf blight resistance genes into rice cultivar RD6	[156]
30.	*Pi54* and *Pi1*	SSR	MAS	Blast resistance	Introgression of blast resistance genes into cold tolerant variety Tellahamsa	[157]
31.	*Pi46* and *Pita*	SSR	MABB	Blast resistance	Blast resistance genes were introgressed into an elite restorer line Hang-Hui-179 (HH179)	[145]
32.	*Pi2* and *Xa23*	SSR	MAS	Blast and Bacterial blight resistance	Pyramiding of broad-spectrum disease resistance genes into GZ63-4S elite thermo-sensitive genic male-sterile line in rice	[158]

Sl. No	Gene/ QTL	Marker type	Technique	Trait	Application	Reference
33.	*Xa21, Xa33, Pi2, Rf3* and *Rf4*	SSR	MAS	Blast and Bacterial blight resistance	Marker-assisted improvement of the elite restorer line of rice, RPHR-1005 for resistance against diseases	[142]
34.	*Xa21, xa13,* and *Pi54*	SSR	MAS	Blast and Bacterial blight resistance	Introgression of Major Bacterial Blight And Blast Resistant Genes into Vallabh Basmati 22	[159]
35.	*Xa21* and *Pi54*	SSR	MABB	Blast and Bacterial blight resistance	Introgression of bacterial blight and blast resistance into DRR17B, an elite, fine-grain type maintainer line of rice	[160]
36.	*Pi54, qSBR11-1, qSBR11-2 and qSBR7-1*	SSR and QTLs	MABB	Blast and Sheath blight	Introgression of multiple disease resistance into a maintainer of Basmati rice CMS line	[161]
37.	*Pikh*	SSR	MABB	Blast	Introgression of Blast Resistance Gene into a Malaysian Cultivar, MR264	[162]

Modified table of Ashkani et al. [163].
SSR, Simple sequence repeat; ISSR, Inter simple sequence repeat; SNP, Single nucleotide polymorphism; RFLP, Restriction fragment length polymorphism; RAPD, Randomly Amplified Polymorphic DNA; MAS, Marker Assisted Selection and MABB, Marker Assisted Backcrossing Breeding.

Table 2.
Examples of marker assisted selection (MAS) and marker assisted backcross breeding (MABB) in rice for blast resistance.

the improved rice lines carrying Pi9 and Pi2 were highly resistant to 43 isolates collected from 13 countries and 455 isolates collected from different parts of the Philippines, and 792 isolates from several regions of China, respectively [86, 164]. Because of their high importance, there are continuing efforts to identify additional major blast resistance genes, especially in wild rice species, and transfer them into elite varieties. For example, Pi9 present in indica rice line 75-1-127 [131] was introgressed from *O. minuta* [55]. Amongst the molecularly characterized major leaf-blast R genes, 22 were; namely, *Pi37, Pit, Pi-sh, Pi64, Pi-b, Pi63, Pi9, Pi-2, Piz-t, Pid3, Pigm, Pi25, Pi36, Pi5, Pi-54, Pik-m, Pik, Pik-p, Pik-e, Pi-a, Pi1,* and *Pita*, belong to the largest class of plant R genes that encode proteins with the nucleotide-binding site (NBS). Leucine-rich repeat (LRR) domains whereas one, *Pid2*, encodes serine–threonine–kinase membrane-spanning protein [165]. Rice blast resistance gene, Pi54 provides broad-spectrum resistance against different strains of *M. oryzae*. Understanding the cellular localization of Pi54 protein is an essential step towards deciphering its interaction with the cognate Avr-gene. A study was conducted to investigate the subcellular localization of Pi54 with Green Fluorescent Protein (GFP) as a molecular tag. This is the first detailed report, which emphasizes the cellular and subcellular distribution of the broad-spectrum blast resistance gene Pi54 in rice and the impact of its constitutive expression towards resistance against other fungal and bacterial pathogens of rice [166]. These R genes function in a gene-for-gene fashion, meaning that for every R gene in the host, there is an Avr gene in the pathogen. Therefore, the pathogen can easily break down the host resistance by modifying or deleting its corresponding Avr gene and rendering the resistant variety susceptible after a few years [167]. The quantitative or partial resistance is more suited to low-risk areas as it cannot suppress *M. oryzae* when the environments are conducive for its growth. The quantitative trait loci (QTL), which in the context of disease resistance also referred to as quantitative resistance loci (QRLs) [168], are thought to play an important role in sustainable food production in the years ahead by manifesting durable resistance against many races of the blast fungus [169]. Chromosomal locations of leaf-blast R genes and quantitative trait loci (QTLs) for neck-blast resistance in rice are illustrated in **Figure 2**. Recently, QTL analysis of introgression line (INGR15002) derived from *O. glumaepatula* led to the identification of two major QTL - qBL3 contributing about 34% and 32%

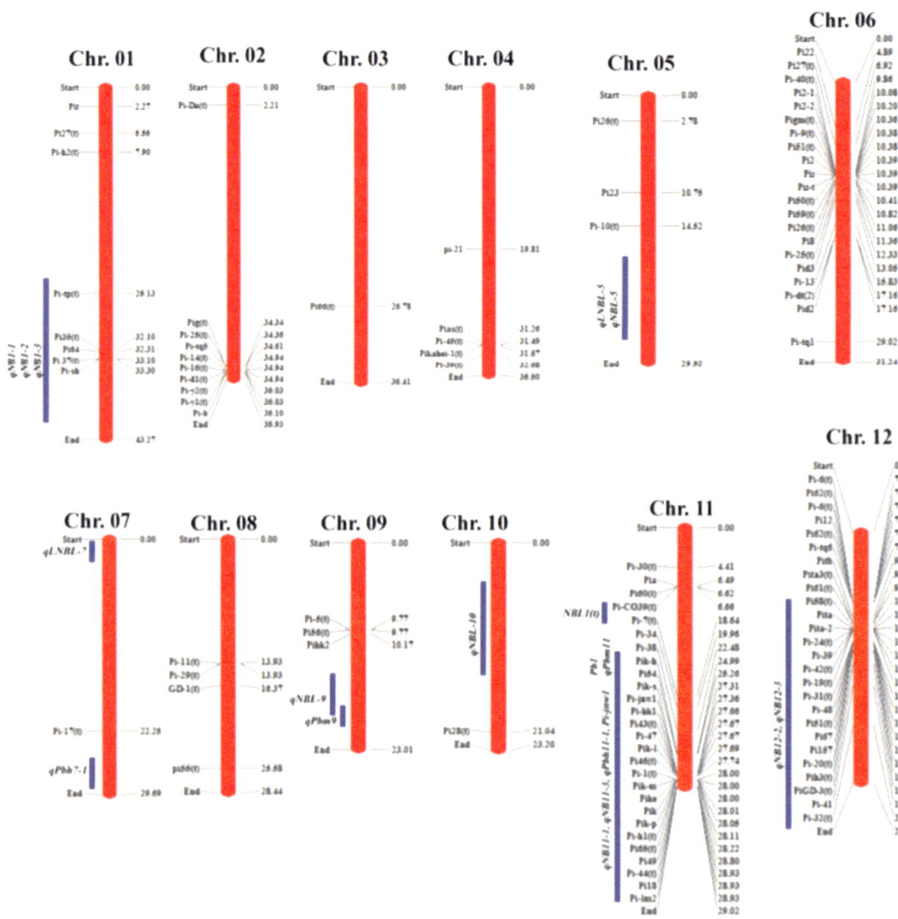

Figure 2.
Chromosomal locations of leaf-blast R genes and quantitative trait loci (QTLs) for neck-blast resistance in rice. The chromosomal locations for R genes and QTLs were deducted by projecting the sequences of closely linked/ flanking markers on the genome sequence of cv. Nipponbare released by International Rice Genome Sequencing Project (http://rapdb.dna.afrc.go.jp) adopted [131].

phenotypic variance towards leaf and neck blast resistance, respectively, and qBL7 contributing about 25% of phenotypic variance for leaf blast [170]. Hence, there are several and continuous attempts to identify QTLs for blast resistance in rice. However, the results of a meta-analysis of QTLs have indicated that the use of QTLs does not offer durable and broad-spectrum resistance compared to that offered by the major genes [171]. Hence, care has to be taken in future breeding programs to effectively combine the major genes and QTLs to achieve durable and long-lasting resistance against several races of the blast fungus.

6. Defence mechanism of rice against blast infection

Rice employs a twolayered innate immune system to defend itself against blast invasion. PAMP-triggered immunity (PTI) forms the first layer of immunity, and is boosted after PAMP recognition by membrane-associated PRR on the host cell membrane [172]. Chitin well-known type of PAMP capable of activating plant immune responses. Chitin from *M. oryzae* is recognized by rice transmembrane LysM receptor-like proteins (LysM-RLPs), including two lysin motif-containing proteins, OsLYP4 and OsLYP6, and a chitin elicitor binding protein (CEBiP). When defending against *M. oryzae*, rice forms a receptor complex called LysM-RLPsOsCERK1. Two rice Receptor like kinases (RLKs), Flagellin Sensing 2 (OsFLS2) and BRI1-Associated receptor Kinase 1 (OsBAK1), are also involved in PTI. To achieve successful infection, virulent *M. oryzae* isolates have evolved a strategy to secrete effectors into the rice cell for subverting PTI, leading to effector-triggered susceptibility (ETS).To combat a blast fungus capable of subverting PTI, rice deploys nucleotide-binding site leucine-rich repeat (NLR) proteins to recognize the effectors named avirulence (AVR) proteins. Several AVR proteins have been cloned, including AVR-Pita, AVR-Pi9, and Avr-Pizt [173, 174]. Recognition of AVR by NLR promotes strong immune responses referred to as effector-triggered immunity (ETI), which arms rice with a second layer of protection in case of disabled PTI [172]. Defence regulators (DR) genes can activate various signaling pathways, such as MAPK cascades and the ubiquitination-mediated pathway, as well as hormonal signaling (**Figure 3**). Upon activation by extracellular stimuli, MAPKs transmit signals from the cell membrane to the nucleus, acting in defense against *M. oryzae* [172]. Transcription factors (TFs) are also involved in defense against infection by *M. oryzae*. of particular interest are broad-spectrum resistance Digu 1 (bsr-d1) and Ideal Plant Architecture 1 (IPA1). Lesion-mimic mutant (LMM) genes are the main DR genes capable of activating immune responses such as ROS bursts. Lesionmimic mutants, including spl30-1, spl33, spl35, lmm24 and spl-D usually show increased disease resistance [172]. Several other DR genes can confer similar blast resistance by initiating ROS bursts. For example, SPL11 cell-Death Suppressor 2 (SDS2) is a ubiquitination substrate of SPL11 (an E3 ubiquitin ligase comprising an armadillo repeat domain and a U-box domain). SDS2 interacts with OsRLCK118/176 and phosphorylates OsRbohB, and then induces a ROS burst, resulting in increased resistance to *M. oryzae* [172]. Hormones are another class of regulators involved in rice blast defense response. Suppressor of Salicylic acid Insensitivity-2 (OsSSI2), OsSec3a (a principal subunit of the exocyst complex in rice), OsAAA-ATPase 1 all mediate resistance by modulating salicylic acid (SA) signaling [175]. JA-resistant 1 (OsJAR1) and JAresponsive MYB (OsJAMyb) are associated with jasmonic acid (JA) signaling, and determine rice blast disease resistance. In the early stage of infection by pathogens, rice accumulates antimicrobial compounds as a defense response. For example, cyanide contributes to rice resistance by restricting fungal growth [176]. Bayogenin 3-O-cellobioside confers cultivar-nonspecific defense against the rice blast fungus [177]. Diterpenoids are a major group of antimicrobial phytoalexins in rice, and their

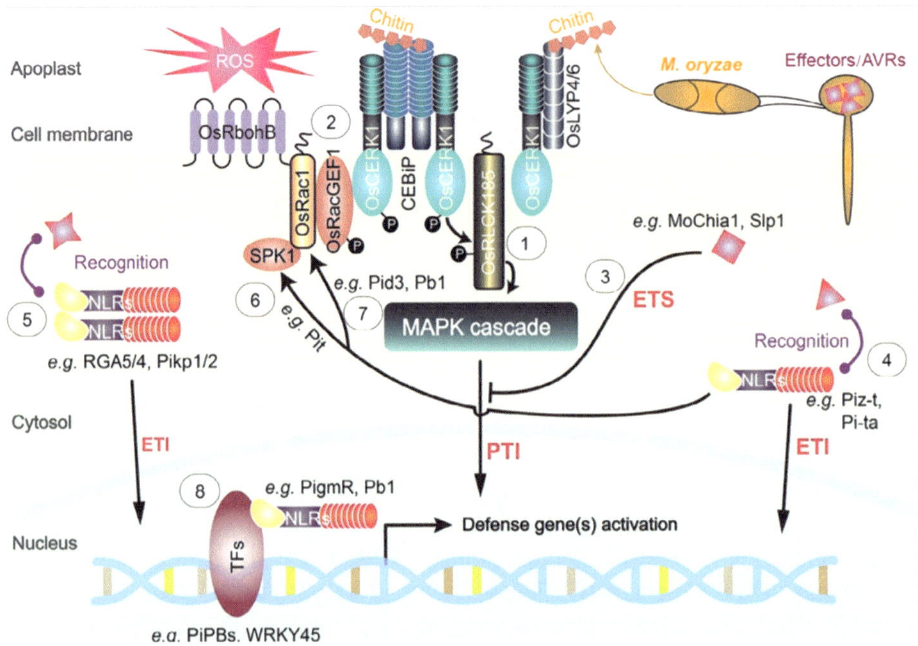

Figure 3.
Rice innate immunity signalling pathways triggered by M. oryzae. (1) On the rice cell membrane (2) Upon chitin perception, the LysM-RLP-OsCERK1 complex also elicits the OsRacGEF1-dependent pathway. (3) The rice blast fungus secretes effectors, such as Chitinase 1 (MoChia1) and Secreted a LysM protein 1 (Slp 1), into the rice cell to subvert PTI, resulting in the emergence of effector-triggered susceptibility (ETS). (4) Alternatively, secreted effectors called avirulence (AVR) proteins are recognized by rice nucleotide-binding site leucine-rich repeat (NLR) proteins, leading to a particularly strong immune response referred to as effector-triggered immunity (ETI). (5) The integrated decoy model (6) Pit employs its CC domain to bind to OsSPK1 for activating OsRac1 and induction of cell death. (7) In ETI, OsRac1 is required for Pb1, Pid3-mediated blast resistance. (8) NLR proteins mediate defense response by direct interaction with transcription factors [172].

role in rice disease resistance has been indicated by functional analysis of a diterpenoid gene cluster (DGC7) located on rice chromosome 7 [178]. Excessive or deficient supply of nutrients, such as nitrogen, phosphate, potassium, and silica, affects stress response and can potentially influence rice disease resistance. For instance, nitrogen partially breaks down rice blast resistance triggered by the Pi1 gene. Potassium is also associated with rice blast resistance. For example, *M. oryzae* can disrupt rice immune response by regulating host K^+ channels. Silicon nutrition can mitigate various biotic stresses [172]. On one hand, silicon acts as a physical barrier against plant disease. On the other hand, silicon boosts the plant's defense by functioning as a biological inducer. Silicon-induced defense response and cell silicification of leaves both contribute to rice blast disease resistance. In *M. oryzae*-infected rice plants, silicon-enhanced blast disease resistance is also associated with an increase in photochemical efficiency and adjustment of mineral nutrient absorption.

7. Strategies for breeding blast resistance in Rice

Although few cultural practices such as nutrient and water management, planting time, spacing, and application of fungicides are employed in managing blast disease, it has not been possible for the farming community to effectively and efficiently offset the blast disease [179]. This is mainly because of the complex etiology of the pathogen *M. oryzae* that includes infection of almost all parts of the rice plant, at all stages starting from seedling stage to maturity. Hence, breeding for durable resistance and resistant cultivars has been a proven ecologically viable and crucial option for addressing the infection by rice blast fungus [155, 180, 181]. Breeding for blast resistance in rice can be broadly categorized into four classes, including conventional breeding methods, marker-dependent breeding methods, breeding approaches requiring genetic transformation, and genome editing.

7.1 Conventional breeding approaches

The conventional breeding approaches, including the pedigree method, backcrossing, recurrent selection, and mutation breeding, have been widely used in developing blast-resistant varieties in rice [182]. The pedigree method of breeding is the most commonly used breeding method for improving pest and disease resistance in rice. The pedigree method is the quick method employed to develop the resistance for one or more pests or diseases governed by major R genes. Backcross breeding improves an agronomically superior and high-yielding elite variety for resistance against insect pests and disease-causing pathogens. The major advantage of backcross breeding is avoiding the undesired genes from the donor parent due to linkage drag [183]. Backcross breeding has been used in South and Southeast Asia to improve blast resistance of several rice varieties including, KDML 105, Basmati, and Manawthukha [184]. With the advantages of shorter breeding cycles, control of genetic gains, and developing a broad range of genetic diversity in breeding lines, the recurrent selection breeding method is another choice of conventional breeding method to improve disease resistance in rice [185]. Mutation breeding is the method of choice when all the alleles available in the germplasm are exhausted, and there is a need to develop novel alleles. Mutation breeding has been effectively used to complement the other conventional methods of breeding. Although there are no breakthroughs achieved using mutation breeding, several examples of the use of this method exist that include the development of blast-resistant varieties RD6, KDML 105, Ratna, and R917 [186, 187]. However, the major limitation of the mutation breeding is the low efficiency, generation of recessive alleles, tracking of the mutated gene, and exposure of the personnel for mutagenic agents.

Further, associated markers have been effectively used to tag the mutated gene and follow them up in the subsequent generations [188]. The blast resistance genes that have been deployed in different rice varieties to address the incidence of the blast pathogen by using the above-mentioned conventional breeding methods include Pib, Pita, Pia, Pi1, Pikh, Pi2(t), and Pi4(t) [130]. Despite several rice varieties with high yield and grain quality in the previous few decades, the conventional breeding methods suffer several limitations, including high cost, labor-intensive, more time consuming, less reliability, difficulties in the appropriate genotypic selection, and linkage drag. Therefore, these limitations have necessitated the development of modern molecular breeding methods, which have overcome the limitations of conventional breeding methods.

7.2 Marker-based breeding methods

The main problem of traditional breeding methods is the selection of a genotype based on the phenotype. For instance, in disease resistance breeding, the resistant genotype is selected on their manifestation of resistance to the disease. However, a particular genotype without any R gene may be selected as resistant in the absence of a minimum level of disease pressure. Therefore, molecular markers associated with specific R genes have been widely employed to make the selection procedure more reliable, effective, and less time-consuming. Modern sequencing technologies have led to the identification of a large number of different DNA markers such as simple sequence repeats (SSRs), single-nucleotide polymorphisms (SNPs), small insertions/deletions (InDels), amplified fragment length polymorphisms (AFLPs), random amplified polymorphic DNAs (RAPDs), restriction fragment length polymorphisms (RFLPs) associated with blast resistance genes that are effectively used in the selection of a genotype during handling of segregating generations [163, 179].

A panel of 80 released varieties from National Rice Research Institute, Cuttack, India, was genotyped with 36 molecular markers that were linked to 36 different blast resistance genes, to investigate the varietal genetic diversity and molecular marker-trait association with blast resistance. The cluster analysis and population structure categorized the 80 National Rice Research Institute released varieties (NRVs) into three major genetic groups. The principal co-ordinate analysis displays the distribution of resistant and moderately resistant NRVs into different groups. Analysis of molecular variance result demonstrated maximum (97%) diversity within populations and minimum (3%) diversity between populations. Among tested markers, two markers (RM7364 and pi21_79-3) corresponding to the blast resistance genes (*Pi56(t)* and *pi21*) were significantly associated and explained a phenotypic variance of 4.9 to 5.1% with the blast resistance [189]. In another research article, molecular analysis of variance of landraces originated from nine diverse rice ecologies of India showed maximum (93%) diversity within the population and least (7%) between populations. Five markers like K3957, Pikh, Pi2–i, RM212and RM302 were strongly associated with blast disease with the phenotypic variance of 1.4% to 7.6% [190].

7.2.1 Marker Assisted Selection (MAS)

In MAS, the resistant phenotype of a variety is selected based on the presence of an R gene linked or R gene-based molecular marker. This selection method is more dependable and time-saving, does not require proper disease favoring environmental conditions, and selects the resistant genotypes even in the absence of the pathogen or disease. Hence, many present-day breeders are resorting to MAS in the developing blast-resistant varieties [163, 179, 191]. A set of well-characterized PCR-based markers such as SSR markers linked to blast R genes have been established currently used in the MAS programs worldwide. Similar to the rice breeders in the rest of the world, Indian rice breeders working on the improvement of blast resistance are not left behind in the use of MAS. The pioneering work of MAS in rice blast improvement began with Hittalmani *et al.* in 2000 [192]. Since then, several rice breeders in India have efficiently used MAS to incorporate different blast resistance genes, resulting in blast-resistant varieties. In China, rice lines were recently bred for blast resistance with four broad-spectrum resistance genes viz., *Pi9*, *Pi47*, *Pi48*, and *Pi49* [193]. A list of blast resistance breeding programs in rice using MAS in India is furnished in **Table 2**.

7.2.2 Marker-Assisted Backcross Breeding (MABB)

Like MAS, the MABB is also dependent upon DNA markers such as SSRs or SNPs. However, the main difference between the MAS and MABB lies in the type of the parent variety in which the improvement is sought and in recurrent parent genome recovery. While the MAS is used to introduce the blast resistance gene into any genotype, the MABB is employed to improve blast resistance in otherwise high-yielding elite varieties or genotypes. Therefore, the end product of the MABB is the same as the original rice variety except with improved blast resistance. Further, the ill effects of the unwanted genes from the donor are avoided by using a set of polymorphic markers for the recovery of the recurrent parent genome. Hence, MABB involves two stages of selection: foreground selection using markers linked to blast resistance genes and background selection using polymorphic markers spread randomly throughout the rice genome. It is reported that short-grained landrace *Mushk Budji* was crossed to a triple-gene donor line, DHMAS 70Q 164-1b, and followed through marker-assisted foreground and background selection in first and second backcross generations that helped to incorporate blast resistance genes *Pi54, Pi1* and *Pita*. Marker-assisted background selection was carried out using 78 SSR and STS markers [194]. Several elite varieties such as MTU1010, IR-64, and Swarna have been improved for their blast resistance in India. A list of all other rice varieties improved for blast resistance in India following MABB is listed in **Table 2**.

7.2.3 Gene pyramiding

A major R gene confers the durable resistance to blast pathogen *M. oryzae* in rice. There are several major genes identified to govern the blast resistance in rice which were discussed earlier. Despite, presence of an R gene, a resistant rice variety becomes susceptible mainly due to the breakdown of the resistance by the evolution of new races of the blast pathogen. The evolution and breakdown of resistance are facilitated by the much longer life cycle of the crop than the quicker life cycle of the pathogen. Further, most of the paddy-growing regions of India are characterized by the presence of a mixture of races of *M. oryzae*. Because of these reasons, the mere presence of a single R gene is not enough for durable blast resistance over the long run. Hence, the deployment of more than one R gene into a single genotype, called gene pyramiding, is most essential. Therefore, gene pyramiding can be described as adding more than one desired gene into a single variety or cultivar. For example, in a recent study, gene pyramided lines were evaluated for key agro-morphological traits, quality, and resistance against blast at different hotspot locations. Two ICF_3 genes, pyramided lines viz., TH-625-159, and TH-625-491 possessing *Pi54* and *Pi1* genes, exhibited a high level of resistance to blast [195]. Often gene pyramiding, ably supported by MAS, involves assembling more than one gene for different insect pests or diseases or a combination of both. This helps to achieve multiple goals in a single effort and reduces the duration of the crop improvement programs. Gene pyramiding has been successfully used for accumulating different blast resistance genes such as Pi1, Pi2, and Pi33 [136], Pib and Pita [196], Pish and Pib [197], Piz5 and Pi54 [138], Piz and Pi5 [198] and Pi21, Pi34 and Pi35 [199].

7.3 Genome editing approaches

During evolution, plants have acquired the ability to recognize the invading pathogen and fight against it. In general, the surface proteins of the plants are called pattern recognition receptors to recognize the pathogen-associated molecular patterns (PAMPs) and trigger an array of defense reactions. While this way of recognizing the pathogens is the first step in the plant defense system, several downstream proteins and plant hormones also play an important role in mediating the plant's fight against the pathogen. Like any gene regulatory system, these mediators of plant defense system may impact the defense system. Recently, the availability of sequence-specific nucleases (SSNs) based genome editing tools, including zinc finger nucleases (ZFNs), transcription activator-like effector nucleases (TALENs), and clustered regularly interspaced short palindromic repeats (CRISPR)/CRISPR-associated (Cas) 9 (CRISPR/Cas9) have made the possibilities of genome editing to regulate gene expression without any genetic modification. Hence, these SSNs-based genome editing technologies may be used in the years ahead to alter the expression of the genes involved in the plant immune system and achieve resistance against the invading pathogen. In a recent study involving

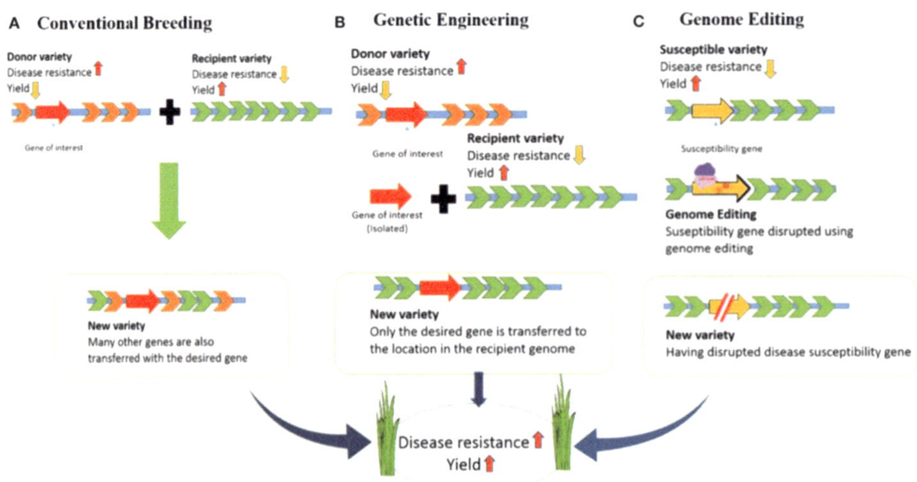

Figure 4.
Developing disease-resistant rice: Comparison of conventional breeding, genetic engineering and genome editing [194].

CRISPR/Cas9, resistance against *M. oryzae* was achieved by knocking out the expression of OsERF922, a plant ethylene-responsive factors (ERF) gene and a key negative regulator of plant immunity [200]. Similarly, *OsSEC3A* gene has been disrupted using CRISPR/Cas9 SSN to explore its role in plant immunity [201]. Proline-rich motif of *Pi*21 was edited to induce resistance against *M. oryzae* [202]. Invention of CRISPR/Cas9 technology in Indian scenario to edit genes mapped or identified against *M. oryzae* is in infancy stage. However, couple of researchers have begun to edit negative regulators of blast resistance genes and other defense related genes identified through RNA sequencing [203]. Rice and other crops genome editing using TALENs produced disease resistance against diverse pathogens [204, 205]. CRISPR-Cas9 is generally limited to perform genome editing at sites with canonical NGG PAMs. Much effort has focused on overcoming this restriction. Numerous Cas9 orthologs have been developed with altered PAM specificities, such as *Staphylococcus aureus* Cas9 (SaCas9) and Cas9-VQR (D1135V/R1335Q/T1337R). The CRISPR-SaCas9 toolset was recently re-optimized by introducing three key mutations, and its activity was analyzed in rice. The newly optimized system performed genome editing with a mutagenesis efficiency of up to 77%. Other versions of Cas9 have also been tested in rice, including expanded PAM SpCas9 (xCas9) and Cas9 that can recognize relaxed NG PAMs (Cas9-NG) [206]. Comparison of conventional breeding, genetic engineering, and genome editing is illustrated in **Figure 4**.

8. Future perspectives and strategies

In the recent past, rapid development in biotechnology and genomics has aided deep understanding of both host and pathogen. In this view, there are a handful of innovative tools and strategies available for developing rice varieties with effective and durable resistance against several races.

8.1 Allele mining

Allele mining is the most widely used method in identifying naturally occurring novel alleles or allelic variants of a gene in a set of germplasm. The allele mining mainly involves two different approaches, *i.e.*, EcoTilling and sequence-based or PCR-based allele mining. Compared to EcoTilling, sequence-based allele mining is reported to be simple and cheaper [163, 207]. The sequence-based allele mining involves PCR amplification of a particular gene and sequencing the PCR product to look for different gene versions. Hence, the host's sequencing information is an essential prerequisite of any allele mining program. The alleles arise in a population due to natural mutations such as transition, transversion, and InDels. Hence, allele mining has to be regularly performed to identify any valuable alleles originating in the germplasm. In a recent report, sequence-based allele mining was performed to amplify and sequence the allelic variants of the major rice blast resistance genes at the *Pi2*/*Pi9* locus of chromosome 6 from the 361 blast-resistant varieties. Thirteen novel *Pi9* alleles (named *Pi9*-Type1 to *Pi9*-Type13) were identified in these 107 varieties. These could potentially serve as a genetic resource for molecular breeding resistance to rice blast [176].

8.2 Identification of SNPs for fast-tracking of MAS

Owing to the evolution of next-generation sequencing technology, genome-wide association mapping (GWAS) has found its way as an efficient tool for mapping genes. Using this method, several QTLs and loci have been identified to be associated with a set of different traits of agronomical importance [208–210]. Similarly, GWAS can identify functional SNPs associated with resistance to rice blast fungus, and MAS can be made much faster and robust. Further, GWAS can also be used to fast-track the background selection of a MAS program by collecting SNPs distributed evenly on the whole genome. The use of these high-density markers and high-resolution genome scans can identify the genomic content contributed by each parent in a breeding program involving multiple parents [181].

8.3 Host induced gene silencing

Induction of host resistance to several pathogenic fungi following the expression of the fungal genes in the host plant has been demonstrated in several cases [211]. Similarly, genes encoding a set of proteins that are very crucial in the initial establishment of *M. oryzae* in rice can be silenced by their expression in the host system. This approach holds considerable potential in breeding the next-generation rice varieties and seeks more research. However, one of the main drawbacks of this approach is that it requires genetic transformation and expression of the foreign genes in the plants.

8.4 Modification of host genes targeted by blast pathogen

The infection of the host by the rice blast pathogen requires recognition of some host proteins for establishing the infection. Hence, articulating the host target proteins by genome editing technologies fails the pathogen to recognize the host targets, limiting the infection. This approach is novel and different as the focus is on articulating the host susceptibility genes rather than R genes. A susceptibility gene refers to genes that render the host susceptible to a pathogen. This approach is now facilitated and made practical with the availability of biotechnology tools such as TALENs and CRISPR/Cas9 technologies. In this direction, the proof of concept has been demonstrated by modifying a specific target gene recognized by *Xanthomonas oryzae pv. oryzae* using TALEN technology [205].

Here, we report the identification and functional characterization of a new member of the miR812 family in rice (named as miR812w) involved in disease resistance. miR812w is present in cultivated Oryza species, both japonica and indica subspecies, and wild rice species within the Oryza genus, but not in dicotyledonous species. miR812w is a 24 nt-long that requires DCL3 for its biogenesis and is loaded into AGO4 proteins. Whereas overexpression of miR812w increased resistance to infection by the rice blast fungus Magnaporthe oryzae, CRISPR/Cas9-mediated MIR812w editing enhances disease susceptibility, supporting that miR812w plays a role in blast resistance.

One recent report showed the identification and functional characterization of a new member of the miR812 family in rice (named as miR812w) involved in disease resistance. miR812w is present in cultivated *Oryza* species, both japonica and indica subspecies. miR812w is a 24 nt-long that requires DCL3 for its biogenesis and is loaded into AGO4 proteins. Overexpression of miR812w in rice increased resistance to infection by *M. oryzae*, CRISPR/Cas9-mediated MIR812w editing enhances disease susceptibility, supporting that miR812w plays a role in blast resistance [212].

8.5 Race dependent deployment of R genes

The success of any resistance breeding program mainly depends on the precise identification of the Avr genes prevailing in a particular location. This challenge can be met by identification and characterization of different races of *M. oryzae* of a location. While the irrational deployment of R genes to address blast disease incidence will not lead to the expected outcome, it adds additional burden to the host in expressing a specific R gene for which the Avr gene is absent and finally results in yield penalty. Furthermore, this exercise has to be continued due to the shift of the avirulence composition in *M. oryzae* populations. To date, a significant number of Avr genes have been identified and cloned. Hence, a simple PCR can be used to ascertain the frequencies of Avr genes and further planning of the breeding program.

9. Conclusion

The management of rice blast fungus is complex due to the continuous evolution of new pathotypes worldwide and India. Although several fungicides, cultural and biological control measures of blast disease are employed at the field level, the use of durable host plant resistance has shown great potential. In addition to being cost-effective, resistance breeding is environmentally friendly and demands less attention and intervention by illiterate farmers. Most of the resistance breeding programs in India were primarily based on single-gene resistance through conventional breeding approaches. However, blast pathogen has successfully overcome the single-gene resistance in a short period and rendered several varieties susceptible to blast, which was otherwise intended to be resistant. Some of the blast endemic areas of India are characterized by the existence of a mixture of more than one race of the blast pathogen, making the situation more challenging. However, the recent technological advancements, including genomics, gene editing, and pyramiding of more than one resistance gene assisted by genetic markers, hold huge promise in counteracting *M. grisea*. Hence, future resistance breeding programs should exploit the modern biotechnology tools and conventional breeding approaches in developing durable blast resistance varieties harboring multiple R genes. The harmonious blending of the bio-control approaches, cultural management practices, and modern breeding methods is the key to successfully addressing blast disease in rice cultivation ecosystems. Further, the effectiveness of the blast-resistant varieties developed for a location can only be achieved when the gene deployed is based on the Avr genes prevalent in that area. Therefore, more efforts are needed to conduct the basic research pertaining to a specific location.

Acknowledgements

We profusely thank Ayeesha Munawery for the discussion and support.

Conflict of interest

The authors have no conflict of interest associated with this work.

Author details

Deepak Chikkaballi Annegowda[1*], Mothukapalli Krishnareddy Prasannakumar*, Hirehally Basavarajegowda Mahesh[3], Chethana Bangera Siddabasappa[1], Pramesh Devanna[4], Sahana Nagaraj Banakar[2], Haniyambadi Basavegowda Manojkumar[5] and Siddegowda Rajendra Prasad[6]

1 Division of Rice Breeding, AICRP (Rice), Zonal Agricultural Research Station, V. C. Farm, Mandya, University of Agricultural Sciences, Bangalore, Karnataka, India

2 Department of Plant Pathology, College of Agriculture, Gandhi KrishiVijnana Kendra, University of Agricultural Sciences Bangalore, Karnataka, India

3 Department of Genetics and Plant Breeding, College of Agriculture, V. C. Farm, Mandya, University of Agricultural Sciences, Bangalore, Karnataka, India

4 Rice Pathology Laboratory, AICRP-Rice, University of Agricultural Sciences, Raichur, Karnataka, India

5 Department of Plant Biotechnology, College of Agriculture, Gandhi KrishiVijnana Kendra, University of Agricultural Sciences, Bangalore, Karnataka, India

6 University of Agricultural Sciences, Bangalore, Karnataka, India

*Address all correspondence to: deepuca@googlemail.com
and babu_prasanna@rediffmail.com

IntechOpen

© 2021 The Author(s). Licensee IntechOpen. This chapter is distributed under the terms of the Creative Commons Attribution License (http://creativecommons.org/licenses/by/3.0), which permits unrestricted use, distribution, and reproduction in any medium, provided the original work is properly cited.

References

[1] Maclean JL, Dawe DC, Hardy B, Hettel GP. (ed.) Rice Almanac, Los Baños (Philippines): International Rice Research Institute, Bouaké (Côte d'Ivoire): West Africa Rice Development Association, Cali (Colombia): International Centre for Tropical Agriculture, Rome (Italy): Food and Agriculture Organization 2002.

[2] GoI (Government of India). First advance estimates of production of foodgrains for 2019-20. Directorate of Economics and Statistics. Department of Agriculture and Cooperation, Ministry of Agriculture, Government of India, New Delhi. 2020. Available from: http://agricoop.gov.in/sites/default/files/Time-Series-1st-Adv-Estimate-2019-20- Final-Press.pdf.

[3] FAO. The State of the World's Land and Water Resources for Food and Agriculture (SOLAW) Managing Systems at Risk. 2011. Available from: http://faostat.fao.org

[4] Wilson RA, Talbot NJ. Under pressure: investigatingthe biology of plant infection by *Magnaporthe oryzae*. Nature Reviews Microbiology.2009; 7: 185-195. DOI:10.1038/nrmicro2032

[5] Padmanabhan SY. Recent advances in the study of blast disease of rice. Madras Agriculture Journal. 1965; 564-583. DOI:10.1080/05331846809432314

[6] Padmanabhan SY, Chakravarti NK, Mathur SC, Veeraraghavan J. Identification of pathogenic races of *Pyricularia oryzae* in India. Phytopathology. 1970; 60(11): 1574-1577. DOI: 10.1094/Phyto-60-1574

[7] Rathour R, Singh BM, Sharma TR, Chauhan RS. Population structure of *Magnaporthe grisea* from the north-western Himalayas and its implications for blast resistance breeding of rice. Journal of Phytopathology. 2004; 152(5): 304-312. DOI:10.1111/j.1439-0434.2004.00846.x

[8] Parker DA, Beckmann M, Enot DP. Rice blast infection of *Brachypodium distachyonas* a model system to study dynamic host/ pathogen interactions. Nature Protocols. 2008;3: 435-444. DOI:10.1038/nprot.2007.499

[9] Teng PS, Torries CQ, Nuque FL, Calvero SB. Current knowledge on crop losses in tropical rice. In: IRRI (ed) Crop loss assessment in rice. International Rice Research Institute, Los Banos. 1990; 39-54.

[10] Shim HS, Hong SJ, Yeh WH, Han SS, Sung JM. Damage analysis of rice panicle blast on disease occurrence time and severity. The Plant Pathology Journal. 2005; 21 (2): 87-92. DOI:10.5423/PPJ.2005.21.2.087

[11] Lavanya B, Gnanamanickam SS. Molecular tools for characterization of rice blast pathogen (*Magnaporthe grisea*) population and molecular marker-assisted breeding for disease resistance. Current Science. 2000; 78 (3): 248-257.

[12] Rajarajeswari NVL, Muralidharan K. Assessments of farm yield and district production loss from bacterial leaf blight epidemics in rice. Crop Protection.2006; 25 (3): 244-252. DOI:10.1016/j.cropro.2005.04.013

[13] Krause RA, Webster RK. Morphology, taxonomy, and sexuality of rice stem rot fungus, *Magnaporthe salvinii* (*Leptosphaeria salvinii*). Mycologia, 1972; 64: 103-114. DOI: 10.2307/3758018

[14] Valent B. Rice blast as a model system for plant pathology. Phytopathology.1990; 80: 33-36. **DOI** :10.1094/Phyto-80-33

[15] Choi J, Park SY, Kim BR, Roh JH, Oh IS. Comparative analysis of pathogenicity and phylogenetic

relationship in *Magnaporthe grisea* species complex. PLoS ONE. 2013; 8(2): e57196. DOI:org/10.1371/journal.pone.0057196

[16] Taylor JW, Jacobson DJ, Kroken S, Kasuga T, Geiser DM. Phylogenetic species recognition and species concepts in fungi. Fungal Genetics and Biology. 2000; 31: 21-32. DOI: 10.1006/fgbi.2000.1228

[17] Couch BC, Kohn LMA. Multilocus gene genealogy concordant with host preference indicates segregation of a new species *Magnaporthe oryzae* from *M. grisea*. Mycologia. 2002;94:683-693 https://www.ncbi.nlm.nih.gov/pubmed/21156541

[18] Taylor JW, Fisher, MC. Fungal multilocus sequence typing – it's not just for bacteria. Current Opinion in Microbiology. 2003; 6: 351-356. DOI:10.1016/s1369-5274(03)00088-2

[19] Hirata K, Kusaba M, Chuma I, Osue J, Nakayashiki H. Speciation in Pyricularia inferred from multilocus phylogenetic analysis. Mycology Research. 2007; 111: 799-808. DOI.org/10.1016/j.mycres.2007.05.014

[20] Ou SH. Rice Diseases. Wallingford, UK: Commonwealth Agricultural Bureaux. 1985.

[21] Farr DF, Rossman AY. Fungal Databases. Systematic Mycology and Microbiology Laboratory, ARS, USDA. 2009

[22] Jong DJC, McCormack BJ, Smirnoff N, Talbot NJ. Glycerol generates turgor in rice blast. Nature, 1997; 389, 471-483. DOI:10.1038/38418

[23] Chumley FG, Valent B. Genetic analysis of melanin deficient, non-pathogenic mutants of *Magnaporthe grisea*. Molecualr Plant Microbe Interaction, 1990; 3: 135-143.DOI: 10.1094/MPMI-3-1

[24] Talbot NJ. On the trail of a cereal killer: Exploring the biology of *Magnaporthe grisea*. Annu. Rev. Microbiol. 2003;57: 177-202. DOI:10.1146/annurev.micro.57.030502.090957.

[25] Prasad MS, Madhav MS, Laha GS, Lakshmi DL, Krishnaveni D, Mangrauthia SK, Balachandran SM, Sundaram RM, Arunakanthi B, Mohan KM, Madhavi KR, Kumar V, Virakthamat BC. Technical Bulletin No 57. Directorate of Rice Research, ICAR, Rajendranagar, Hyderabad, India 2011; pp-52.

[26] Anwar A, Teli MA, Bhat GN, Parray GA, Wani. Status of rice blast (*Pyricularia grisea*), cultivar reaction and races of its causal fungus in temperate agro- ecosystem of Kashmir, India. SAARC Journal Agriculture XII (2). 2009;pp. 25-37.

[27] Jagadeesh D, Prasanna Kumar MK, Devaki NS. Population analysis of *Magnaporthe oryzae* by using endogenous repetitive DNA sequences and mating-type alleles in different districts of Karnataka, India. J. Appl. Genetics. 2018;59: 365-375.

[28] Mahesh HB, Meghana S, Shailaja H, Prasannakumar MK, Mahadevu P, Channabyregowda MV Malali G. Acquisition of the grasshopper retro transposon by rice *Magnaporthe* isolates indicates a dynamic gene flow between rice and non-rice *Magnaporthe* population. J Pathol & Microbiol. 2016;1(2): 1011

[29] Priyadarshinee P , Yadav MK, Prabhukarthikeyan SR, Raghu S, Keerthana U, Ngangkham U, Pramesh D. Morpho-molecular characterization of Indian population of Magnaporthe oryzae, an incitant of blast disease in rice. Research Journal of Biotechnology.2021;16(1) https://www.researchgate.net/publication/347910076

[30] Dean RA, Talbot NJ, Ebbole DJ, Farman ML, Mitchell TK, Orbach MJ,

Thon M, Kulkarni R, Xu J-R, Pan H. The genome sequence of the rice blast fungus *Magnaporthe grisea*. Nature.2005; 434: 980-986. DOI: 10.1038/nature03449

[31] Yoshida K, Saitoh H, Fujisawa S, Kanzaki H, Matsumura H, Yoshida K, Tosa Y, Chuma I, Takano Y, Win J. Association genetics reveals three novel avirulence genes from the rice blast fungal pathogen *Magnaporthe oryzae*. The Plant Cell Online. 2009; 21: 1573-1591. DOI: 10.1105/tpc.109.066324

[32] Xue M, Yang J, Li Z, Hu S, Yao N, Dean RA, Zhao W, Shen M, Zhang H, Li C. Comparative analysis of the genomes of two field isolates of the rice blast fungus *Magnaporthe oryzae*. PLoS genetics.2012;8: e1002869. DOI:10.1371/journal.pgen.1002869

[33] Chen C, Lian B, Hu J, Zhai H, Wang X, Venu R, Liu E, Wang Z, Chen M, Wang B. Genome comparison of two *Magnaporthe oryzae* field isolates reveals genome variations and potential virulence effectors. BMC genomics.2013; 14: 887. DOI: 10.1186/1471-2164-14-887

[34] Gowda M, Shirke MD, Mahesh HB, Chandarana P, Rajamani A, Chattoo BB. Genome analysis of rice-blast fungus *Magnaporthe oryzae* field isolates from southern India. Genomics Data.2015; 5: 284-291. DOI: 10.1016/j.gdata.2015.06.018

[35] Shirke MD, Mahesh H, Gowda M. Genome-Wide Comparison of Magnaporthe Species Reveals a Host-Specific Pattern of Secretory Proteins and Transposable Elements. PloS one.2016;11: e0162458. DOI: 10.1371/journal.pone.0162458

[36] Kachroo P, Leong SA, Chattoo BB. Mg-SINE: a short interspersed nuclear element from the rice blast fungus, *Magnaporthe grisea*. Proceedings of the National Academy of Sciences.1995; 92: 11125-11129. PMCID: PMC40584

[37] Kumar A, Sheoran N, Prakash G, Ghosh A, Chikara SK, Rajashekara H, Singh UD, Aggarwal R, Jain RK. Genome sequence of a unique Magnaporthe oryzae RMg-Dl isolate from India that causes blast disease in diverse cereal crops, obtained using PacBio singlemolecule and Illumina HiSeq2500 sequencing. Genome Announc. 2017;5:e01570-16. https:// doi.org/10.1128/genomeA.01570-16.

[38] Stephen AG, Darrell Ricke, Tien-Hung Lan, Gernot Presting, Ronglin Wang, Molly Dunn, Jane Glazebrook, Allen Sessions, Paul Oeller, Hemant Varma, David Hadley, Don Hutchison, Chris Martin, Fumiaki Katagiri, B. Markus Lange, Todd Moughamer, Yu Xia, Paul Budworth, Jingping Zhong, Trini Miguel, Uta Paszkowski, Shiping Zhang, Michelle Colbert, Wei-lin Sun, Lili Chen, Bret Cooper, Sylvia Park, Todd Charles Wood, Long Mao, Peter Quail, Rod Wing, Ralph Dean, Yeisoo Yu, Andrey Zharkikh, Richard Shen, Sudhir Sahasrabudhe, Alun Thomas, Rob Cannings, Alexander Gutin, Dmitry Pruss, Julia Reid, Sean Tavtigian, Jeff Mitchell, Glenn Eldredge, Terri Scholl, Rose Mary Miller, Satish Bhatnagar, Nils Adey, Todd Rubano, Nadeem Tusneem, Rosann Robinson, Jane Feldhaus, Teresita Macalma, Arnold Oliphant a, Briggs S. A Draft Sequence of the Rice Genome (*Oryza sativa* L. ssp. *japonica*. Science. 2002; 296: 92-100. DOI: 10.1126/science.1068275

[39] Yu J, Hu S, Wang J, Wong GK, Li S, Liu B, Deng Y, Dai L, Zhou Y, Zhang X, Cao M, Liu J, Sun J, Tang J, Chen Y, Huang X, Lin W, Ye C, Tong W, Cong L, Geng J, Han Y, Li L, Li W, Hu G, Huang X, Li W, Li J, Liu Z, Li L, Liu J, Qi Q, Liu J, Li L, Li T, Wang X, Lu H, Wu T, Zhu M, Ni P, Han H, Dong W, Ren X, Feng X, Cui P, Li X, Wang H, Xu X, Zhai W, Xu Z, Zhang J, He S, Zhang J, Xu J, Zhang K, Zheng X, Dong J, Zeng W, Tao L, Ye J, Tan J, Ren X, Chen X, He J, Liu D, Tian W, Tian C, Xia H, Bao Q, Li G, Gao H, Cao T, Wang J, Zhao W, Li P, Chen W, Wang X, Zhang Y, Hu J, Wang J,

Liu S, Yang J, Zhang G, Xiong Y, Li Z, Mao L, Zhou C, Zhu Z, Chen R, Hao B, Zheng W, Chen S, Guo W, Li G, Liu S, Tao M, Wang J, Zhu L, Yuan L, Yang H. A draft sequence of the rice genome (*Oryza sativa* L. ssp. *indica*). Science. 2002; 296: 79-92. DOI: 10.1126/science.1068037

[40] Schatz MC, Maron LG, Stein JC, Wences AH, Gurtowski J, Biggers E, Lee H, Kramer M, Antoniou E, Ghiban E, Wright MH, Chia J-m, Ware D, McCouch SR, McCombie WR. Whole genome de novo assemblies of three divergent strains of rice, *Oryza sativa*, document novel gene space of aus and indica. Genome Biology.2014; 15: 506. DOI: 10.1186/PREACCEPT-2784872521277375

[41] Sakai H, Kanamori H, Arai-Kichise Y, Shibata-Hatta M, Ebana K, Oono Y, Kurita K, Fujisawa H, Katagiri S, Mukai Y: Construction of pseudomolecule sequences of the aus rice cultivar kasalath for comparative genomics of Asian cultivated rice. DNA research: dsu006. 2014; DOI: 10.1093/dnares/dsu006

[42] Mahesh HB, Shirke MD, Singh S, Rajamani A, Hittalmani S, Wang G-L, Gowda *M. indica* rice genome assembly, annotation and mining of blast disease resistance genes. BMC genomics.2016;17: 1. DOI: 10.1186/s 12864-016-2523-7

[43] Li J-Y, Wang J, Zeigler RS. The 3,000 rice genomes project: new opportunities and challenges for future rice research. GigaScience. 2014; 3: 1-3. DOI: 10.1186/2047-217x-3-8

[44] Wang Q, Li J, Lu L, He C, Li C. Novel variation and evolution of avrpiz-t of *Magnaporthe oryzae* in field isolates. Front. Genet. 2020;11:746. doi: 10.3389/fgene.2020.00746

[45] Alexandrov N, Tai S, Wang W, Mansueto L, Palis K, Fuentes RR, Ulat VJ, Chebotarov D, Zhang G, Li Z. SNP-Seek database of SNPs derived from 3000 rice genomes. Nucleic acids research .2014; 43: D1023-D1027. DOI: 10.1093/nar/gku1039

[46] Wang GL, Mackill DJ, Bonman JM, Mccouch SR, Champoux MC, Nelson RJ. RFLP mapping of genes conferring complete and partial resistance to blast in a durably resistant rice cultivar. *Genetics* .1994;136: 1421-1434. https://www.ncbi.nlm.nih.gov/pmc/articles/PMC1205921/

[47] Goto I. Genetic studies on resistance of rice plant to blast fungusII. Difference in resistance to the blast disease between Fukunishiki and it parental cultivar, Zenith. Ann Phytopathol Soc Jap. 1976; 42:253-260. DOI: 10.3186/jjphytopath.42.253.

[48] Fujii K, Hayano-Saito Y, Sugiura N, Hayashi N, Saka N, Tooyama T, Izawa T, Shumiya A. Gene analysis of panicle blast resistance in rice cultivars with rice stripe resistance. Breed Res. 1999; 1:203-210 doi.org/10.1270/jsbbr.1.203

[49] Hayashi N, Inoue H, Kato T, Funao T, Shirota M, Shimizu T, Kanamori H, Yamane H, Hayano-Saito Y,Matsumoto T, Yano M, Takatsuji H. Durable panicle blast-resistance gene Pb1 encodes an atypical CC-NBS-LRR protein and was generated by acquiring a promoter through local genome duplication. Plant J. 2010; 64:498-510. DOI: 10.1111/j.1365-313X.2010.04348.x.

[50] Fujii K, Hayano-Saito Y, Shumiya A, Inoue M. Genetical mapping based on the RFLP analysis for the panicle blast resistance derived from a rice parental line St. No. 1. Breed Sci.1995; 45:209.

[51] Causse MA, Fulton TM, Cho YG, Ahn SN, Chunwongse J, Wu K, Xiao J, Yu ZH, Ronald PC, Harrington SE, Second G, McCouch SR, Tanksley SD. Saturated molecular map of the rice genome based on an interspecific

backcross population. Genetics. 1994; 138:1251-1274. PMCID:PMC1206261

[52] Yu ZH, Mackill DJ, Bonman JM, Tanksley SD. Tagging genes for blast resistance in rice via linkage to RFLP markers. Theor Appl Genet. 1991; 81:471-476. doi.org/10.1007/BF00219436.

[53] Naqvi NI, Chattoo BB. Development of a sequence characterized amplified region (SCAR) based indirect selection method for a dominant blast-resistance gene in rice. Genome. 1996; 39:26-30. doi.org/10.1139/g96-004.

[54] Inukai T, Nelson RJ, Zeigler RS, Sarkarung S, Mackill DJ, Bonman JM, Takamure I, Kinoshita T . Genetic analysis of blast resistance in tropical rice cultivars using near-isogenic lines. In: Khush GS (ed) Rice genetics III,Proceeding of the third international rice genetics symposium, International Rice Research Institute, Manila,Philippines, 1996; 447-455.

[55] Amante-Bordeos A, Stitch LA, Nelson R, Dalmacio RD, Oliva NP, Aswidinnoor H, Leung H. Transfer of bacterial blight and blast resistance from the tetraploid wild rice *Oryza minuta* to cultivated rice *Oryza sativa*. Theor Appl Genet. 1992;84:345-354. doi: 10.1007/BF00229493.

[56] Pan HQ, Tanisaka T, Ikehashi H. Studies on the genetics and breeding of blast resistance in rice VI. Gene analysis of the blast resistance of two Yunnan native cultivars GA20 and GA25. Breed Sci. 1996; 46(Suppl 2):70.

[57] Pan QH, Wang L, Ikehashi H, Tanisaka T. Identification of a new blast resistance gene in the indica rice cultivar Kasalath using Japanese differential cultivars and isozyme markers. Phytopathology.1996; 86:1071-1075. DOI: 10.1094/Phyto-86-1071.

[58] Pan QH, Wang L, Ikehashi H, Yamagata H, Tanisaka T. Identification of two new genes conferring resistance to rice blast in the Chinese native cultivar 'Maowangu'. Plant Breed.1998; 117:27-31. Doi: 10.1111/j.1439-0523.1998.tb01443.x.

[59] Pan QH, Hu ZD, Tanisaka T, Wang L. Fine mapping of the blast resistance gene Pi15, linked to Pii, on rice chromosome 9. Acta Bot Sinica.2003; 45:871-877.

[60] Pan QH, Wang L, Tanisaka T. A new blast resistance gene identified in the Indian native rice cultivar Aus373 through allelism and linkage tests. Plant Pathol. 1999; 48:288-293. Doi: 10.1046/j.1365-3059.1999.00337.x.

[61] Iwata N. Registration of new gene symbols. Rice Genet Newsl.1996; 13:12-18.

[62] Pan QH, Tanisaka T, Ikehashi H. Studies on the genetics and breeding of blast resistance in rice IV Gene analysis for the blast resistance of a indica variety Kasalath. Breed Sci.1995; 45(Suppl. 2):170

[63] Ahn S-N, Kim Y-K, Hong H-C, Han S-S, Kwon S-J, Choi H-C, Moon H-P, McCouch S. Molecular mappingof a new gene for resistance to rice blast. Euphytica. 2000; 116:17-22. Doi: 10.1023/A%3A1004045812357.

[64] Hayashi N, Ando I, Naito H. Gene analysis of a new blast resistance in the paddy rice variety, Aichi Asahi. Breed Sci. 1996; 46(Suppl. 2):168.

[65] Imbe T, Oba S, Yanoria MJT, Tsunematsu H. A new gene for blast resistance in rice cultivar, IR24. Rice Genet Newsl. 1997; 14:60-62. https://shigen.nig.ac.jp/rice/oryzabase/asset/rgn/vol14/v14p60.html

[66] Fukuoka S, Okuno K. QTL analysis and mapping of pi21, a recessive gene for field resistance to rice blast in Japanese upland rice. Theor Appl Genet.

2001; 103:185-190. Doi: 10.1007/s001220100611.

[67] Ahn SN, Kim YK, Hong HC, Han SS, Kwon SJ, Choi HC, Moon HP, McCouch SR. Molecular mapping of a gene for resistance to a Korean isolate of rice blast. Rice Genet Newsl. 1996; 13:74. https://shigen.nig.ac.jp/rice/oryzabase/asset/rgn/vol13/v13p74.html

[68] Sallaud C, Lorieux M, Roumen E, Tharreau D, Berruyer R, Svestasrani P, Garsmeur O, Ghesquiere A, Notteghem JL. Identification of five new blast resistance genes in the highly blast-resistant rice variety IR64 using a QTL mapping strategy. Theor Appl Genet. 2003; 106:794-803. Doi: 10.1007/s00122-002-1088-9.

[69] Zhuang JY, Wu JL, Fan YY, Rao ZM, Zheng KL. Genetic drag between a blast resistance gene and QTL conditioning yield trait detected in a recombinant inbred line population in rice. Rice Genetics Newsl.2001;18:69-70. https://shigen.nig.ac.jp/rice/oryzabase/asset/rgn/vol18/c30.html

[70] Wu JL, Fan YY, Li DB, Zheng KL, Leung H, Zhuang JY. Genetic control of rice blast resistance in thedurably resistant cultivar Gumei 2 against multiple isolates. Theor Appl Genet. 2005; 111:50-56. Doi: 10.1007/s00122-005-1971-2.

[71] Mackill DJ, Bonman JM. Inheritance of blast resistance in near-isogenic lines of rice. Phytopathology. 1992; 82:746-749. Doi: 10.1094/Phyto-82-746.

[72] Zenbayashi-Sawata K, Fukuoka S, Katagiri S, Fujisawa M, Matsumoto T, Ashizawa T, Koizumi S.Mapping of the QTL (quantitative trait locus) conferring partial resistance to leaf blast in rice cultivar Chubu 32.Theor Appl Genet. 2002; 104:547-552. DOI: 10.1007/s00122-001-0779-y

[73] Nguyen TTT, Koizumi S, La TN, Zenbayashi KS, Ashizawa T, Yasuda N, Imazaki I, Miyasaka A. Pi35(t), a new gene conferring partial resistance to leaf blast in the rice cultivar Hokkai 188. Theor Appl Genet.2006; 113:697-704. Doi: 10.1007/s00122-006-0337-8

[74] Liu XQ, Wang L, Chen S, Lin F, Pan QH. Genetic and physical mapping of Pi36(t), a novel rice blast resistance gene located on rice chromosome 8. Mol Gen Genomics.2005; 274:394-401. Doi: 10.1007/s00438-005-0032-5.

[75] Lin F, Chen S, Que Z, Wang L, Liu X. The blast resistance gene Pi37 encodes an NBS-LRR protein and is a member of a resistance gene cluster on rice chromosome 1. Genetics. 2007; 177:1871-1880. Doi:10.1534/genetics.107.080648.

[76] Li W, Wang B, Wu J, Lu G, Hu Y, Zhang X, Zhang Z, Zhao Q, Feng Q, Zhang H, Wang Z, Wang G, Han B, Wang Z, Zhou B. The *Magnaporthe oryzae* avirulence gene AvrPiz-t encodes a predicted secreted protein that triggers the immunity in rice mediated by the blast resistance gene Piz-t. Mol Plant Microbe Interact.2009; 22:411-420. Doi:10.1094/MPMI-22-4-0411

[77] Gowda M, Roy-Barman S, Chattoo BB. Molecular mapping of a novel blast resistance gene Pi38 in rice using SSLP and AFLP markers. Plant Breed. 2006; 125:596-599. Doi: 10.1111/j.1439-0523.2006.01248.x.

[78] Liu X, Yang Q, Lin F, Hua L, Wang C, Wang L, Pan Q. Identification and fine mapping of Pi39(t), a major gene conferring the broad-spectrum resistance to *Magnaporthe oryzae*. Mol Gen Genomics.2007; 278:403-410. Doi: 10.1007/s00438-007-0258-5.\

[79] Jeung JU, Kim BR, Cho YC, Han SS, Moon HP, Lee YT, Jena KK. A novel gene, Pi40 (t), linked to the DNA markers derived from NBS-LRR motifs confers broad spectrum of blast resistance in rice. Theor Appl Genet. 2007; 115:1163-1177. Doi: 10.1007/s00122-007-0642-x.

[80] Yang Q, Lin F, Wang L, Pan Q. Identification and mapping of Pi41, a major gene conferring resistance to rice blast in the *Oryza sativa subsp. indica* reference cultivar, Theor Appl Genet.2009; 118:1027-1034. Doi: 10.1007/s00122-008-0959-0.

[81] Kumar P, Pathania S, Katoch P, Sharma TR, Plaha P, Rathour R. Genetic and physical mapping of blast resistance gene Pi-42(t) on the short arm of rice chromosome 12. Mol Breed.2010; 25:217-228. Doi: 10.1007/s11032-009-9320-9

[82] Chen DH, Dela Vina M, Inukai T, Mackill DJ, Ronald PC, Nelson RJ. Molecular mapping of the blast-resistance gene, Pi44(t), in a line derived from a durably resistant rice cultivar. Theor Appl Genet. 1999; 98:1046-1053. Doi: 10.1007/s001220051166.

[83] Huang H, Huang L, Feng G, Wang S, Wang Y, Liu J, Jiang N, Yan W, Xu L, Sun P, Li Z, Pan S, Liu X, Xiao Y, Liu E, Dai L, Wang GL. Molecular mapping of the new blast resistance genes Pi47 and Pi48 in the durably resistant local rice cultivar Xiangzi 3150. Phytopathology. 2011; 101(5):620-626. Doi:10.1094/PHYTO-08-10-0209.

[84] Jeon JS, Chen D, Yi GH, Wang GL, Ronald PC. Genetic and physical mapping of Pi5(t), a locus associated with broad-spectrum resistance to rice blast. Mol Gen Genomics.2003; 269:280-289. DOI: 10.1007/s00438-003-0834-2

[85] McCouch SR, Nelson RJ, Tohme J, Zeigler RS. Mapping of blast resistance genes in rice. In: Zeigler RS, Leong SA, Teng PS (eds) Rice blast disease. C.A.B. International, Wallingford UK. 1994; pp 167-186.

[86] Qu S, Liu G, Zhou B, Bellizzi M, Zeng L, Dai L, Han B, Wang G. The broad-spectrum blast resistance gene Pi9 encodes a nucleotide-binding site-leucine-rich repeat protein and is a member of a multigene family in rice. *Genetics.* 2006; 172: 1901-1914. DOI: 10.1534/genetics.105.044891

[87] Goto I, Jaw YL, Baluch AA. Genetic studies on resistance of rice plant to blast fungus IV. Linkage analysisof four genes, Pi-a, Pi-k, Pi-z and Pi-i. Ann Phytopathol Soc Jap.1981; 47:252-254. Doi: 10.3186/jjphytopath.47.252.

[88] Okuyama Y, Kanzaki H, Abe A, Yoshida K, Tamiru M, Saitoh H, Fujibe T, Matsumura H, Shenton M, Galam DC, Undan J, Ito A, Sone T, Terauchi R (2011) A multifaceted genomics approach allows the isolation of the rice Pia blast resistance gene consisting of two adjacent NBS-LRR protein genes. Plant J. 2011; 66:467-479. Doi: 10.1111/j.1365-313X.2011.04502.x.

[89] Wang Z-X, Yano M, Yamanouchi U, Iwamoto M, Monna L, Hayasaka H, Katayose Y, Sasaki Takuji .The Pib gene for rice blast resistance belongs to the nucleotide binding and leucine-rich repeat class of plant disease resistance genes. Plant J. 1999; 19:55-64. Doi:10.1046/j.1365-313X.1999.00498.x.

[90] Yokoo M, Kikuchu F, Fujimaki H, Nagai K. Breeding of blast resistance lines (BL1 to 7) from indica-japonica crosses of rice. Jap J Breed. 1978; 28:359-385.

[91] Tabien RE, Pinson SRM, Marchetti MA, Li Z, Park WD, Paterson AH, Stansel JW. Blast resistance genes from Teqing and Lemont. In: Khush GS (ed) Rice genetics III. Proceedings of Third International Rice Genetics Symposium, Oct 16-20. Manila, The Philippines. 1996; pp 451-455. Doi:10.1142/9789812814289_0046.

[92] Chauhan RS, Farman ML, Zhang HB, Leong SA. Genetic and physical mapping of a rice blast resistance locus, Pi-CO39(t), that corresponds to the avirulence gene AVR1-CO39 of *Magnaporthe grisea*. Mol Gen Genomics. 2002; 267:603-612. Doi: 10.1007/s00438-002-0691-4.

[93] Chen XW, Li SG, Ma YQ, Li HY, Zhou KD, Zhu LH. Marker-assisted selection and pyramiding for three blast resistance genes, Pi-d(t)1, Pi-b, Pi-ta2, in rice. Sheng Wu Gong Cheng Xue Bao .2004; 20:708-714. PMID:15973995.

[94] Chen XW, Shang J, Chen D, Lei C, Zou Y, Zhai W, Liu G, Xu J, Ling Z, Cao G, Ma B, Wang Y, Zhao X, Li S, Zhu L. A B-lectin receptor kinase gene conferring rice blast resistance. Plant J.2006; 46:794-804. Doi:10.1111/j.1365-313X.2006.02739.x.

[95] Shinoda H, Toriyama K, Yunoki T, Ezuka A, Sakurai Y. Studies on the varietal resistance of rice to blast. VI. Linkage relationship of blast resistance genes. Bull Chugoku Nat Agric Exp Stat A.1971; 20:1-25.

[96] Zhou JH, Wang JL, Xu JC, Lei CL, Ling ZZ. Identification and mapping of a rice blast resistance gene Pi-g(t) in the cultivar Guangchangzhan. Plant Pathol.2004; 53:191-196. Doi: 10.1111/j.0032-0862.2004.00986.x.

[97] Liu B, Zhang S, Zhu X, Yang Q, Wu S, Mei M, Mauleon R, Leach J, Mew T, Leung H. Candidate defense genes as predictors of quantitative blast resistance in rice. Mol Plant Microbe Interact.2004; 17:1146-1152. DOI/pdf/10.1094/MPMI.2004.17.10.1146

[98] Deng Y, Zhu X, Shen Y, He Z. Genetic characterization and fine mapping of the blast resistance locus Pigm(t) tightly linked to Pi2 and Pi9 in a broad-spectrum resistant Chinese variety. Theor Appl Genet. 2006; 113:705- 713. Doi:10.1007/s00122-006-0338-7.

[99] Ise K. Linkage analysis of some blast resistance gene in rice, *Oryza sativa* L. Jap J Breed.1991; 42(Suppl. 2):388-389

[100] Kinoshita T, Kiyosawa S. Some considerations on linkage relationships between Pii and Piz in the blast resistanceof rice. Rice Genetics Newsl.1997;14:57-59.

[101] Goto I. Genetic studies on the resistance of rice plant to the blast fungus I. Inheritance of resistance in crossesSensho x H-79 and Imochishirazu x H-79. Ann Phytopathol Soc Jap.1970; 36:304-312. Doi:10.3186/jjphytopath.36.304.

[102] Hayasaka H, Miyao A, Yano M, Matsunaga K, Sasaki T. RFLP mapping of a rice blast resistance gene Pi-k. Breed Sci 46. 1996; (Suppl. 2):68. https://shigen.nig.ac.jp/rice/oryzabase/gene/detail/937

[103] Zhai C, Lin F, Dong Z, He X, Yuan B, Zeng X, Wang L, Pan Q. The isolation and characterization of Pik, a rice blast resistance gene which emerged after rice domestication. New Phytol. 2011; 189:321-334. Doi:10.1111/j.1469-8137.2010.03462.x.

[104] Wang L, Xu XK, Lin F, Pan QH. Characterization of rice blast resistance genes in the Pik cluster and fine mapping of the Pik-p locus. Phytopathology. 2009; 99:900-905. Doi:10.1094/PHYTO-99-8-0900.

[105] Sharma TR, Shanker P, Singh BK, Jana TK, Madhav MS, Gaikwad G, Singh NK, Plaha P, Rathour R. Molecular mapping of rice blast resistance gene Pi-k h in rice variety Tetep. J Plant Biochem Biotechnol.2005; 14:127-133. DOI.org/10.1007/BF03263240

[106] Kaji R, Ogawa T. RFLP mapping of a blast resistance gene, Pi-km, in rice. Breed Sci 46. 1996; (Suppl. 2):70

[107] Ashikawa I, Hayashi N, Yamane H, Kanamori H, Wu J. Two adjacent nucleotide-binding site-leucine-rich repeat class genes are required to confer pikm-specific rice blast resistance. Genetics. 2008; 180:2267-2276. Doi:10.1534/genetics.108.095034.

[108] Hayashi K, Yoshida H, Ashikawa I. Development of PCR-based allele specific and InDel marker sets for nine

riceblast resistance genes. Theor Appl Genet.2006; 113:251-260. Doi:10.1007/s00122-006-0290-6.

[109] Fjellstrom RG, Conaway-Bormans CA, McClung AM, Marchetti MA, Shank AR, Park WD. Development of DNA markers suitable for marker assisted selection of three Pi genes conferring resistance to multiple *Pyricularia grisea* pathotypes. Crop Sci.2004; 44:1790-1798. Doi:10.2135/cropsci2004.1790.

[110] Goto I. Genetic studies on resistance of rice plant to blast fungus (VII). Blast resistance genes of Kuroka. Ann Phytopathol Soc Jap.1988; 54:460-465. Doi:10.3186/jjphytopath.54.460.

[111] Tabien RE, Li Z, Paterson AH, Marchetti MA, Stansel JW, Pinson SRM. Mapping of four rice blast resistance genes from 'Lemont' and 'Teqing' and evaluation of their combinatorial effect for field resistance Theor Appl Genet. 2000; 101:1215-1225. Doi:10.1007/s001220051600.

[112] Dwinita WU, Sugiono M, Hajrial A, Asep S, Ida H. Blast resistance genes in wild rice *Oryza rufipogon* andrice cultivar IR64. Ind J Agric.2008; 1:71-76.

[113] Hayashi K, Yasuda N, Fujita Y, Koizumi S, Yoshida H. Identification of the blast resistance gene Pit in rice cultivars using functional markers. Theor Appl Genet.2010; 121(7):1357-1367. Doi:10.1007/s00122-010-1393-7.

[114] Bryan GT, Wu K-S, Farrall L, Jia Y, Hershey HP, McAdams SA, Faulk KN, Donaldson GK, Tarchini R, Valent B. A single amino acid difference distinguishes resistant and susceptible alleles of the rice blast resistance gene Pi-ta. Plant Cell. 2000;12:2033-2046. Doi:10.1105/tpc.12.11.2033.

[115] Kiyosawa S. Genetic studies on host-pathogen relationship in the rice blast disease. Proc. Symp. Rice diseases and their control by growing resistant varieties and other measures. Tokyo.1967; p. 137-153.

[116] Nakamura S, Asakawa S, Ohmido N, Fukui K, Shimizu N, Kawasaki S. Construction of an 800-kb contig in the near-centromeric region of the rice blast resistance gene Pi-ta2using a highly representive rice BAC library. Mol Gen Genomics. 1997; 254:611-620. https://www.ncbi.nlm.nih.gov/pubmed/9202377

[117] Barman SR, Gowda M, Venu RC, Chattoo BB. Identification of a major blast resistance gene in the rice cultivar Tetep. Plant Breeding.2004;123(3): 300-302. DOI: 10.1111/j.1439-0523.2009.00982.x

[118] Lei CL, Huang DY, Lil WJ, Wang L, Liu ZL, Wang XT, Shi K, Cheng ZJ, Zhang X, Ling ZZ, Wan JM. Molecular mapping of a blast resistance gene in an indica rice cultivar Yanxian No. 1. Rice Genet Newsl.2005; 22:76-77. https://shigen.nig.ac.jp/rice/oryzabase/asset/rgn/vol22/v22.html

[119] Zhou B, Qu S, Liu G, Dolan M, Sakai H. The eight amino-acid differences within three leucine-rich repeats between Pi2 and Piz-t resistance proteins determine the resistance specificity to *Magnaporthe grisea*. Mol Plant Microbe Interact.2006; 19:1216-1228. Doi:10.1094/MPMI-19-1216.

[120] Jiang N, Li Z, Wu J, Wang Y, Wu L, Wang S, Wang D, Wen T, Liang Y, Sun P, Liu J, Dai L, Wang Z, Wang C, Luo M, Liu X, Liang Wan G. Molecular mapping of the Pi2/9 allelic gene Pi2-2 conferring broad-spectrum resistance to *Magnaporthe oryzae* in the rice cultivar Jefferson. Rice. 2012; 5:29. DOI:10.1186/1939-8433-5-29.

[121] Shang J, Tao Y, Chen X, Zou Y, Lei C, Wang J, Li X, Zhao X, Zhang M, Lu Z, Xu J, Cheng Z, Wan J, Zhu L. Identification of a new rice blast resistance gene, Pid3, by genomewide comparison of paired nucleotide-binding site--leucine-rich repeat genes

and their pseudogene alleles between the two sequenced rice genomes. Genetics Society of America. 2009; 182: 1303-1311. DOI: 10.1534/genetics.109.102871

[122] Joshi BK, Bimb HP, Parajuli G, Chaudhary B. Molecular tagging, allele mining and marker aided breeding for blast resistance in rice. BSN e-Bulletin 2009; 1: 1-23.

[123] Hayashi K, Hashimoto N, Daigen M, Ashikawa I. Development of PCR-based SNP markers for rice blast resistance genes at the *Piz* locus. Theor Appl Genet. 2004; 108: 1212-1220. https://doi.org/10.1007/s00122-003-1553-0.

[124] Lee SK, Song MY, Seo YS, Kim HK, Ko S, Cao PJ, Suh JP, Yi G, Roh JH, Lee S, An G, Hahn TR, Wang GL, Ronald P, Jeon JS. Rice Pi5-mediated resistance to *Magnaporthe oryzae* requires the presence of two coiled-coil-nucleotide-binding-leucine-rich repeat genes. Genetics. 2009; 181(4):1627-38. DOI: 10.1534/genetics.108.099226.

[125] Inoue H, Nakamura M, Mizubayashi T, Takahashi A, Sugano S, Fukuoka S, Hayashi N. Panicle blast 1 (Pb1) resistance is dependent on at least four QTLs in the rice genome. Rice. 2017; 10(1):36. DOI: 10.1186/s12284-017-0175-0.

[126] Yang S, Tingting GU, Chunyu Pan, ZhumeiFeng, Jing DingYueyu Hang, Jian-Qun Chen, DachengTian. Genetic variation of NBS-LRR class resistance genes in rice lines. Theoretical and Applied Genetics. 2008; 116: 165. DOI.org/10.1007/s00122-007-0656-4.

[127] Das A, Soubam D, Singh PK, Thakur S, Singh NK, Sharma TR. A novel blast resistance gene, *Pi54rh* cloned from wild species of rice, *Oryza rhizomatis* confers broad spectrum resistance to *Magnaporthe oryzae*. Functional Integrative Genomics. 2012; 12: 215. DOI.org/10.1007/s10142-012-0284-1.

[128] Devanna NB, Vijayan J, Sharma TR. The blast resistance gene *Pi54of* cloned from *Oryza officinalis* interacts with *Avr-Pi54* through its novel non-LRR domains. Plos One. 2014;9(8): e104840. DOI: 10.1371/journal.pone.0104840

[129] Sharma TR, Madhav MS, Singh BK, Shanker P, Jana TK, Dalal V, Pandit A, Singh A, Gaikwad K, Upreti HC, Singh NK. High resolution mapping, cloning and molecular characterization of the *Pi-kh*gene of rice, which confers resistance to *M. grisea*. Mol Genet Genom. 2005; 274(6): 569-578. DOI: 10.1007/s00438-005-0035-2

[130] Korinsak S, Sirithunya P, Meakwatanakarn P, Sarkarung S, Vanavichit A, Toojinda T. Changing allele frequencies associated with specific resistance genes to leaf blast in backcross introgression lines of KhaoDawk Mali 105 developed from a conventional selection program. Field Crops Res.2011;122(1): 32-39. DOI:10.1016/j.fcr.2011.02.005

[131] Liu SP, Li X, Wang CY, Li XH, He YQ. Improvement of resistance to rice blast in Zhenshan 97 by molecular marker aided selection. Acta Bot Sin. *2002;*45: 1346-1350.

[132] Narayanan NN, Baisakh N, Cruz CMV, Gnanamanickam SS, Datta K, Datta SK. Molecular breeding for the development of blast and bacterial blight resistance in rice cv. IR50. Crop Sci. 200; 42(6): 2072-2079. DOI: 10.2135/cropsci2002.2072

[133] Fjellstrom R, Mc-Clung AM, Shank AR. SSR markers closely linked to the *Pi-z* locus are useful for selection of blast resistance in a broad array of rice germplasm. Mol Breeding, 2006. 17(2): 149-157. DOI 10.1007/s11032-005-4735-4

[134] Singh A, Singh VK, Singh SP, Pandian RTP, Ellur RK, Singh D, Bhowmick PK, Gopala Krishnan S, Nagarajan M, Vinod KK, Singh UD,

Prabhu KV, Sharma TR, Mohapatra T, Singh AK. Molecular breeding for the development of multiple disease resistance in Basmati rice. AoB Plants, 2012; pls029. DOI: 10.1093/aobpla/pls029

[135] Wang Z, Jia Y, Rutger JN, Xia Y: Rapid survey for presence of a blast resistance gene *Pi-ta* in rice cultivars using the dominant DNA markers derived from portions of the *Pi-ta* gene. Plant Breeding.2007; 126(1): 36-42. DOI: 10.1111/j.1439-0523.2007.01304.x

[136] Chen HQ, Chen ZX, Ni S, Zuo SM, Pan XB, Zhu XD. Pyramiding three genes with resistance to blast by marker assisted selection to improve rice blast resistance of Jin23B. Chin J Rice Sci. 2008; 22: 23-27. http://www.ricesci.cn/EN/abstract/abstract1698.shtml

[137] Fu CY, Wu T, Liu WG, Wang F, Li JH, Zhu XY, Huang HJ, Liu ZR, Liao YL, Zhu MS, Chen JW, Huang YJ: Genetic improvement of resistance to blast and bacterial blight of the elite maintainer line Rongfeng B in hybrid rice (*Oryza sativa* L.) by using marker-assisted selection. Afr J Biotechnol. 2012; 11(67): 13104-13114. DOI: 10.5897/AJB12.1465

[138] Singh VK, Singh A, Singh SP, Ellur RK, Singh D, Gopala KS, Bhowmick PK, Nagarajan M, Vinod KK, Singh UD, Rathore R, Prasanthi SK, Agrawal PK, Bhatt JC, Mohapatra T, Prabhu KV, Singh AK. Marker-assisted simultaneous but step wise back cross breeding for pyramiding blast resistance genes *Piz5* and *Pi54* into an elite Basmati rice restorer line 'PRR78'. Plant Breeding. 2013; 132: 486-495. DOI: 10.1111/pbr.12077

[139] Wen SS, Gao BJ: Introgressing blast resistant gene *Pi-9*(t) into elite rice restorer Luhui 17 by marker assisted selection. Rice Genom Genet. 2011; 2:31-36. DOI: 10.5376/rgg.2011.02.0004

[140] Gouda PK, Saikumar S, Varma CMK, Nagesh K, Thippeswamy S, Shenoy V, Ramesha MS, Shashidhar HE: Marker-assisted breeding of *Pi-1* and *Piz-5* genes imparting resistance to rice blast in PRR78 restorer line of Pusa RH-10 Basmati rice hybrid. Plant Breeding.2013; 132(1): 61-69. DOI: 10.1111/pbr.12017

[141] Hua LX, Liang LQ, He XY, Wang L, Zhang WS, Liu W, Liu XQ, Lin F: Development of a marker specific for the rice blast resistance gene *Pi39* in the Chinese cultivar Q15 and its use in genetic improvement. Biotech Biotechnol Equip.2015; 29(3): 448-456. DOI:10.1080/13102818.2015.1011894

[142] Kumar VA, Balachiranjeevi CH, Naik SB, Rambabu R, Rekha G, Harika G, Hajira SK, Pranathi K, Vijay S, Anila M, Mahadevaswamy HK, Kousik M, Yugander A, Aruna J, Hari Prasad MS, Madhav MS, Laha GS, Balachandran SM, Prasad MS, Ravindra Babu V, Sundaram RM: Marker-assisted improvement of the elite restorer line of rice, RPHR-1005 for resistance against bacterial blight and blast diseases. J Genet.2016; 95(4): 895-903. DOI:10.1007/s1204-016-0711-5

[143] Beser N, Del Valle MM, Kim SM, Vinarao BR, Surek H, Jena KK: Marker-assisted introgression of a broad-spectrum resistance gene, *Pi40* improved blast resistance of two elite rice (*Oryza sativa* L.). Mol Plant Breeding. 2016; 7(33): 1-15. DOI:10.5376/mpb.2016.07.0033

[144] Hegde SS, Prashanthi SK: Identification of polymorphic markers and introgression of *Pi1* and *Pi2* genes for blast resistance in rice. J Farm Sci.2016; 29: 327-331

[145] Xiao WM, Luo LX, Wang H, Guo T, Liu YZ, Zhou JY, Zhu XY, Yang QY, Chen ZQ: Pyramiding of *Pi46* and *Pita* to improve blast resistance and to evaluate the resistance effect of the two R genes. J Integr Agric.2016; 15: 2290-2298. DOI: 10.1016/s2095-3119(16)61415-6

[146] Luo WL, Huang M, Guo T, Xiao WM, Wang JF, Yang GL, Liu YZ,

Wang H, Chen ZH, Zhuang CX: Marker-assisted selection for rice blast resistance genes *Pi2* and *Pi9* through high-resolution melting of a gene-targeted amplicon. Plant Breeding.2016; 136: 67-73. DOI:10.1111/pbr.12447

[147] Wu YY, Yu L, Pan CH, Dai ZY, Li YH, Xiao N, Zhang XX, Ji HJ, Huang NS, Zhao BH, Zhou CH, Liu GQ, Liu XJ, Pan XB, Liang CZ, Li AH. Development of near isogenic lines with different alleles of *Piz* locus and analysis of their breeding effect under Yangdao 6 background. Mol Breeding. 2016; 36: 1-12. doi: 10.1007/s11032-016-0433-7

[148] Usatov AV, Kostylev PV, Azarin KV, Markin KV, Markareno MS, Khachumov VA, Bibov MY: Introgression of rice blast resistance gene *Pi1*, *Pi2* and *Pi33* into Russian rice varieties by marker assisted selection. Ind J Gent Plant Breeding. 2016; 76(1): 18-23. DOI: 10.5958/0975-6906.2016.00003.1

[149] Tanweer FA, Rafii MY, Sijam K, Rahim HA, Ahmed F, Ashkani S, Latif M: Introgression of blast resistance genes (putative *Pi-b* and *Pi-kh*) into elite rice cultivar MR219 through marker-assisted selection. Front Plant Sci.2015; 6: 1-11. DOI: 10.3389/fpls.2015.01002

[150] Araujo LD, Prabhu AS, Arraes Pereira PA, and da Silva GB: Marker-assisted selection for the rice blast resistance gene *Pi-ar* in a backcross population. Crop Breeding and Applied Biotechnology, 2010; 10: 23-31.

[151] Ellur RK, Apurva Khanna Ashutosh Yadav, Sandeep Pathania H, Rajashekara Vikas K, Singh S, Gopalakrishnan Prolay K, Bhowmick M, Nagarajan KK, Vinod G, Prakash Kalyan K, Mondal Nagendra K, Singh Ashok K, Singh: Improvement of Basmati rice varieties for resistance to blast and bacterial blight diseases using marker assisted backcross breeding. Plant Science. 2015;3: 14-20.

[152] Singh VK, Singh A, Singh SP, Ellur RK, Choudhary V, Sarkel S, Singh D, Gopala Krishnan S, Nagarajan M, Vinod KK, Singh UD, Rathore R, Prasanthi SK, Agrawal PK, Bhatt JC, Mohapatra T, Prabhu KV, Singh AK: Incorporation of blast resistance gene in elite Basmati rice restorer line PRR78, using marker assisted selection. Field Crops Res.2012; 128: 8-16. DOI: 10.1016/j.fcr.2011.12.003

[153] Divya B, Robin S, Rabindran R, Senthil S, Raveendran M, Joel AJ: Marker assisted backcross breeding approach to improve blast resistance in Indian rice (*Oryza sativa*) variety ADT43. Euphytica.2014; 200(1): 61-77. DOI: 10.1007/s10681-014-1146-9

[154] Madhavi KR, Rambabu R, Abhilash Kumar V, Vijay Kumar S, Aruna J, Ramesh S, Sundaram RM, Laha GS, Sheshu Madhav M, Ravindra babu V, Srinivas Prasad M: Marker assisted introgression of blast (Pi-2 and Pi-54) genes in to the genetic background of elite, bacterial blight resistant indica rice variety, Improved Samba Mahsuri.2016; 1-13. DOI 10.1007/s10681-016-1784-1.

[155] Aruna Kumari K, Durgarani CV, Satturu V, Sarikonda K R, Chittoor PDR, Vutukuri B, Laha GS, Nelli APK, Gattu S, Jamal M, Prasad Babu A, Hajira S, Sundaram RM. Marker-Assisted pyramiding of genes conferring resistance against bacterial blight and blast diseases into Indian rice variety MTU1010. Rice Science. 2016; 23(6): 306–316. DOI: 10.1016/j.rsci.2016.04.005

[156] Pinta WT, Toojinda P, Thummabenjapone J, Sanitchon: Pyramiding of blast and bacterial leaf blight resistance genes into rice cultivar RD6 using marker assisted selection. African Journal of Biotechnology.2013; 12 (28): 4432-4438. DOI: 10.5897/AJB12.2028

[157] Oddin J, Durga Rani, H.V. Swathi, G., Anuradha, C.H. Vanisri, S., Sheshu Madhav, M. Sundaram, R. M., Arun Prem Kumar, N. 2015.Introgression of blast resistance genes *Pi-54* and *Pi1* into cold

tolerant variety Tellahamsa, by marker assisted selection. International Journal of Current Research; 7(11): 22348-22353.

[158] Jiefeng Jiang, Dabing Yang, Jauhar Ali, Tongmin Mou, Molecular marker-assisted pyramiding of broad-spectrum disease resistance genes, *Pi2* and *Xa23*, into GZ63-4S, an elite thermo-sensitive genic male-sterile line in rice. Molecular Breeding. 2015; 35: 83-96. DOI. org/10.1007/s1103

[159] Srikanth. S, Manish K Pandey, Bala Chiranjeevi. CH, Hajira. SK, Vijay Kumar. S, MBVN Kousik, Bhadana VP, Sheshu Madhav. M, Suneetha. K, Subba Rao LV, Kumaraswamy. M, Archana Giri, Laxmi Narasu. B, Shobha Rani. N, and Sundaram RM. Introgression of major bacterial blight and blast resistant genes into Vallabh Basmati 22, an elite Basmati variety. International Journal of Development Research. 2016; 6(7): 8366-8370.

[160] Balachiranjeevi CH, Naik S, Bhaskar V, Abhilas HS. Akanksha BC, Viraktamath MS, Madhav AS, Hariprasad GS, Laha MS, Prasad SM, Balachandran CN, Neeraja M, Satendra Kumar P, Senguttuvel K B, Kemparaju VP, Bhadana. Marker-assisted introgression of bacterial blight and blast resistance into DRR17B, an elite, fine-grain type maintainer line of rice. Molecular Breeding. 2015; 35: 151. DOI.ORG/10.1007/S1103

[161] Singh Ashok K, Singh Vikas K, Atul Singh, Ranjith K. Ellur, Pandian RTP, Gopala Krishnan S, Singh UD, Nagarajan M, Vinod KK, -Prabhu KV. Introgression of multiple disease resistance into a maintainer of Basmati rice CMS line by marker assisted backcross breeding. Euphytica. 2015; 35(1): 97-107. DOI.org/10.1007/s1068.

[162] Hasan NA, Mohd Y, Rafii, Harun A, Rahim, Nusaibah Syd Ali, Norida Mazlan and Shamsiah Abdullah. Introgression of *Pi-kh* resistance gene into a Malaysian cultivar, MR264 using Marker-Assisted Backcrossing (MABC). International Journal of Agriculture and Biology. 2015;17: 1172-117. DOI: 10.17957/IJAB/15.0040.

[163] Ashkani S, Rafi MY, Shabanimofrad M, Miah G, Sahebi M, Azizi P, Tanveer FA, Akhtar MS, Nasehi, A. Molecular breeding strategy and challenges towards improvement of blast disease resistance in rice crop. Front Plant Sci. 2015; 6: 886. DOI: 10.3389/fpls.2015.00886

[164] Chen DH, Zeigler S, Ahn SW, Nelson RJ. Phenotypic characterization of the rice blast resistance gene Pi-2(t). Plant Dis.1996;80: 52-56.

[165] Kalia S, Rathour R. Current status on mapping of genes for resistance to leaf- and neck-blast disease in rice. 3 Biotech. 2019;9, 209. https://doi. org/10.1007/s13205-019-1738-0

[166] Singh J, Gupta SK, Devanna BN, Singh S, Upadhyay A, Sharma TR. Blast resistance gene Pi54 over-expressed in rice to understand its cellular and sub-cellular localization and response to different pathogens. Scientific reports. 2020;10(1):1-13.

[167] Chen H, Wang S, Xing Y, Xu C, Hayes PM, Zhang Q. Comparative analyses of genomic locations and race specificities of loci for quantitative resistance to *Pyricularia grisea* in rice and barley. *Proc. Natl. Acad. Sci. U.S.A. 2003;* 100: 2544-2549. DOI: 10.1073/ pnas.0437898100

[168] Young ND. QTL mapping and quantitative disease resistance in plants. *Annu. Rev. Phytopathol.* 1996;34:479-501. DOI: 10.1146/annurev.phyto.34.1.479

[169] Song FM, Goodman RM. Molecular biology of disease resistance in rice. Physiol Mol Plant Pathol.2001;59(1): 1- 11.DOI: 10.1006/pmpp.2001.0353

[170] Devi SJSR, Singh K, Umakanth B. Identification and characterization of a

large effect qtl from *Oryza glumaepatula* revealed *Pi68(t)* as putative candidate gene for rice blast resistance. Rice. 2020;13, 17 https://doi.org/10.1186/s12284-020-00378-4

[171] Ballini E, Morel JB, Droc G, Price A, Courtois B, Notteghem JL. A genome-wide meta-analysis of rice blast resistance genes and quantitative trait loci provides new insights into partial and complete resistance. Mol. Plant Microbe Interact. 2008; 21: 859– 868. DOI: 10.1094/MPMJ-21-7-0859

[172] Yin J, Zou L, Zhu X. et al. Fighting the enemy: How rice survives the blast pathogen's attack. The Crop Journal. 2021; https://doi.org/ 10.1016/j.cj.2021.03.009

[173] Yang C, Yu Y, Huang J, Meng F, Pang J, Zhao Q, Islam M, Xu N, Tian Y, Liu J. *Magnaporthe oryzae* binding of the chitinase MoChia1 by a rice tetratricopeptide repeat protein allows free chitin to trigger immune responses. Plant Cell. 2019; 31:172-188.

[174] Kim S, Kim C, Park C, Kim K, Jeon J, Chung H, Choi G, Kwon S, Choi J, Jeon J, Jeon J, Khang C, Kang S, Lee Y. Two nuclear effectors of the rice blast fungus modulate host immunity via transcriptional reprogramming, Nature Communication. 2020; 11:5845.

[175] Liu X, Inoue H, Tang X, Tan Y, Xu X, Wang C, Jiang C. Rice OsAAA-ATPase1 is induced during blast infection in a salicylic acid-dependent manner, and promotes blast fungus resistance, Int. J. Mol. Sci. 2020; 21:1443.

[176] Zhou Y, Lei F, Wang Q. Identification of Novel Alleles of the Rice Blast-Resistance Gene *Pi9* through Sequence-Based Allele Mining. Rice. 2020;13, 80. https://doi.org/10.1186/s12284-020-00442-z

[177] Norvienyeku J, Lin L, Waheed A, Chen X, Bao J, Aliyu S, Lin L, Shabbir A, Batool W, Zhong Z, Zhou J, Lu G, Wang Z. Bayogenin 3-O-cellobioside confers non-cultivar-specific defence against the rice blast fungus Pyricularia oryzae. Plant Biotechnol. J. 2020: https://doi.org/10.1111/pbi.13488

[178] Zhan C, Lei L, Liu Z, Zhou S, Yang C, Zhu X, Guo H, Zhang F, Peng M, Zhang M, Li Y, Yang Z, Sun Y, Shi Y, Li K, Liu L, Shen S, Wang X, Shao J, Jing X, Wang Z, Li Y, Czechowski T, Hasegewa H, Graham I, Tohge T, Qu L, Liu X, Fernie A, Chen L, Yuan M, Luo J. Selection of a subspecies-specific diterpene gene cluster implicated in rice disease resistance, Nat. Plants. 2020; 6:1447– 1454.

[179] Deepti S, MdShamim, Mahesh K, Anurag M, Pramila P,Deepak K, Prashant Y, Mohammed HS, Kapildeo NS. Current status of conventional and molecular interventions for blast resistance in rice. Rice Science. 2017; 24(6): 299:321.

[180] Bonman J. Durable resistance to rice blast disease-environmental influences. *Euphytica.1992;* 63:115-123. DOI: 10.1007/BF00023917

[181] Mi J, Yang D, Chen Y, Jiang J, Mou H, Huang J, Ouyang Y and Mou T. Accelerated molecular breeding of a novel P/TGMS line with broad-spectrum resistance to rice blast and bacterial blight in two-line hybrid rice. Rice. 2018;11:11 doi.org/10.1186/s12284-018-0203-8

[182] Miah G, Rafii MY, Ismail MR, Puteh AB, Rahim HA, Asfaliza R, Latif MA. Blast resistance in rice: A review of conventional breeding to molecular approaches. MolBiol Rep. 2013; 40: 2369-2388.DOI: 10.1007/s11033-012-2318-0

[183] Xi ZY, He FH, Zeng RZ, Zhang ZM, Ding XH, Li WT, ZhangGQ. Development of a wide population of chromosome single-segment substitution lines in the genetic background of an elite cultivar of rice (*Oryza sativa*l.). Genome. 2008; 49(5): 476-484. DOI: 101139/g06-005

[184] Sreewongchai T, Toojinda T, Thanintorn N, Kosawang C, Vanavichit A, Tharreau D, Sirithunya P. Development of elite indica rice lines with wide spectrum of resistance to Thai blast isolates by pyramiding multiple resistance QTLs. Plant Breed. 2010;129:176-180. DOI/abs/10.1111/j.1439-0523.2009.01669.x

[185] Fujimski H. Recurrent selection by using male sterility for rice improvement. Japan Agric Res Quart. 1979;13:153-156.

[186] Kaur S, Padmanabhan SY, Rao M. Induction of resistance to blast disease (*Pyricularia oryzae*) in the high yielding variety, Ratna (IRE 9 TKM 6). *In*: Proceedings of the IAEA research coordination. Geoling, Ames, Iowa. 1975; 141-145.

[187] Zhang MX, Xu JL, Luo RT, Shi D, Li ZK. Genetic analysis and breeding use of blast resistance in a japonica rice mutant *R917*. Euphytica. 2003;130: 71-76. DOI: 10.1023/A:102238062

[188] Shu QY. Induced plant mutations in the genomics era. Food and Agriculture Organization of the United Nations. Rome. 2009; pp 425-427.

[189] Yadav MK, Aravindan S, Ngangkham U. Candidate screening of blast resistance donors for rice breeding. J Genet. 2019;98:73 https://doi.org/10.1007/s12041-019-1116-z

[190] Yadav MK, Aravindan S, Ngangkham U, Raghu S, Prabhukarthikeyan SR, Keerthana U. Blast resistance in Indian rice landraces: Genetic dissection by gene specific markers. PLoS ONE. 2019;14(1):e0211061. https://doi.org/10.1371/journal.pone.0211061

[191] Koide Y, Kobayashi N, Xu D, Fukuta Y. Resistance genes and selection DNA markers for blast disease in rice (*Oryzasativa* L.). *Japan Agric. Res. Q.* 2009; 43: 255-280. DOI: 10.1016/j.crvi.2014.11.003

[192] Hittalmani S, Parco A, Mew TV, Zeigler RS, Huang N. Fine mapping and DNA marker-assisted pyramiding of the three major genes for blast resistance in rice. Theor Appl Genet.2000; 100(7): 1121-1128. **DOI:** 10.1007/s001220051395

[193] Chen Q, Zeng G, Hao M. Improvement of rice blast and brown plant hopper resistance of PTGMS line C815S in two-line hybrid rice through marker-assisted selection. Mol Breeding. 2020;40, 21 https://doi.org/10.1007/s11032-020-1098-9

[194] Khan GH, Shikari AB, Vaishnavi R. Marker-assisted introgression of three dominant blast resistance genes into an aromatic rice cultivar *Mushk Budji*. Sci Rep. 2018;**8**, 4091 https://doi.org/10.1038/s41598-018-22246-4

[195] Jamaloddin M, Durga Rani CV, Swathi G, Anuradha C, Vanisri S, Rajan CPD. Marker Assisted Gene Pyramiding (MAGP) for bacterial blight and blast resistance into mega rice variety "Tellahamsa". PLoS ONE. 2020;15(6): e0234088. https://doi.org/10.1371/journal.pone.0234088

[196] Fujita D, Ebron LA, Kobayashi N, Fukuta Y. DNA marker analysis of blast resistance genes Pib and Pita in IRRI-bred rice varieties comparision with gene estimation by a differential system. Advances in Genetics, Genomics and Control of Rice Blast Disease. 2009; 315-324. DOI: 10.1007/978-1-4020-9500-9_31

[197] Koide Y, Kawasaki A, Telebancoâaaa-Yanoria M, Hairmansis A, Nguyet N, Bigirimana J. Development of pyramided lines with two resistance genes, Pish and Pib, for blast disease (*Magnaporthe oryzae* B. Couch) in rice (*Oryza sativa* L.). *Plant Breed.* 2010; 129: 670-675. DOI: 10.1111/j.1439-0523.2010.01781.x

[198] Urso S, Orasen G, Perrini R, Tacconi G, Delfanti S, Biselli C. Pyramiding of Pi resistance genes to

increase blast resistance in Italian rice varieties using marker-assisted selection approaches. Proceedings of the 57th Italian Society of Agricultural Genetics Annual Congress, 16th-19th September, Foggia. 2013;

[199] Yasuda N, Mitsunaga T, Hayashi K, Koizumi S, Fujita Y. Effects of pyramiding quantitative resistance genes pi21, Pi34, and Pi35 on rice leaf blast disease. *Plant Dis.* 2014; 99: 904-909. DOI: 10.1094/PDIS-02-14-0214-RE

[200] Wang F, Wang C, Liu P, Lei C, Hao W, Gao Y. Enhanced rice blast resistance by CRISPR/Cas9-Targeted mutagenesis of the ERF transcription factor gene *OsERF922*. PLoS ONE. 2016; 11(4): e0154027. DOI: 10.1371/journal.pone.0154027

[201] Ma J, Chen J, Wang M, Ren YL, Wang S, Lei CL, Cheng ZJ. Disruption of *OsSEC3A* increases the content of salicylic acid and induces plant defense responses in rice. J Exp Bot. 2018; 69:(5)1051-1064.

[202] Li SF, Shen L, Hu P, Liu Q, Zhu XD, Qian Q, Wang KJ, Wang YX. Developing disease-resistant thermo sensitive male sterile rice by multiplex gene editing. J Integr Plant Biol. 2019; 61 (12):1201-1205.

[203] Mahesh HB, Shirke MD, Wang GL & Gowda M. In planta transcriptome analysis reveals tissue-specific expression of pathogenicity genes and microRNAs during rice-*Magnaporthe* interactions. Genomics. 2021; 113(1): 265-275.

[204] Wang Y, Cheng X, Shan Q, Zhang Y, Liu J, Gao C. Simultaneous editing of three homoeoalleles in hexaploid bread wheat confers heritable resistance to powdery mildew. Nat Biotechnol. 2014; 32(9): 947-951. DOI: 10.1038/nbt.2969

[205] Li T, Liu B, Spalding MH, Weeks DP and Yang B. High-efficiency TALEN-based gene editing produces disease-resistant rice. Nat Biotechnol. 2012;30: 390-392. DOI:10.1038/nbt.2199

[206] Zafar K, Sedeek KEM, Rao GS, Khan MZ, Amin I, Kamel R, Mukhtar Z, Zafar M, Mansoor S, Mahfouz MM. Genome Editing Technologies for Rice Improvement: Progress, Prospects, and Safety Concerns. Front. Genome Ed. 2020;2:5. doi: 10.3389/fgeed.2020.00005

[207] Ramkumar G, Biswal AK, Mohan KM, Sakthivel K, Sivaranjani AKP, Neeraja CM, Rani NS, Viraktamath BC, and Madhav, M.S. Identifying novel alleles of rice blast resistance genes Pikh and Pita through allele mining. Int. Rice Res. Notes 35 1-6.

[208] Huang X, Wei X, Sang T, Zhao Q, Feng Q, Zhao Y, Li C, Zhu C, Lu T, Zhang Z, Li M, Fan D, Guo Y, Wang A, Wang L, Deng L, Li W, Lu Y, Weng Q, Liu K, Huang T, Zhou T, Jing Y, Lin Z, BucklerES, Qian Q, Zhang QF, Li J and Han B. Genome-wide association studies of 14 agronomic traits in rice landraces. Nat Genet. 2010; 42: 961-967. DOI: 10.1038/ng.695

[209] Huang X, Zhao Y, Wei X, Li C, Wang A, Zhao Q, Li W, Guo Y, Deng L, Zhu C, Fan D, Lu Y, Weng Q, Liu K, Zhou T, Jing Y, Si L, Dong G, Huang T, Lu T, Feng Q, Qian Q, Li J, Han B. Genome-wide association study of flowering time and grain yield traits in a worldwide collection of rice germplasm. Nat Genet. 2012;44: 32-39. DOI: 10.1038/ng.1018

[210] Zhao K, Tung CW, Eizenga GC, Wright MH, Ali ML, Price AH, Norton GJ, Islam MR, Reynolds A, Mezey J, McClung AM, Bustamante CD, McCouch SR. Genome-wide association mapping reveals a rich genetic architecture of complex traits in *Oryza sativa*. Nat Commun. 2011;2: 467. DOI: 10.1038/ncomms1467

[211] Nunes CC, Dean RA. Host-induced gene silencing: a tool for understanding fungal host interaction and for developing novel disease control strategies. Mol Plant Pathol. 2012; 13: 519-529. DOI:10.1111/j.1364-3703.2011.00766.x

[212] Campo SS, anchez-Sanuy F, Camargo-Ramırez R, Gomez-Ariza J, Baldrich P, Campos-Soriano L, SotoSuarez M, San Segundo B. A novel Transposable element-derived microRNA participates in plant immunity to rice blast disease. Plant Biotechnol J. 2021;https://doi.org/10.1111/ pbi.13592

Chapter 10

Emerging Minor Diseases of Rice in India: Losses and Management Strategies

Raghu Shivappa, Devanna B. Navadagi, Mathew Seikholen Baite, Manoj Kumar Yadav, Prabhukarthikeyan S. Rathinam, Keerthana Umapathy, Prajna Pati and Prakash Chandra Rath

Abstract

Rice (*Oryza sativa* L) being one of the imperative food crops of the word contributes immensely to the food and nutritional security of India. The cultivation of rice is changed over the decades from a simple cultivation practices to the advanced cultivation to increase yield. Increased in rice yields especially after 1960s is mainly due to the introduction of high yielding semi-dwarf varieties which requires more inputs like chemical fertilizers, water and other resources. As a result, India achieved self sufficiency in rice and currently producing more than 115 MT of rice to meet country's demand. Now India is exporting rice to other nations and earning foreign returns. With the change in rice cultivation practices, problems also aroused side by side. A number of biotic and abiotic stresses emerged as major constraints for rice cultivation in diverse agro-climatic conditions and growing ecologies. Diseases are the major biotic constraints to rice which can reduce the yields by 20–100% based on severity. Major diseases like blast, brown spot, bacterial blight, sheath blight and tungro still causing more damage and new minor diseases like bakanae, false smut, grain discoloration, early seedling blight, narrow brown spot, sheath rot have emerged as major problems. The losses due to these diseases may 1–100% based on the growing conditions, varietal susceptibility etc.., At present no significant source of resistance available for any of the above emerging diseases. But looking into the severity of these diseases, it is very important to address them by following integrated management practices like cultural, mechanical, biological and finally chemical control. But more emphasis has to be given to screen gerrmplasm against these diseases and identify stable source of resistance. Finally utilizing these sources in resistance breeding program by employing molecular breeding tools like marker assisted selection (MAS), marker assisted back cross breeding (MABB), gene pyramiding and transgenic tools. The present chapter discusses the importance of these emerging minor diseases of rice, the losses and possible management measures including resistance breeding.

Keywords: sheath rot, bakanae, false smut, narrow brown spot, early seedling blight, narrow brown spot, disease management, resistance breeding, molecular tools

IntechOpen

1. Introduction

Indian agriculture is considered as "Gambling with monsoon" which means the production and productivity of Indian agriculture is mainly depends on the quantity and distribution of monsoon rains. Among them, South West monsoons are very important which covers major area of the country and directly indicates the country's production. Agriculture sector plays a vital role in country's economy and is the backbone of the country. Country produces almost all the agricultural commodities starting with cereals, pulses, oilseeds, commercial crops, fruits, vegetables and plantation crops [1]. Among the commodities cereals contribute to the major proportion which is the means of daily staple food. Major cereals like Rice, Wheat, Maize and minor cereals like sorghum, bajra, ragi (finger millet), etc.., contribute significantly to the daily calories of the people [2].

Rice (*Oryza sativa* L) is considered as one of the imperative food crops of the globe. Rice is being cultivated since ancient times to feed the population. The crop is the staple food source feeding more than half of the world's ever-growing population [3]. The crop is grown by more than one hundred countries of which more than 70% of the production comes from Asian countries. The crop also cultivated in small to medium scale in Africa, Europe and American countries [4]. India grows rice in 43 M ha with production of 112 million tons (Mt) of milled rice and average productivity of 2.6 t^{-1} ha. The crop is grown in highly diverse conditions ranging from hills to coasts. Primarily a *kharif* crop, it is cultivated round the year in one or the other parts of the country [5]. Rice production in India has made tremendous progress over the years. However, it is facing unprecedented challenges of environmental degradation and climate change in recent years [6–8]. Low and uncertain income, degraded natural resource base, growing labour and energy shortages and threats of climate change are making Indian agriculture highly vulnerable and unsustainable [9–12].

Though green revolution brought tremendous changes in production and productivity of food grains, it also led to some of the related problems. There has been a constant increase in the number of insect pests and diseases, some of the non-insect pests, nematodes over the years [13]. The country witnessed a number of epidemics due to these biotic stresses. The level of incidence, quantum of damage has changed over the years [14]. These pests and diseases showed a concomitant shifts from their minor status to major status/intensity depending on the region, varieties grown, environmental conditions and cultivation practices [15]. Major diseases of rice such as blast, brown spot, sheath blight, bacterial leaf blight and tungro have become more severe over the years, and a number of minor diseases like sheath rot, bakanae, false smut, grain discoloration, early seedling blight and narrow brown spot have emerged as major problems [16]. Similarly, among the insect pests, gall midge, white backed planthopper, gundy bug, leaf folder has emerged as major problems to the rice cultivation [4].

There may be a number of factors responsible for this changed scenario of pest status like intensified rice cultivation of high yielding varieties, cultivation of varieties lacking the resistance to major pests which provide rapid multiplication of pests, imbalanced application of chemical fertilizers, particularly the nitrogenous fertilizers, non-judicious application of chemical pesticides which leads to resistance development and changes in climatic conditions [13, 17]. Present review provides the information on important emerging diseases of rice, nature of damage, yield losses and possible management strategies to mitigate the losses due to these new emerging problems.

2. Important minor diseases, biology and the losses

2.1 Sheath rot

2.1.1 Distribution

Sheath rot has become a common problem in almost all the rice growing ecologies [18]. It is most common almost all the rice varieties with dense planting, nitrogen-responsive semi dwarf and tall varieties growing in India [19]. In eastern Indian states like Odisha, West Bengal, Jharkhand, North Eastern states, the famous varieties become highly susceptible [20]. The disease is emerging as major problem in North and South Indian states also causing both qualitative and quantitative losses. High yield losses due to the disease have been reported from all Asian countries also [21]. Sheath rot was first identified by Sawada in 1922 in Japan where it became a common problem [22]. Later the disease has been reported from low to moderate incidence from South and South East Asia [23], Kenya, Nigeria [24], USA [25] and Brazil [26, 27] Reported that, the disease is a major problem in Upland rice in West Africa where the rice cultivars were introduced from Asia. The yield losses due to the disease may vary from 20 to 85% depending on the growing conditions, cultivars and environmental conditions.

2.1.2 Identification symptoms

The disease is very important emerging problem caused by a number of pathogens. Among them, *Sarocladium oryzae* is the major fungal pathogen. The initial symptoms start as discoloration of the flag leaf. Rotting occurs on the on the leaf sheath which encloses young panicle. Irregular lesions on leaf sheath which are having gray centre with reddish brown margin. The lesions coalesce each other and covers entire sheath. Under severe infestation, the developing grains also infected and turn to dirty discolored having white fungal mass. The grains ultimately rot and gets chaffy. The panicle will not emerge out of sheath. The plants may stunt and dies prematurely [22]. The disease symptoms are provided in **Figures 1–4**.

2.1.3 Pathogen

Rice sheath rot is a complex disease that can be caused by various fungal and bacterial phytopathogens. Major phytopathogens associated with the disease are *Sarocladium oryzae* and *Fusarium fujikuroi* species complex (*Fusarium*

Figure 1.
Infected leaf sheath.

Figure 2.
Severe infection.

Figure 3.
Infected grains.

Figure 4.
The panicle turned chaffy.

fujikuroi, Fusarium verticilloides and other *Fusarium* spp.). Bacterial pathogen like *Pseudomonas fuscovaginae* is associated with the disease. Among these pathogens, *Sarocladium oryzae* is the major pathogen which was originally identified as *Acrolyndrium oryzae* when it was first reported from Taiwan in 1922 [28]. Later in 1975, the genus *Sarocladium* was established [29] and the pathogen was renamed as *Sarocladium oryzae*. The genus Sarocladium currently comprises of 16 species including plant pathogens, saprophytic microbes, endophytes, mycoparasites and some of the potential human pathogens [30]. The pathogen grows slowly on Potato Dextrose Agar (PDA) at 2.5 mm per day at 28°C. It produces sparsely branched white mycelium with yellow pigmentation underside. It also produces numerous microsclerotia which are round in shape and orange colored. The pathogen produces simple or branched conidiophores. The conidia are cylindrical, septate./aseptate and hyaline measuring 4–7 X 1–2 μm in size (**Table 1**).

2.1.4 Epidemiology and integrated disease management

The disease is reported from moderate to severe form in almost all the rice growing countries. The disease is very common and severe in Monsoon/rainy season and reported in low to moderate form in summer months [31]. Sharma et al. [32] observed that, the disease incidence in Nepal was found below 1250 M altitude at a temperature of 20-30°C and relative humidity of 65–85%. Similar observation found by [33]. The sheath rot pathogen survives in infected seeds (having seed borne nature), plant residues like straw, stubble, chaffy grains. It can also found in soil, water when environmental conditions become favorable. The pathogen may attack the plants at various growth stages. The fungus enters through stomata, wounds and found to be most destructive during booting and grain developmental stages [34]. The nature of entry and extent of damage also facilitated by insect and mites damage by weakening of the plants [34]. Secondary transmission of the pathogen may be by wind and rain splash.

2.1.5 Integrated management

Though the disease can be managed through number of measures, breeding for resistance is the best option for its management. Breeding for disease resistance is bit difficult and challenging as the disease is caused by a number of pathogens. Many researchers have identified few resistant varieties. Hemalatha et al. [35] developed method of screening for resistance against. *S oryzae* based on a crude toxin preparation. Pathogen variability, its virulence pattern, geographical location and cultivars growing are the points to be considered while breeding for diseases resistance. Select suitable resistant varieties for sowing: Jalmagna, Latisali, Rasi, Pankaj etc., Apart from resistance breeding other management practices should also be followed [36]. Use of healthy seeds, limiting insect pest population, avoiding densely planting, balanced application of chemical fertilizers especially nitrogen and increased application of potassic fertilizers. Similarly adoption of field sanitation, weed control, crop residue management is the some of the recommended cultural practices. Treat the seeds with carbendazim 50% WP @ 2 g/kg or biological control agents like *Trichoderma* or *Pseudomonas* talc based formulations @ 8–10 g/kg of seeds. Soil application of *Pseudomonas fluorescens* @ 2.5 kg/ha after 30 days of transplanting mixing with 50 kg FYM. Removal and destruction of weeds, infected stubbles should be done at critical periods. Application of potash at tillering stage is helpful in disease reduction. At booting stage, spray fungicides like carbendazim 50% WP @ 500 g/ha or mancozeb 75%WP @ 1 kg/ha or Iprobenphos 48EC @ 1 kg/ha or Thiophanate methyl 70%WP @ 500 g/ha or Isoprothiolane 40%EC @ 750 ml/ha. During grain maturity stage, spray systemic fungicides like Ediphenphos @ 500 g/ha or a combination of Tridemorph 80% EC (fungicide) + phosphamidon 40% SL (insecticide) to give better control of sheath rot. Foliar spray of calcium sulphate and zinc sulphate is effective against sheath rot.

Sl. No	Pathogen	Survival	Host range	Most susceptible plant stage	Dissemination	Reproduction	Relevant metabolites
1	*Sarocladium oryzae*	Seeds, plant residues, soil, water	Weeds, bamboo, sedge	After booting stage	Wind, rain, insects, mites	Aseptate conidia	Helvolic acid, cerulenin
2	*Fusarium fujikuroi*	Seeds, plant residues, soil		All stages	Wind, rain	Macro- and microconidia, no chlamydospores	Fumonisins (low levels in some strains), gibberellins, moniliformin
3	*Fusarium proliferatum*	Seeds, plant residues, soil	Wide host range	All stages	Wind, rain	Macro- and microconidia, no chlamydospores	Fumonisins (high levels), moniliformin
4	*Fusarium verticillioides*	Seeds, plant residues, soil	Wide host range	All stages	Wind, rain	Macro- and microconidia, no chlamydospores	Fumonisins (high levels)
5	*Pseudomonas fuscovaginae*	Seeds, epiphytically and endophytically on rice	Wild and cultivated Gramineae	Seedling and booting stages	Wind, rain	Bacterial cells	Fuscopeptin, syringotoxin

Source: Bigirimana et al. [21].

Table 1.
Pathogens associated with rice sheath rot disease and their characteristics.

2.2 Bakanae/foot rot (foolish seedling disease)

2.2.1 Distribution

Bakanae or foot rot is one of the important emerging diseases of rice, caused by *Fusarium fujikuroi* (Nirenberg) [teleomorph: *Gibberella fujikuroi* (Sawada) Ito]. The disaese has become a major concern in basmati growing tracts of north India during last few years [37, 38]. The northern western states like Punjab, Haryana, eastern UP, Uttarakhand and Delhi are facing serious problem of disease especially in basmati growing regions. But recently the disease is emerging as a major problem in eastern and north eastern states like Odisha, West Bengal and Assam leading to the susceptibility of popular varieties in this region [39]. Bakanae disease induces grain sterility resulting in a considerable loss of grain yield [40]. Reports have shown that the disease can cause even 70% yield loss and quality under field conditions [41]. In eastern and north eastern states of India, popular varieties like Pooja, Swarna and Abhishek became highly susceptible to the disease. Earlier reports indicates that, the disease occurs sporadically in Asia. The term 'Bakanae' is of Japanese origin meaning 'bad', 'naughty' or 'foolish' seedling, indicating the unusual early elongation of seedlings due to the production of gibberellins on infection process. The fungus produces both gibberellins which causes seedling elongation and fusaric acid, attributed to seedling death [42].

2.2.2 Identification symptoms

Initial symptoms appear as pale green and lanky seedling sporadically in the field. The seedling later shows abnormal elongation which is much taller than normal plant. The intermodal length will be more and production of fibrous roots seen from each nodes. White powdery mass of the pathogen produces in each node later covers entire plant. The entire plant gets killed without producing any grains. Death of the plants is called as foot-rot. The pathogen produces two toxins, (1) Gibberellic acid which is a growth hormone leads to elongation symptoms and (2) Fusaric acid which lead to death of the young seedlings.

The symptoms and disease severity is mainly depends on the quantity of these two metabolites produced in response to pathogen infection and environmental conditions. The pathogen is both seed-borne and soil-borne, so the infection may occur either by sowing the infested seeds in non-infested fields or by sowing the healthy seeds in infested fields. The inoculums of the pathogen may build up in soil if a susceptible variety grown in the same field year after year. Seed-borne inoculum plays a major role in secondary transmission of the disease under favorable environmental conditions by producing numerous conidia and infects fresh plant [43]. The pathogen infects rice grains during field and carried to storage, the contaminated seeds after sowing in field will results in disease incidence by the colonization of pathogen in seedlings [44]. *F. fujikuroi*-infected seedlings show morphological and colure abnormalities (**Figures 5–8**). The abnormal symptoms include elongation, stunting, large angle between leaf and stem, production of roots from each node and yellowish-green leaves [45]. Because of the different kinds of symptoms, bakanae disease is a complex and contradictory (e.g., elongation and stunting symptoms), and depends on varietal response also.

2.2.3 Epidemiology and Integrated management

Management of the disease is very challenging as the pathogen is seed borne. Once the disease establish in field, it is very difficult to manage. Most commonly

Figure 5.
Death of the seedling after transplanting.

Figure 6.
Abnormal elongation of seedlings.

Figure 7.
Production of fibrous roots from nodes.

used management practice was hot water treatment or fungicide seed treatment [45]. But it was found ineffective as the thermal effect is not efficiently transmitted to the pericarp layer of the seeds. Several fungicides also found in effective as seed treatment chemicals due to development of resistance and as a result fungal spores will not destroy [46]. But several researcher shown that, some of the fungicides as seed treatment and seedling dip treatment found effective [47]. There are also reports of management of the disease with Biocontrol agents such as *Bacillus* spp. (), *Pseudomonas* spp. A combination antagonistic yeasts and thermotherapy was

Figure 8.
White powdery mass of the pathogen at the base of stem.

found to be efficient in managing the disease [48]. Presently the disease is emerging as an alarming state in almost all the rice growing areas of India and worldwide [38]. Looking for alternative management practices other than usage of chemical fungicides is the need of the hour. Alternative management measures such as usage of bio-control agents, chemical elicitor compounds which induce resistance are promising. Along with this, identification of rice bakanae resistant cultivars is more promising and to be taken up in priority [49]. Many researchers have screened numerous accessions against rice bakanae disease to identiofy genes/QTLs responsible for resistance [50, 51]. Select suitable resistant varieties for sowing. Varieties like Chandan, Improved Tapaswini, Sarasa, Kshira, CR Dhan-311, Wifa-10, Improved Lalat, Palguni, Saket-4, Maudamani, Kalinga-II, Naveen, Kalinga-I, CR Dhan-305, Khitish, Satabdi, CR Dhan-310, CR-29-83, IR-29, Udaya, Padma. Treat the seeds with carbendazim 50% WP @ 2 g/kg or biological control agents like *Trichoderma* or *Pseudomonas* talc based formulations @ 8–10 g/kg of seeds. Remove and destroy infected plants from main field. Do not used farmer saved seeds for sowing in the next season. Foliar spray with combination fungicide, Trifloxystrobin 25% + Tebuconazole 50% WG (75%WG) @ 1 g/lit will give the protection to some extent.

2.3 False smut

2.3.1 Distribution

False smut or green smut is a common disease of rice caused by *Ustilaginoidea virens* in rice growing regions of India. Epidemics of false smut disease of rice were reported in Tamil Nadu in India and later in many countries of world [52]. Pannu et al., [53] also reported losses up to 44 per cent in Punjab. In Uttar Pradesh, yield losses up to 44 per cent were observed by Singh and Dube. In some rice growing districts of Bihar, 15–50 percent losses occurs due to false smut of rice when comes as medium to severe form [54]. The fungus overwinters in soil by means of sclerotia and chlamydospores. Sclerotia produces ascospores, which are primary source of infection to rice plants, whereas secondary infection may come from air-borne chlamydospores [55, 56]. Sclerotia can survive in the field for several months. Infection starts in grains of rice before flowering. Infection results in one or more kernels on mature heads of plants being replaced by globose, yellowish-green, velvety smut balls. When smut balls burst open, powdery dark green spores are released [57]. The infection of *U. virens* is favored by high relative humidity (>90%) and temperatures between 25 and 30°C [58]. Rainfall, high humidity, and soils with high nitrogen content during flowering also favors disease development [59]. Reports on the effect of rainfall are conflicting, high disease intensity has been attributed to rainfall at heading, but the opposite (low rainfall favoring the disease) has also been reported [60]. The fungus attacks some of the weed species that commonly occur in rice fields and may also serve as sources of inoculums [57].

The pathogen specifically infects rice flowers and later transforms the grains into smut balls. Initially the balls are slightly flattened, yellow in color covered with a thin membrane. As the smut ball matures, it increases in size converting yellowish-green to green or greenish-black. At final maturity, the entire smut ball looks dark black in color with rough surface. Very few grains in the panicle or all the grains may convert into smut balls (**Figures 9–12**).

2.3.2 Integrated management

Select seeds which are free of smut balls for sowing. Avoid excess nitrogenous fertilizer application. Follow alternate wetting and drying of the field to avoid moisture build-up which helps in disease incidence. Remove and destroy infected panicles, crop debris after harvest. Identification of the stable and durable sources of resistance is always been the better option in disease management [56]. Large number of varieties have been screened and identified for their resistance or tolerance to false smut disease under artificial/natural inoculation conditions [60].

Figure 9.
Yellow ball covered with thin membrane.

Figure 10.
The ball converting to greenish-yellow.

Figure 11.
The ball converted to black sori.

Figure 12.
Heavily infected panicle.

Artificial inoculation is still not dependable which is the major obstacle in screening large number of varieties under artificial conditions [56, 61]. Phenotyping for false smut resistance has been taken by many researchers in India, Philippines, China, Bangladesh, Pakistan and other countries by following standard evaluation system (SES) scale of IRRI (2002). In India, Kaur et al. [62] identified some of the hybrids namely VNR-211, GK-5025, HRI-140, IRH-74, PRSH-9018, KPH-467, RH-10428, 27P64 and KRH-4 showing complete resistance to false smut. Screening and identification of genotypes and QTL mapping is been carried out by pioneer rice research institutes in India such as National Rice Research Institute, Cuttack (Odisha), Indian Institute of Rice Research (IIRR), Hyderabad (Telangana), Punjab Agricultural Universty (PAU), Ludhiana (Punjab) and Indian Agricultural Research Institute (IARI), New Delhi. Identification of quantitative trait loci (QTL) and utilization in resistance breeding program is atmost priority in management of this disease [63]. Other management options like Seed treatment should be followed strictly: use carbendazim 50% WP @ 2 g/kg or biological control agents like *Trichoderma* or *Pseudomonas* talc based formulations @ 8–10 g/kg of seeds. Fungicidal spray with Propiconazole 25EC @ 500 ml/ha or copper hydroxide 77%WP @ 1.25 kg/ha at boot leaf initiation stage. Repeat the above spray at 50% flowering stage. Biological control of false smut disease has been successful with the strains of *Bacillus subtilis* in solution of Validamycine [64, 65].

2.4 Grain discoloration/dirty panicle disease

Pathogen: *Drechslera Oryzae, Sarocladium oryzae, Alternaria padwickii, Curvularia spp., Epicoccum sp., Fusarium moniliforme, Aspergillus spp.*

2.4.1 Identification symptoms

Seed discoloration has time and again proved to be a major recurring issue in the Indian coastal regions haltering the levels of desired grain production. Discoloration, though a minor disease in nature, has assumed a greater importance in agriculture owing to effect of global warming in production due to unpredictable hailstorms, delayed or low levels of rainfall, higher temperature and humidity levels after fowering stage. This has further led to relaxation in the procurement norms by the agencies. Grain discoloration serves as a visible indicator of seeds having lower quality in association with [66] microorganisms. This malady has been a prime retardant in the post-harvesting of paddy grains. The minimum threshold for the procurement of discolored kernels of paddy crop is kept at 3%, and all the samples above that are rejected. Several biotic and abiotic factors are responsible for discoloration of rice seeds. Glume discoloration is term for the alteration in color of mature seed from its original color, and seed possesses series of problems in seed certification programme. This disease though minor in nature reduces the vigor and yield of the crop and causes grain discoloration at maturity, thus reducing the economical and marketable value of the crop. The prevalence of monoculture and year around growth of only economical crops like paddy and wheat have led to seed discoloration gradually turning out to be a major problem [67]. This is also due to the many pests and pathogen that are common to both the crops. Since rice is a crop of immense importance, there is an increased need for research and development on various fungal fora that can afect the vigor, yield, morphology and constitution of the newly introduced higher yielding and aromatic rice cultivars. The reason for such discoloration whether pathological and/or non-pathological is not always clearly understood. In most of the cases, discolored rice seeds are frequently associated with micro-organisms, mainly fungi, though sometimes it also occurs due to insect bite and physiological or genetic reasons. Attempts have been made in the past to identify causal agents causing seed discoloration and also to control them by the use of chemicals. Grain discoloration is the early indication of poor yield and quality which leads to reduced market value. The discolored grain suffers more infection during storage also due to the development and infection of many storage fungi [68, 69]. Storage fungi such as *Aspergillus* spp. and *Fusarium* spp. produce deadly micotoxins which are detrimental to the human and animal health. Fungal pathogens such as *Alternaria alternata* (cause ashy gray discoloration) and *Helminthosporium oryzae* (cause black dicolration with dark brown spots on seeds) found mostly on the seed coat and endosperm region of the seeds. *Curvularia geniculata* which caused eye shaped spots, *Fusarium* spp. (*Fusarium oxysporum, F. moniliformae*) are responsible for pink discoloration and *Sarocladium oryzae* causes light brown discoloration. All these fungi found in embryo, seed coat and endosperm of the seeds [70]. Maximum colonization of seed borne fungi was observed in seed coat (0–3.30%) and endosperm (0.1–1.65%) as reported by Halgekar and Giri [71]. The losses due to grain discoloration were estimated approximately about 20–25 percent [72]. In our previous studies we reported that, the grain discoloration incidence ranged from 25 to 92% in different rice genotypes [73]. There may be a number of factors responsible for the disease starting from the varietal susceptibility, changed climatic conditions and agronomic managements which ultimately leads to the increased incidence of the disease with more damage.

The disease is an important emerging problem in all the rice growing seasons. The type of infection may be external contamination by saprophytic pathogens or internally seed born where pathogens infects endosperm. There are different kinds of like ashy gray discoloration, black discoloration, pink discoloration and light brown discoloration based on the pathogens involved (**Figures 13–16**). The infected grains convert into dirty discolored grains which have numerous spots and lesions on them. Infection in field may transfer to storage where the pathogens multiply and produces some of the harmful toxins. The early infection in field leads to complete yield losses due chaffiness of the grains. The severe infection leads to rotting of the panicles leads to qualitative and quantitative losses.

2.4.2 Integrated management

Use disease free seeds. The seeds should also be free from damage, chaffiness and other deformities. Treat the seeds with carbendazim 50% WP @ 2 g/kg or biological control agents like *Trichoderma* or *Pseudomonas* talc based formulations @ 8–10 g/kg of seeds. At tillering and pre-flowering stage, spraying of carbendazim 50% WP @500 g/ha or copper oxychloride 50% WP @ 1 kg/ha will control the disease effectively. At boot leaf initiation stage, spray with Mancozeb 75% WP @ 1 kg/ha or Iprobenphos 48% EC @ 500 ml/ha or Carbendazim 50% WP @250 g/hac.

Figure 13.
Infected panicle.

Figure 14.
Discolored grains.

Figure 15.
Severe infection of the grains.

Figure 16.
Chaffy grains.

During grain maturity stage, spray with Mancozeb 75% WP @ 1 kg/ha or Iprobenphos 48% EC @ 500 ml/ha. Biological control agents like *Trichoderma atroviride* or *Bacillus amyloliquefaceance* talc based formulations @10 g/lit of water at grain maturity stage.

2.5 Early seedling blight

Pathogen: *Sclerotium oryzae, Sclerotium rolfsii.*

2.5.1 Identification symptoms

The disease is one of the important emerging disease generally appears in nursery beds during cold season (winter months). Initial symptoms starts as wilting or drooping of the seedling in the nursery bed. Later, white mycelia mat of fungus developes on soil surface covering the seedlings (**Figures 17** and **18**). The fungal mat covers almost both root and shoots. As the disease progresses, development of white mustard shaped sclerotial bodies seen which later turn to dark brown hard structures. The disease may appear sporadically in nursery or covers entire nursery. Severe infection leads to death of the seedlings.

Figure 17.
Development of white sclerotial bodies of the fungus.

Figure 18.
Drying and death of the seedlings in nursery.

2.5.2 Integrated management

The disease is seasonal and do not have any resistant varieties. Follow clean cultivation: do not allow the nursery to in wet condition, always allow stand in water. Remove and destroy the infected seedlings. Drench with Carbendazim 50% WP @ 3 g/lit or Carbendazim (12%) + Mancozeb (63%) WP@ 2 g/lit of water.

2.6 Narrow brown leaf spot

Pathogen: *Cercospora janseana.*

2.6.1 Identification symptoms

Narrow brown leaf spot symptoms and disease cycle Cercospora janseana causes narrow brown leaf spot of rice [74]. Symptoms of narrow brown leaf spot (NBLS) disease include long cylindrical dark brown spots with dark margins and grayish centers with or without chlorosis. Lesions range from 1 to 10 mm x 1–1.5 mm on leaves and 15–45 x 1–2 mm on mid-ribs and leaf sheaths [74]. Morphology of symptoms varies with the susceptibility of the cultivar. On resistant cultivars, symptoms are long, narrow lesions that sometimes do not develop fully. In susceptible cultivars, spots are broad and necrotic. Initially, dark spots develop on the leaf lamina and later on the leaf mid-vein, leaf sheath, panicle, seed coat and glumes. Symptoms appear late in the season on all leaves regardless of age. NBLS causes a premature ripening of the grains, reduces yield quantity, and grain milling quality [75]. The disease cycle begins when *C. janseana* enters the plant tissues through stomata, establishes beneath the stomata in the parenchyma cells, and spreads longitudinally in intercellular. Upon development, conidiophores emerge through the stomata. Preliminary studies have shown that 30 or more days are required to develop symptoms after inoculation [76]. This long latent period may be the probable reason of late appearance of symptoms during the season even though infection occurs at early plant developmental stages. The initial source of inoculum appears to be from *C. janseana* that has survived on residues of previous rice crops, infected seeds, and seasonal weeds [77].

The disease is emerging as a major problem in wet periods. The disease causes severe necrosis of the leaf tissue and drying and death of the leaves. Initially, short, linear, brown lesions appears on leaf blades later extends to leaf sheath, pedicels and glumes. The lesions are 2–10 mm long and 1 mm wide in size (**Figures 19** and **20**). The size of the lesions depends on varietal susceptibility. The resistant cultivars have narrower and darker lesions whereas; the susceptible cultivars have wider and light brown. The lesions coalesce each other giving dried and burning appearance to field [78].

2.6.2 Integrated management

Use of resistant varieties should be the top priority in managing this disease. Mechanical control measures like removal and destruction of weeds, balanced nutrients especially Potasium is important. Foliar spray of Carbendazim 50% WP @ 500 g or Mancozeb 75% WP @ 2 kg or Hexaconazole 5% SC @ 500 ml/ha.

Figure 19.
Elongated, broad and light brown lesions on susceptible variety.

Figure 20.
Severely infected field.

3. Conclusion

Rice cultivation is subject to various constraints including environmental conditions, socio-economic status of the farmers, adequate resources availability. Change in cultivation practices like practicing intensive agricultural practices to enhance yield, the crop is suffering from a number of biotic constraints such as pest and diseases. In order to feed the ever-growing global population, it is necessary to manage the diseases to avoid losses. Over the decades, disease and pest management in rice relied primarily and most importantly on resistant varieties and chemical pesticides. But continuous cultivation of a single variety resulted in the familiar "Bom and Burst" cycle where most popular and high yielding variety will become highly susceptible to certain pest and diseases. This is mainly due to the shift in pathogen/pest dynamics, virulence level and population build-up. In several cases continuous use of chemical pesticides had lead to development of resistance, pest resurgence etc..,. on the other hand, a number of minor diseases have emerged as major problems which have created new challenges to rice cultivation. So, it is most important to address these emerging issues urgently by integrated management practices. Proper agronomic management measures, nutrient management is most essential. Developing resistance should be prioritized in tackling these emerging problems. Use of molecular tools like MAS, gene pyramiding, RNAi, Gene silencing, QTL mapping and CRISPR Cas9 technologies should be widely adopted. Used of biological control agents and their products is ecofriedly approach. At last, chemical pesticides should be an option not priority.

Author details

Raghu Shivappa[1*], Devanna B. Navadagi[2], Mathew Seikholen Baite[1], Manoj Kumar Yadav[1], Prabhukarthikeyan S. Rathinam[1], Keerthana Umapathy[1], Prajna Pati[3] and Prakash Chandra Rath[1]

1 Crop Protection Division, ICAR-National Rice Research Institute (NRRI), Cuttack, Odisha, India

2 Crop Improvement Division, ICAR-National Rice Research Institute (NRRI), Cuttack, Odisha, India

3 Department of Entomology, Institute of Agricultural Sciences, SOA Deemed to be University, Bhubaneswar, Odisha, India

*Address all correspondence to: raghurm531@gmail.com; s.raghu@icar.gov.in

IntechOpen

© 2021 The Author(s). Licensee IntechOpen. This chapter is distributed under the terms of the Creative Commons Attribution License (http://creativecommons.org/licenses/by/3.0), which permits unrestricted use, distribution, and reproduction in any medium, provided the original work is properly cited.

References

[1] Agricultural Statistics at a Glance. Government of India Ministry of Agriculture & Farmers Welfare Department of Agriculture. Cooperation & Farmers Welfare Directorate of Economics and: Statistics; 2020

[2] Sekhara K. Trends in Area, Production and Productivity of Paddy Crop: an Overview. International Journal of Humanities and Social Science Invention. 2019;**8**(1):50-58

[3] Bhattacharjee P, Singhal RS, Kulkarni PR. Basmati rice: A review. International Journal of Food Science and Technology. 2002;**37**(1):1-2

[4] Jena M, Pandi GGP, Adak T, Rath PC, Gowda B, Patil NKB, et al. Paradigm shift of insect pests in rice ecosystem and their management. Oryza. 2018;**55**(Special Issue):82-89

[5] Pathak H, Nayak AK, Jena M, Singh ON, Samal P and Sharma HG. Rice Research for Enhancing Productivity, Profitability and Climate Resilience. 2018. Pp.552

[6] Wassmann R, Jagadish SVK, Heuer S, Ismail A, Redona E, Serraj R, et al. Climate Change Affecting Rice Production: The Physiological and Agronomic Basis for Possible Adaptation Strategies. Adv. Agron. 2009a;**101**:59-122

[7] Wassmann R, Jagadish SVK, Sumleth K, Pathak H, Howell G, Ismail A, et al. Regional vulnerability of climate change impacts on Asian rice production and scope for adaptation. Adv. Agron. 2009b;**102**:91-133

[8] Pathak H. Greenhouse gas emission from Indian agriculture: Trends, drivers and mitigation strategies. Proc. Indian Natn. Sci. Acad. 2015;**81**(5):1133-1149

[9] Pathak H. Samal P and Shahid MRevitalizing rice-systems for enhancing productivity, proitability and climate resilience. In: Pathak H et al., editors. Rice Research for Enhancing Productivity. ICAR-National Rice Research Institute, Cuttack, Odisha: Proitability and Climate Resilience; 2018b. pp. 1-17

[10] Pathak H, Nayak AK, Maiti D, Kumar GAK, Reddy JN, Rath PC, Swain P and Bhagawati R (Eds.) National Rice Research Institute: Activities, Achievements and Aspirations. ICAR-National Rice Research Institute, Cuttack, Odisha, (2019a.) p264 + viii, ISBN: 81-88409- 08-1

[11] Pathak H, Pradhan SK, Mondal B, Jambhulkar NN, Parameswaran C, Tripathi R, et al. Assessing area, production and return with rice varieties of NRRI. Cuttack. Oryza. 2019b;**56**:169-173

[12] Pathak H, Voleti SR, Meera Shaik N, Tripathi R, Sailaja B, Nayak AK, Subba Rao LV, Mondal B, Reddy JN and Mohapatra T. Reorientation of All India Coordinated Crop Improvement Projects: The Case of Rice.NRRI Research Bulletin No. 18, ICAR-National Rice Research Institute, Cuttack 753006, Odisha, India. (2019c) pp 20+viii.

[13] Prakash A, Rao J, Berliner J, Mukherjee AK, Adak T, Lenka S, et al. Emerging pest scenario in rice in India. Journal of Applied Zoological Researchers. 2014;**25**(2):179-181

[14] Guru-Pirasanna-Pandi Govindharaj, Basana Gowda, R. Sendhil, Totan Adak, S. Raghu, Naveenkumar Patil, Annamalai Mahendiran, Prakash Chandra Rath, G.A.K. Kumar, Christos A. Damalas. Determinants of rice farmers' intention to use pesticides in eastern India: Application of an extended version of the planned

behavior theory, Sustainable Production and Consumption. (2021). 26: 814-823.

[15] Aashish Kumar Anant, Govindharaj Guru-Pirasanna-Pandi, Mayabini Jena, Gajendra Chandrakar, Parameshwaran Chidambaranathan, Raghu S, G Basana Gowda, Mahendiran Annamalai, Naveenkumar Patil, Totan Adak, Naveenkumar Ramasamy, Prakash Chandra Rath. Genetic dissection and identification of candidate genes for brown planthopper, *Nilaparvata lugens* (Delphacidae: Hemiptera) resistance in farmers' varieties of rice in Odisha, Crop Protection. (2021). 144:105600.

[16] Raghu S, Yadav MK, Prabhukarthikeyan SR, Baite MS, Lenka S, Jena M. Occurrence pathogenicity characterization of *Fusarium fujikuroi* causing rice bakanae disease from Odisha and *in vitro* management Oryza. (2018).55: 214-223, doi 10.5958/2249-5266.2018.00025.5

[17] Sain M, Prakash A. Major insect pests of rice and their changing scenario. AZRA, India: In Rice Pest management; 2008. pp. 7-17

[18] Yadav MK, Aravindan S, Mukherjee AK, Lenka S. Sheath rot: Emerging threat to rice production. Everyman's Science. 2015:286-288

[19] Phookan AK, Hazarika DK. Distribution of sheath rot (ShR) in six agroclimatic zones of Assa m. India. IRRN. 1992;**17**:16

[20] Singh M and Das BC. Screening of Aromatic rice (Joha) genotype against Sheath rot disease of Rice and its management under Field condition. International Journal of Advanced Research (2016). 4(4): 922-925.

[21] Bigirimana VP, Hua GKH, Nyamangyoku OI, Höfte M. Rice Sheath Rot: An Emerging Ubiquitous Destructive Disease Complex. Front. Plant Sci. 2015;**6**:1066. DOI: 10.3389/fpls.2015.01066

[22] Sakthivel N and Gnanamanickam SS. Appl. Environ. Microbiol. 1987;**53**:2056-2059

[23] CABI; EPPO, Sarocladium oryzae. [Distribution map]. Distribution Maps of Plant Diseases, 1997.December (Edition 2). Wallingford, UK: CAB International, Map 619.

[24] Alagarsamy G, Bhaskaran, R. Sheath rot of rice. In: Kannaiyan S, ed. Advances in Rice Pathology. (1987). Coimbatore, India: Tamil Nadu Agricultural Univiversity, 185-209.

[25] Shahjahan AKM, Harahap Z, Rush MC. Sheath rot of rice caused by Acrocylindrium oryzae in Louisiana. Plant Disease Reporter. 1977;**61**(4):307-310

[26] Mathur SC. Observations on diseases of dryland rice in Brazil, March 1981. International Rice Research Newsletter. 1981;**6**(6):11-12

[27] Ou SH. Rice diseases. Wallingford, UK; CAB International, (1985). 380 pp.

[28] MewTW and Gonzales PA. *HandbookofRiceSeedborneFungi*. (2002). Enfield, NH:SciencePublishers.

[29] Gams W, Hawksworth DL. Identity of *Acrocylindrium oryzae* Sawada and a similar fungus causing sheath-rot of rice. Kavaka. 1975;**3**:57-61

[30] Giraldo A, Gené J, Sutton DA, Madrid H, DeHoog GS and Cano J etal. Phylogeny of *Sarocladium* (Hypocreales). Persoonia (2015).34, 10-24.doi: 10.3767/003158515X685364.

[31] Mew TW and Gonzale P. Handbook of Rice Seedborne Fungi. (2002). Enfield. NH: Science Publishers.

[32] Sharma S, Sthapit B, Pradhanag B and Joshi K. Bacterial Sheath Brown Rot of Rice Caused by *Pseudomonas fuscovaginae* in Nepal,"in *Rice Cultivation*

in Highland Areas, (edsC.PoissonandJ. Rakotoarisoa) Proceedings of the CIRAD Conference, Antananarivo, Madagascar, 107-112.

[33] Sakthivel N. Sheath rot disease of rice:current status and control strategies,"in *Major Fungal Diseases of Rice:Recent Advances*, edsS. Sreenivasaprasad and R. Johnson (Dordrecht:Springer), (2001).271-283.

[34] Pearce DA, Bridge PD and Hawksworth DL. Species concept in *Sarocladium*, the causal agent in sheath rot in rice and bamboo blight,"in *Major Fungal Diseases of Rice:RecentAdvances*, edsS.Sreenivasaprasadand R. Johnson(Dordrecht:Springer) (2001). 285-292.

[35] Hemalatha R, Jebaraj S, Raja J, Raguchander T, Ramanathan A, Samiyappan R. et al. Employing a crude toxin preparation from *Sarocladium oryzae* as a molecular sieve to select sheath rot- resistant somaclones of rice. *J. Plant Biochem.Biotechnol.* 8, (1999).75-80.doi: 10.1007/BF03263062.

[36] Deepmala K, Rakesh Kumar B, Tiwari PK. Screening of Aromatic Rice Entries against Sheath Rot of Rice Under Field Condition. Trends in Biosciences. 2015;**8**(2):0974-0978 462-46

[37] Bashyal BM and Aggarwal R Molecular identification of *Fusarium* spp. associated with bakanae disease of rice in India. Indian Journal of Agricultural Sciences. (2013). 83:72-77.

[38] Bashyal BM, Aggarwal R, Sharma S, Gupta S, Rawat K, Singh D and Singh AK, Gopalakrishnan S. Occurrence, identification and pathogenicity of Fusarium species associated with bakanae disease of basmati rice in India.Eur. J.Plant Pathol. (2014)144:457-466. doi:10.1007/s10658-015-0783-8.

[39] Raghu S, Yadav MK, Prabhukarthikeyan SR, Baite MS, Lenka S and Jena M Occurrence, pathogenicity, characterization of *Fusarium fujikuroi* causing rice bakanae disease from Odisha and *in vitro* management. *Oryza* (2018). 55:214-223. DOI 10.5958/2249-5266.2018.00025.5.

[40] Zainudin NAIM. Razak AA and Salleh B Bakanae disease of rice in Malaysia and Indonesia: etiology of the causal agent based on morphological, physiological and pathogenicity characteristics. Journal of Plant Protection Research. 2008;**48**:475-485

[41] Fiyaz RA, Gopala Krishnan S, Rajashekara H, Yadav AK, Bashyal BM, Bhowmick PK, Singh NK, Prabhu KV and Singh AK. Development of high throughput screening protocol and identification of novel sources of resistance against bakanae disease in rice (*Oryza sativa* L.) Indian J. Genet. (2014). 74: 414-422. DOI: 10.5958/0975-6906.2014.00864.5.

[42] Singh R, Sunder S. Foot rot and bakanae of rice: an overview. Rev. Plant Pathol. 2012;**5**:565-604

[43] Rosales AM, Mew TW. Suppression of *Fusarium moniliforme* in Rice by Rice-Associated Antagonistic Bacteria. Plant Dis. 1997;**81**:49-52. DOI: 10.1094/pdis. 1997.81.1.49

[44] Amatulli MT, Spadaro D, Gullino ML and Garibaldi A. Molecular identification of *Fusarium* spp. associated with bakanae disease of rice in Italy and assessment of their pathogenicity. Plant Pathol.(2010). 839-844.

[45] Hayasaka T, Ishiguro K, Shibutani K and Namai T (2001) Seed disinfection using hot water immersion to control several seed-borne diseases of rice plants. J Phytopathol 67:26-32. https://doi. org/10.3186/jjphytopath.67.26.

[46] Lee YH, Lee MJ, Choi HW, Kim ST, Park JW, Myung IS, et al. Development

of in vitro seedling screening method for selection of resistant rice against bakanae disease. Res Plant Dis. 2011;**17**:288-294

[47] Singh R, Sunder S. Foot rot and bakanae of rice: an overview. Rev. Plant Pathol. 2012;**5**:565-604

[48] Matic S, Spadaro D, Garibaldi A, Gullino ML. Antagonistic yeasts and thermotherapy as seed treatments to control *Fusarium fujikuroi* on rice. Biol Control. 2014;**73**:59-67. DOI: 10.1016/j.biocontrol.2014.03.008

[49] Volante A, Tondelli A, Aragona M, Valente MA, Biselli C, Desiderio F, et al. Identification of bakanae disease resistance loci in japonica rice through genome wide association study. Rice. 2017;**10**:29. DOI: 10.1186/s12284-017-0168-z

[50] Fiyaz, RA, Yadav AK, Krishnan SG, Ellur RK, Bashyal BM, Grover N, Bhowmick PK, Nagarajan M, Vinod KK, Singh NK, Prabhu KV and Singh AK (2016) Mapping quantitative trait loci responsible for resistance to Bakanae disease in rice **9**:45. DOI: 10.1186/s12284-016-0117-2.

[51] Kim MH, Hur YJ, Lee SB, Kwon TM, Hwang UH, Park SK, et al. Largescale screening of rice accessions to evaluate resistance to bakanae disease. J Gen Plant Pathol. 2014;**80**:408-414

[52] Singh AK, Pophaly DJ. An unusual rice false smut epidemic reported in Raigarh District. Chhattisgarh. International Rice Research Notes. 2010;**35**:1-3

[53] Pannu PPS, Thind TS, Goswami S. Standardization of technique for artificial creation of false smut of rice and its management. Indian Phytopathology. 2010;**63**:234-235

[54] Laha GS, Prasad MS, Krishnaveni D, Ladhalakshmi D, Prakasam V. Production oriented survey. Directorate of Rice Research Rajendranagar. Hyderabad. 2013:45-48

[55] Ashizawa T, Takahashi M, Moriwaki J, Hirayae K. Quantification of the rice false smut pathogen *Ustilaginoidea virens* from soil in Japan using real-time PCR. European Journal of Plant Pathology. 2010;**128**:221-232

[56] Baite MS, Raghu S, Lenka S, Mukherjee AK, Prabhukarthikeyan SR, Jena M. Survey of rice false smut caused by *Ustilaginoidea virens* in Odisha. The Bioscan. 2017;**12**(4):2081-2085

[57] Atia MMM. Rice false smut (*Ustilaginoidea virens*) in Egypt. Journal of Plant Disease and Protection. 2004;**111**:71-82

[58] Bhagat AP, Prasad Y, Effect of irrigation on incidence of false smut of rice.Journal of Applied Biology. (1996); 6:31-32.

[59] Ladhalakshmi D, Laha GS, Singh R, Karthikeyan A, Mangrauthia SK, Sundaram RM. Isolation and characterization of *Ustilaginoidea virens* and survey of false smut disease of rice in India. Phytoparasitica. 2012;**40**:171-176

[60] Dodan DS, Singh SR. False smut of rice: present status. Agricultural Review. 1996;**17**:227-240

[61] Zhang Y, Zhang K, Fang A, Han Y, Yang J, Xue M. Specific adaptation of Ustilaginoidea virens in occupying host florets revealed by comparative and functional genomics. Nat. Commun. 2014;**5**:3849

[62] Kaur Y, Lore JS, Pannu PPS. Evaluation of rice genotypes for resistance against false smut. Plant Dis. Res. 2015;**30**(1):46-49

[63] Andargie M, Lia L, Feng A, Zhub X, Lia J. Mapping of the quantitative trait

locus (QTL) conferring resistance to rice false smut disease. Current Plant Biology. 2018;**15**:38-43

[64] Mohiddin FA, Bhat FA, Gupta V, Gupta D, Kalha CS. Integrated disease management of false smut of rice caused by Ustilaginoidea virens. Trends in Biosciences. 2012;**5**(4):301-302

[65] Brooks SA, Anders MM, Yeater KM. Effect of cultural management practices on the severity of false smut and kernel smut of rice. Plant Disease. 2009;**93**:1202-1208

[66] Chandramani B, Awadhiya GK. Assessment of percent grain discoloration in important rice varieties. Int J Curr Res Biosci Plant Biol. 2014;**1**(4):61-64

[67] Bala A, Pannu PPS. Status of seed discoloration of rice, mycoflora associated and its impact on seed health and quality of farmer saved seeds. Seed Res. 2017;**45**(2):131-135

[68] Bag MK. Mycoflora causing grain discoloration of rice and its effect on some yield components. J Mycol Pathol Res. 2010;**48**:149-152

[69] Raghu S, Baite MS, Patil NB, Sanghamitra P, Yadav MK, Prabhukarthikeyan SR, et al. Grain discoloration in popular rice varieties (Oryza sativa L) in eastern India, associated mycoflora, quality losses and management using selected biocontrol agents. Journal of Stored Products Research. 2020;**88**(2020):101682

[70] Sachan IP, Agrawal VK. Seed discoloration of rice, location of inoculum and influence on nutritional value. Indian Phytopathol. 1995;**48**:14e20

[71] Halgekar NY and Giri GK. Detection of seed borne fungi in rice. Environ. Ecol. (2015); 33.1599e1603.

[72] Ghose RLM, Ghatge M and Subramanyam V. Rice in India, second ed. ICAR, New Delhi. (1960); 474pp.

[73] Baite MS, Raghu S, Prabhukarthikeyan SR, Keerthana U, Jambulkar NN, Rath PC. Disease incidence and yield loss in rice due to grain discoloration. J. Plant Dis. Prot. 2019. DOI: https://doi.org/10.1007/s41348-019-00268-y

[74] Groth DE, Hollier CA, NBLS of Rice LSU AgCenter Pub 3105, (2010). Available at: http://www.lsuagcenter.com/NR/rdonlyres/BA9CBCB0-085B-4211-A422-A788382DE25C/74889/pub3105NarrowBrownLeafSpotHIGHRES.

[75] Zhou, X.G., Jo, Y.K., Disease management. In: Way, M.O., McCauley, G.M., Zhou, X.G., Wilson, L.T., Brandy, M. (Eds.), 2014 Texas Rice Production Guidelines. Texas Rice Research Foundation, (2014). pp. 44-57. Available at: https://beaumont.tamu.edu/eLibrary/Bulletins/2014_Rice_Production_Guidelines.

[76] Sah DN, Rush MC. Physiological races of Cercospora oryzae in the southern United States. Plant Dis. 1988;**72**:262-264

[77] Uppala S, Zhou XG. Field efficacy of fungicides for management of sheath blight and narrow brown leaf spot of rice. Crop Protection. 2018;**104**:72-77

[78] Mani KK. Study of Factors Affecting Growth and Development of Narrow Brown Leaf Spot of Rice Caused by Cercospora janseana (Racib.) O. Const. LSU Doctoral Dissertations. 2015:3568

Chapter 11

Potential Role of Rice Plant Growth Promoting Phylloplane and Rhizospheric Bacteria in Controlling *Xanthomonas oryzae* pv. *oryzae*

Md. Mahfujur Rahman, Md. Mostafa Masud, Muhammad Iqbal Hossain, Nur-E-Tajkia Islam, Md. Zahangir Alam, Md. Mamunur Rashid, Mohammad Ashik Iqbal Khan, Md. Abdul Latif, Krishna Pada Halder and Md. Rashidul Islam

Abstract

Rice is an important cereal worldwide and it occupies the top position among the cereals in Bangladesh. Rice plant suffers from around 32 diseases of which ten are major in Bangladesh at present. Among the diseases, Bacterial Blight (BB) caused by *X. oryzae* pv. *oryzae* (*Xoo*) considered as a most destructive disease occurs in both rainfed and irrigated seasons of Bangladesh. BB causes considerable yield loss varies from 30 to 50% depending on the outbreak. It is also an important disease in most of the South and Southeast Asian countries. To develop environment-friendly sustainable management approach against BB of rice, in total sixty three plant growth promoting bacteria were identified from rice phylloplane and rhizosphere that are antagonistic to *X. oryzae* pv. *oryzae* during boro and aman seasons 2018 and 2019.These bacterial species inhibited the growth of *X. oryzae* pv. *oryzae in vitro* by 20.83 to 76.19%. These bacterial isolates were identified by sequencing of PCR products of 16SrDNA belonging to the genera mostly *Pseudomonas*, *Bacillus* and *Serratia*. Out of these bacterial species, 48 bacterial species were found as positive for IAA production, all 63 bacterial species were found positive for siderophore production and 48 were found capable to solubilize insoluble phosphate. Based on growth inhibition of *X. oryzae* pv. *oryzae* in *in vitro*, thirty two bacterial species were selected for plant growth promotion assessment and evaluation of net house and field efficacy in controlling BB of rice. These bacterial species were formulated using talcum powder which was viable for at least three months post formulation. Assessment of plant growth promoting determinants revealed that all 32 bacterial species identified in this study enhance the growth of rice plants as measured by root and shoot length and and reduced the BB severity in susceptible rice cultivar significantly as compared with untreated control.

Keywords: Rice, Plant growth promoting phylloplane and rhizospheric bacteria, control, *X. oryzae* pv. *oryzae*

1. Introduction

Rice (*O. sativa* L.) suffers from 32 diseases of which in Bangladesh 10 has been known as dreadful diseases [1]. Among the diseases three bacterial diseases are frequently occurred in Bangladesh. Among these three diseases, Bacterial Blight (BB) caused by *X. oryzae* pv. *oryzae* (*Xoo*) considered as a most destructive disease occurs in all Agro Ecological Zones (AEZs) of Bangladesh and mostly in two rice growing seasons namely viz. raifed and irrigated [2–4] and cause severe yield loss. In Japan, India and Bangladesh due to this devastating disease around 50%, 60% and 30% yield loss was observed [5], respectively in the highly infected rice fields. It is also a crucial disease in most of the South and Southeast Asian countries [6]. Bacterial blight (BB) is disease associated with several growth phases of rice plant showing either "Kresek" (acute wilting of young plants) symptoms and "leaf blight" (straw color blighted area with weavy margin) symtoms [7]. Excess amount of nitrogenous fertilizer in rice varieties (HYV) facilitates the emergence of this disease and its severity in the field [8–12]. In Bangladesh different pathogenic [13, 14] and genetic variability [15] have been detected and those were excessively perilous for rice [16].

Chemical fungicides (copper compounds, other chemicals and antibiotics) are not effective in controlling this disease [17]. However, control measures are including chemical, cultural, host resistance, genetic modification methods, among them cultural practices are not also effective in all circumstances as well as no fruitful chemical control and commercial product was found in this tropical climatic area which can be suppressed this disease nicely [18, 19]. Moreover, using antibiotics, toxic residues and chemicals have several limitations against BB of rice [20]. Apart from that, the uses of host resistance genes are used, in case of breeding single gene (Xa4) are manifested ineffective BLB management due to sub-populations [21].

Thus, biological control alleviates costs and it also serves as an environment friendly approach to mitigate this devastating threat [22], besides, the application of biological strains of PGPB would be the fullest alternative way of minimizing chemical pesticides, fertilizer and environmental pollution [23]. PGPB plays a crucial role in developing immunization in plants body, ISR is triggered by PGPB which is a signaling pathway while SAR mainly dependent on salicylic acid triggering a induced resistance by a particular infection, However, it is observed that ISR requires salicylic acid (SA) and ISR demands ethylene (ET) and jasmonic acid (JA) signal pathways [24] and both of these are triggered latent resistance mechanism subsequently after inoculation [25]. In recent years, application of PGPB in the field has been evaluated as an inducer showing systematic resistance [26, 27, 38]. Due to fruitful leaf colonization, quick growth, normal application procedure of *L. antibioticus* have been utilized as a bio control agents against Xoo [28]. *Bacillus* spp. also found effective in quelling BLB of rice under greenhouse condition [29]. According to [30], *S. globisporus* have been effective against rice blast. Sheath blight disease was alleviated by using a few biofilm and surfactant delivering strains of *Bacillus subtilis* [31]. Amalgamation of *B. subtilis* and *Streptomyces philanthi* were biologically effective againstrice sheath blight adding with chemical fungicides [32]. HCN (Hydrogen cyanide) played an effective role inhibiting the surges of *M. oryzae* as well as developing its bio control agents against blast of rice [33]. These antagonistic bacteria have the ability to subvert plant pathogens by releasing chemicals such as glucanases, proteases and chitinases, siderophores [34]. Rice disease can be controlled by the antagonistic strains of *Bacillus* and *Pseudomonas* spp. up to 90% based on what kind of strains are used [35]. When systemic resistance is exposed is called as ISR, and conversely, by other phenomenon is called SAR [36]. No necrosis manifested while ISR developed by PGPB [36]. Last few decades, PGPB have been showing as a systematic resistance in the field [26, 27, 37, 38]. ISR demands three systematic pathway which are jasmonic acid (JA), ethylene (ET), salicylic acid (SA) signaling pathways [24]. PGPB can induce priming by the release of volatiles. For instance, *Bacillus subtilis* GBO3 induces a signaling pathway that is independent of salicylic acid (SA), jasmonic acid (JA) and the Npr1 gene (SA insensitive or nonexpresser of PR genes), yet it requires ethylene [39]. Priming offers an energy cost efficient strategy, enabling the plant to react more effectively to any invader encountered by boosting infection induced cellular defense responses [40, 41]. The increased levels of defense related enzymes during ISR are known to play a crucial role in host resistance [42, 43], reported that *Pseudomonas fluorescens* have been used as a bacterial antagonists against BLB of rice. A plentiful of bacterial strains *B. cereus, B. pasteurii, pumilus, Bacillusmycoides, B. amyloliquefaciens, B. sphaericus, B. pumilus, B. cereus have been effective in reducing disease resistance upon using* ASM (acibenzolar-S-methyl) [39, 43].

Species such as *Bacillus* spp. which showed ISR are radically linked to plant growth modification promotion [39] and this strains have been manifested resistance activity against a number of plant diseases studied by several researcher [44–50]. In rice, limited number of studies found discussing on induced resistance, the main theme of PGPB also includes production of growth hormones such as IAA and IA (inorganic phosphate) (Khan et al., 1997 and [51]), and zinc solubilization [52], atmospheric nitrogen [53]. Plant health also maintained by PGPB by producing ISR, siderophores and competition [54] as well as mitigate plant pathogens by developing enzymes such as antibiotics, proteases, glucanases and chitinases [34]. In both lab and field conditions PGPB bacteria are significantly reduced plant disease incidence, among them *Bacillus* and *Pseudomonas spp*. suppressed diseases up to 90% based on rice variety and types of pathogens [35]. ISR (Induced systemic resistance) is an environment friendly option for plant disease control because it initiates defense related genes and enzymes in host plant through inoculated bacteria to reduce disease incidence [29]. Bacterial Blight pathogen, however, radical information on rice PGPB which can be used as both biopesticide and biofertilizer is not disclosed in Bangladesh. Besides, more investigation needs to be executed from other dimension to completely minimize this deadly disease.

2. Materials and methods

2.1 Isolation and identification of bacteria from rice phylloplane and rhizosphere

2.1.1 Plant sample collection

To isolate the bacteria from rice phylloplane and rhizosphere, the healthy rice plants with root system and soils of different rice cultivars were collected from 40 districts representing 30 Agroecological Zones (AEZs) of Bangladesh from the vicinity of BB infected rice plants during boro and aman season, 2018 and 2019 at maximum tillering stage to pre-ripening stage. Then the rice plant samples were brought into the laboratory in labeled polybags.

2.1.2 Isolation and purification of bacteria

The phylloplane bacteria were isolated using washing method. Freshly harvested 2nd, 3rd, 4th leaves were vortexed in sterile saline solution for 12 minutes with two or three brief intervals. Then 100 μl solution was placed at the center of Luria

Bartani (LB) or King's B agar plate and the solution was spread with glass spreader. The inoculated plates were incubated for 3–5 days at room temperature. After incubation of the inoculated plates, bacterial colonies appeared with various types of colors. Then the bacterial colonies were selected and isolated depending on their color and were streaked on LB media separately. Again the streaked LB plates were incubated at room temperature for 2 days. For isolation of antagonistic bacteria from rhizosphere, 1 g roots with rhizospheric soils were taken and then it was shaken with 100 ml sterile water for about 10–15 min to obtain soil suspension. Isolation of bacteria were carried out from rhizospheric soil by serial dilution technique up to 10^{-5} to 10^{-6} using LB (Luria Bertani) medium. Then the solution was placed at the center of Luria Bartani (LB) or King's B agar plate and the solution was spread with glass spreader. The inoculated plates were incubated for 3–5 days at room temperature. After incubation of the inoculated plates, bacterial colonies appeared with various types of colors. Then the bacterial colonies were selected and isolated depending on their color and were streaked on LB media separately. Again the streaked LB plates were incubated at room temperature for 2 days.

2.2 Assay of antagonism of bacterial spp. to *X. oryzae* pv. *oryzae* by dual culture method

Antimicrobial activity of antagonistic strains of *Pseudomonas* spp./*Bacillus* spp. were determined by agar diffusion technique method [55] with some modifications. Antagonistic bacterial suspension was spot inoculated (5 μl of 10^8 CFU/ml) at three places on the NBY plates that were prior inoculated with *X. oryzae* pv. *oryzae* cell suspension (10^8CFU/ml $\sim$ optical density: 0.3). The plates were incubated for 7 days post inoculation at 28°C. Then *X. oryzae* pv. *oryzae* growth inhibition by the antagonistic bacterial isolates indicated by clear halo zones were measured with a ruler in mm. The percent growth inhibition of *X. oryzae* pv. *oryzae* by bacterial isolates was calculated as follows:

$$\text{Growth inhibition } (\%) = \left[\text{Total diameter } (\text{Colony diameter} + \text{clear halo zones}) - \text{Colony diameter}\right] \text{ x } 100/\text{Total diameter} \tag{1}$$

2.3 Assessment of plant growth promoting determinants of bacteria antagonistic to *X. oryzae* pv. oryzae

Active isolates with antagonistic potential against *X. oryzae* pv. *oryzae* were further evaluated for their ability to produce plant growth promoting determinants viz. siderophore production, Indole acetic acid (IAA) production and phosphate solubilization capability as follows:

2.3.1 Assay for siderophore production

Siderophore productions by antagonistic bacterial isolates were tested qualitatively as described by Alexander and Zuberer [56]. 5 μl of antagonistic bacterial cell suspension (5 × 108 CFU/mL) was spot inoculated on Chrome azurol S (CAS) agar plate. The plates were then incubated at 30°C for 5 days. Development of yellow-orange halo zone around the bacterial growth was considered as positive for siderophore production. Experiment was performed with a completely randomized design with 3 replications. CAS agar was prepared from 4 solutions. Solution 1 (Fe-CAS indicator solution) was prepared by mixing 10 mL of 1 mmol $L^{-}1$ FeCl3.6H2O (in 10 mmol $L^{-}1$ HCl) with 50 mL of an aqueous solution of CAS (1.21 g $L^{-}1$). The resulting dark purple mixture was added slowly with constant stirring to 40 mL of aqueous solution of hexadecyl trimethyl ammonium bromide (1.821 g $L^{-}1$). The yielded of dark blue solution which was autoclaved, then cooled to 50°C. The entire reagent was freshly prepared for each batch CAS agar. Solution 2 (buffer solution) was prepared by dissolving 30.24 g of piperazine-N, N-bis (2-ethane sufonic acid) (PIPES) in 750 mL of salt solution containing 0.3 g K_2PO_4, 0.5 g NaCl and 1.0 g NH_4Cl. The pH was adjusted to 6.8 with 50% (w/v) KOH, and water was added to bring the volume 800 mL. The solution was autoclaved after adding 15 g of agar then cooled to 50°C. Solution 3 contained 2 g glucose, 2 g mannitol, 493 mg $MgSO_4$.7H2O, 11 mg $CaCl_2$, 1.17 mg $MnSO_4$.2H2O, 1.4 mg H_3BO_3, 0.04 mg $CuSO_4$.5H2O, 1.2 mg $ZnSO_4$.7H2O, 1.0 mg $NaMoO_4$.2H2O in 70 mL water, autoclaved, cooled to 50°C. Solution 4 was 30 mL filter sterilized 10% (w/v) casamino acid. Finally, solution 3 added to solution 2 along with solution 4, solution 1 was added last, with sufficient.

2.3.2 Assay for indole acetic acid (IAA) production

IAA production of antagonistic bacterial isolates were carried out as per the procedure described by Patten and Glick [57]. Every isolate was grown in LB media supplemented with (0.005%) L-tryptophan and incubated in shaker at 30°C with 160 rpm for 48 h. Then bacterial culture was centrifuged at 8000 rpm for 15 min and 1 mL culture filtrate was mixed with 4 mL salkowski's reagent (1.5 mL FeCl3.6H2O 0.5 M solution in 80 mL 60% H2SO4) and the mixture was incubated at room temperature for 30 min, presence of pink color indicate qualitatively that isolate produced IAA. Formation of pink color indicated the presence of indoles [58].

2.3.3 Phosphate solubilization assay by antagonistic bacterial isolates

Phosphate solubilization was determined according to the method of Azman et al. [59]. Sterile filter papers (5.0 mm) were soaked in antagonistic bacterial cell suspension (5 × 108 CFU/mL) was dispensed using pipette onto sterile filter paper (6.0 mm) that was placed on National Botanical Research Institute's phosphate (NBRIP) agar plate (Glucose (10 g/L), Ca3 (PO4)2 (5 g/L), MgCl2.6H2O (5 g/L), MgSO4.H2O (0.25 g/L), KCl (0.2 g/L), (NH4)2SO4 (0.1 g/L), Bacteriological Agar (15 g/L) [60]. The plates were then incubated at 28°C for 7 days. Phosphate solubilization was assessed by observing the clear halo zone. The experiment was performed with a completely randomized design (CRD) with 3 replications.

2.4 Identification of selected plant growth promoting antagonistic bacterial isolates by sequence analyses of 16SrDNA

2.4.1 Extraction of genomic DNA

Bacterial culture from NA media was transferred in LB broth and shaken for 18 h at 28°C. Then genomic DNA of antagonistic bacteria was extracted according to Wizard® Genomic DNA purification Kit (Promega, Madison, USA). Obtaining the DNA pellet was rehydrated by adding 25 μL DNA rehydration solution and kept it overnight at 4°C. Finally the genomic DNA samples of the isolates were preserved at −20°C for further use.

2.4.2 Primers and PCR conditions

To identify the antagonistic bacterial isolates, the primer sets 27F (5′-AGA GTT TGATCM TGG CTC AG-3′) and 1518R (5′-AAG GAG GTG ATC CAN CCR CA-3′) specific to 16SrDNA were used for amplification of 16SrDNA from the prepared genomic DNA template [61]. The PCR conditions were as follows: initial denaturation at 95°C for 5 min, 35 cycles denaturation at 94°C for 1 min, annealing at 55°C for 1 min, extension at 72°C for 2 min and finally a 7 min extension at 72°C. PCR products were visualized by electrophoresis on 1.0% agarose gel containing 0.5% of ethidium bromide using a Gel documentation System after separating the PCR products in the agarose gel for 50 min at 80 volt.

2.4.3 Sequencing of PCR products

A partial nucleotide sequencing of 16SrDNA was performed from amplified PCR products using primers 27F (5′-AGA GTT TGATCM TGG CTC AG-3′) in the Macrogen Lab, South Korea via Biotech Concern Bangladesh. The sequencing was done directly from PCR products according to the standard protocols for the ABI 3730xl DNA genetic analyzer (Applied Biosystems, Foster City, CA, USA) with BigDye® Terminator v1.1 and 3.1 Cycle Sequencing Kits.

2.4.4 Processing of sequence data

The sequencing data were processed and nucleotide sequence data was exported using Chromas software version 2.6.4.The quality of nucleic acid sequences was evaluated using Chromas (Version 2.6) software to avoid the use of low quality bases.

2.4.5 Analyses of nucleotide sequences

The nucleotide sequences were analyzed using online bioinformatics tools. The DNA sequences of 16Sr DNA of the bacterial isolates were compared with 16Sr DNA of the bacterial spp. and the sequences of ITS region of the fungal isolates were compared with ITS region of the fungal spp. that were available in the NCBI database using Basic Local Alignment Search Tool (BLAST) algorithm to identify closely related sequences (https://blast.ncbi.nlm.nih.gov/Blast.cgi).

2.5 Formulation of plant growth promoting antagonistic bacterial species

The pure cultures of thirty two selected potential bacterial antagonists were grown on LB agar medium for 24 hrs. Then the bacterial isolates were transferred in LB broth for about six hours by taking a loopful of bacteria from the LB agar plate. After that the liquid culture was then centrifuged and resuspended the pellet in previously prepared 200 ml peptone broth aimed to fortify the carrier materials. This culture broth was then grown for another two hours with shaking. After that 5 ml of sterile 100% glycerol was added to this 200 ml culture. Then the cultures of the bacterial antagonists (200 ml fortified with 1% peptone and 1% glycerol) were added to the mixture of 500 g talcum powder amended with 5 g carboxy methyl cellulose (CMC) and 7.5 g Calcium carbonate which were autoclaved for two consecutive days at 121°C under 15PSI pressure for 30 min each. The formulations were then being dried overnight in the tray. After that the formulations were powdered with hand, the formulated bacterial antagonists were packed in plastic bags. The formulated bacterial antagonists were then kept at both room and 4-8°C temperature in the refrigerator.

2.6 Assessment of viability of the formulated fungal and bacterial antagonists

The viability of the bacterial and fungal antagonists were checked by drawing 1 g of the formulated products in sterile water in every 30 days after formulation and diluted serially up to 10^{-4} or 10^{-5}. The numbers of viable cells (colony forming unit) were counted per gram formulations kept at both room temperature and 4-8°C temperature in the refrigerator.

2.7 Assessment of plant growth promotion induced by antagonistic bacterial and fungal isolates

Rice seeds (cv.IR24) were surface sterilized and dried. Then the sterilized rice seeds were treated with formulated bacterial and fungal antagonists (10 g/kg seeds) and the treated seeds were left for 1 h under shade. The rice seeds were then sown in the plastic pots previously filled with sterile soils. Fifty seeds were sown in each pot and three replications were maintained. Then the germination of seeds were recorded at 7DAS. The seedlings were uprooted at 7 DAS, 14 DAS and 28 DAS to measure the root length, shoot length and to calculate the vigor index [= (root length + shoot length) × germination percentage] were measured.

2.7.1 Seed priming, raising of seedlings and transplanting

Seeds of IR24 were treated with 32 selected formulated PGP antagonistic bacterial isolates. The treated sees were left for 1 hr. for adherence of the bacterial and fungal isolates with the treated seed surface. The treated seeds were then sown in the plastic pots filled with sterilized soils. One month old seedlings were then transplanted in the plastic pots filled with puddle soils.

2.7.2 Foliar spray of formulated PGP bacterial and fungal isolates

Formulated PGP antagonistic bacterial isolates were sprayed two times (at 50 and 55 DAS) before inoculation and two times after inoculation i.e. 65 and 70 DAS.

2.7.3 Inoculation of the rice plant with X. oryzae *pv.* oryzae

Rice plants were inoculated with a strain of *X. oryzae* pv. *oryzae* by Scissor clip method at 60 DAS.

3. Results

3.1 Isolation and identification of antagonistic bacteria against *X. oryzae* pv. *oryzae*

Rice plant samples were collected from 40 districts of Bangladesh representing 30 AEZs during boro seasons 2018–2019 and aman seasons 2018–2019. In total 300 bacterial isolates and 100 fungal isolates were isolated and purified from rice plant samples during boro season, 2018. Some selected representative bacterial species were shown in **Figure 1**. Out of 300 bacterial isolates, eighteen were identified as antagonist against *X. oryzae* pv. *oryzae* and inhibited the growth of *X. oryzae* pv. *oryzae in vitro* which was ranged by 28.39–76.19% (**Table 1** and **Figure 2**). The maximum (76.14%) growth inhibition of *X. oryzae* pv. *oryzae in vitro* was recorded by BDISOB05P while the minimum (28.59) growth inhibition was exhibited by

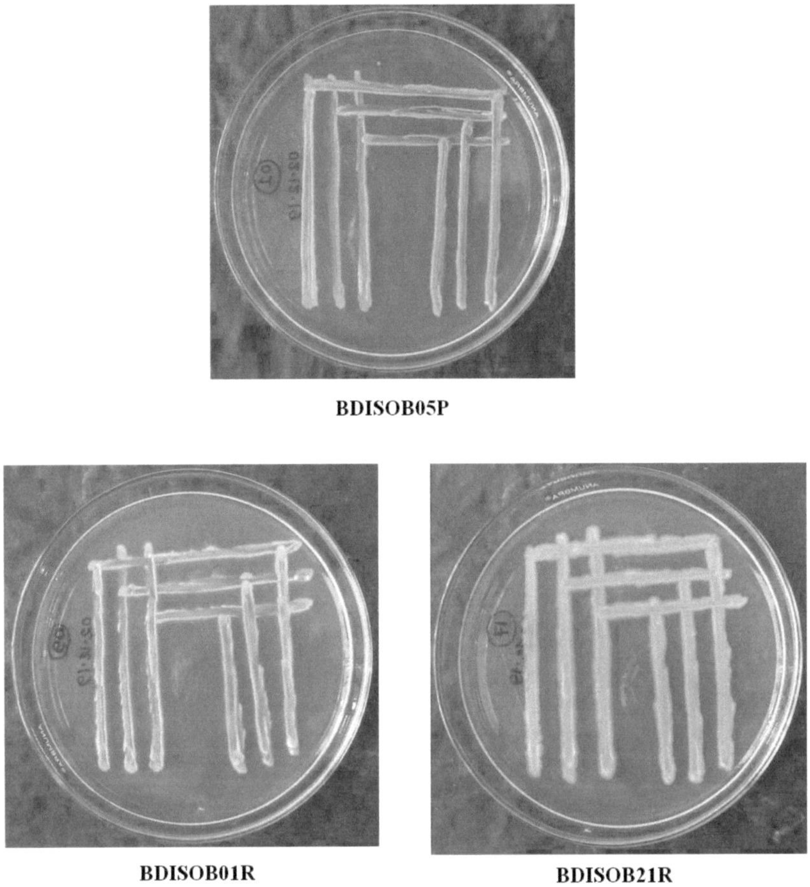

Figure 1.
Representative photographs of purified bacterial isolates obtained from rice phylloplane and rhizosphere. BDISOB05P: an isolate from Mymensingh, BDISOB01R: an isolate from Mymensingh and BDISOB21R: an isolate from Chattagram.

BDISOB272R. These antagonistic bacterial isolates were identified by sequencing of PCR products of 16SrDNA gene (**Figure 3A**). The identified bacterial species were BDISOB04P (*P. putida*), BDISOB05P (*P. putida),* BDISOB98P (*Stenotrophomonas maltophilia*), BDISOB241P (*Burkholderia* sp.), BDISOB242P *(B. gladioli),* BDISOB219R (*P. taiwanensis*), BDISOB220R (*Serratia* sp.), BDISOB221R (*Pseudomonas* sp.), BDISOB222R (*P. plecoglossicida*), BDISOB258R (*P. putida*), BDISOB272R (*Stenotrophomonas maltophilia*), BDISOB275R (*P. putida*), BDISOB186R (*Pseudomonas* sp.), BDISOB283R (*Pseudomonas fluorescens*), BDISOB306R (*P. putida*), BDISOB53R (*P. putida*), BDISOB61R (*Delftia tsuruhatensis*) (**Table 1**). In total 400 bacterial isolates and 40 fungal isolates were isolated and purified from rice plant samples collected in aman season, 2018. Seventeen bacterial isolates were identified as antagonist against *X. oryzae* pv. *oryzae* and inhibited the growth of *X. oryzae* pv. *oryzae in vitro* which was ranged by 38.33–60.66% (**Table 2**). The highest (60.66%) growth inhibition of *X. oryzae* pv. *oryzae* was exhibited by BDISO147Pand the lowest (38.33%) growth inhibition was shown by BDISO135P.These antagonistic bacterial isolates were identified by sequencing of PCR products of 16SrDNA gene (**Figure 3B**). The bacterial species were BDISO04P (*P. putida*), BDISO45P (*Bacillus paramycoides*), BDISO356P (*P. hibiscicola*), BDISO198P (*Serratia plymuthica*), BDISO135P (*Bacillus sp.*), BDISO148P (*Serratia marcescens*), BDISO92P (*Serratia marcescens*), BDISO237P (*Alcaligenes faecalis*), BDISO12P (*Alcaligenes faecalis*),

Isolates	Closest relatives	Accession no.	Alignment	Homology	Growth inhibition of *X. oryzae* pv. *oryzae* (%)
BDISOB04P	*P. putida strain*	MF838698.1	968/1086	89	61.67
BDISOB05P	*P. putida strain*	MH085459.1	931/1140	82	76.14
BDISOB16P	*Bacillus* sp.	MH819972.1	702/738	95	59.94
BDISOB98P	*Stenotrophomonas maltophilia*	AY486381.1	1224/1271	96	33.04
BDISOB241P	*Burkholderia* sp.	GU979224.1	1154/1222	94	63.64
BDISOB242P	*B. gladioli*	MH748602.1	1186/1239	96	51.18
BDISOB219R	*P. taiwanensis*	KC293831.1	913/969	94	63.12
BDISOB220R	*Serratia*sp.	FM875872.1	150/186	81	61.77
BDISOB221R	*Pseudomonas* sp.	MG021242.1	303/341	89	68.33
BDISOB222R	*P. plecoglossicida*	KC864769.1	614/751	82	64.79
BDISOB258R	*P. putida*	MF417798.1	917/1050	87	64.40
BDISOB272R	*Stenotrophomonas maltophilia*	KJ534495.1	794/923	86	28.59
BDISOB275R	*P. putida*	KT984874.1	1201/1229	98	71.86
BDISOB186R	*Pseudomonas* sp.	JQ977022.1	29/29	100	64.43
BDISOB283R	*Pseudomonas fluorescens*	KF010368.1	969/1006	96	66.04
BDISOB306R	*P. putida*	KF030905.1	1298/1374	94	44.97
BDISOB53R	*P. putida*	JQ833720.1	53/60	88	48.19
BDISOB61R	*Delftia tsuruhatensis*	MF353931.1	976/1168	84	38.54

Table 1.
List of antagonistic bacterial isolates identified by homology search ofsequences of 16SrDNA by BLAST program obtained from plant samples collected in boro season 2018.

BDISO196P (*Alcaligenes faecalis*), BDISO145P (*Serratia marcescens*), BDISO09P (*Serratia marcescens*), BDISO21R (*Serratia marcescens*), BDISO154P (*P. taiwanensis*), BDISO154P (*P. taiwanensis*), BDISO147P (*Serratia marcescens*), BDISO158R (*Serratia marcescens*), BDISO0R (*B. amyloliquefaciens*). In boro season 2019, 300 bacterial isolates were isolated and purified. In boro season 2019, out of 400 bacterial isolates fourteen were identified as antagonist against *X. oryzae* pv. *oryzae* and inhibited the growth of *X. oryzae* pv. *oryzae in vitro* which was ranged by 20.83–60.87% (**Table 3** and **Figure 3C**). The maximum (60.87%) growth inhibition of *X. oryzae* pv. *oryzae in vitro* was recorded by BDISOB37R while the minimum (20.83%) growth inhibition was exhibited by BDISOB14R. The bacterial species identified were BDISOB37R [*Pseudochrobactrum asaccharolyticum*], BDISOB16R [*Pseudochrobactrum asaccharolyticum*], BDISOB91R [*Pseudochrobactrum asaccharolyticum*], BDISOB17R [*Limnolyngbyacircumcreta*], BDISOB15R [*Pseudochrobactrum asaccharolyticum*], BDISOB86R [*Enterobacteraerogenes*], BDISOB30R [*Pseudochrobactrum asaccharolyticum*], BDISOB92R [*Pseudomonas fluorescens*], BDISOB178R [*Serratia marcescens*], BDISOB11R [*Pseudochrobactrum asaccharolyticum*], BDISOB21R [*Stenotrophomonas maltophilia*], BDISOB24R [*P. asaccharolyticum*], BDISOB23R [*Pseudochrobactrum asaccharolyticum*] and BDISOB14R [*Pseudochrobactrum asaccharolyticum*] by sequencing of bacterial 16SrDNA. In aman season 2019, 400 bacterial isolates were isolated and purified. In aman season 2019, out of 400 bacterial isolates fourteen were identified as antagonist against *X. oryzae* pv. *oryzae* and inhibited the growth of *X. oryzae* pv. *oryzae in vitro*which was ranged

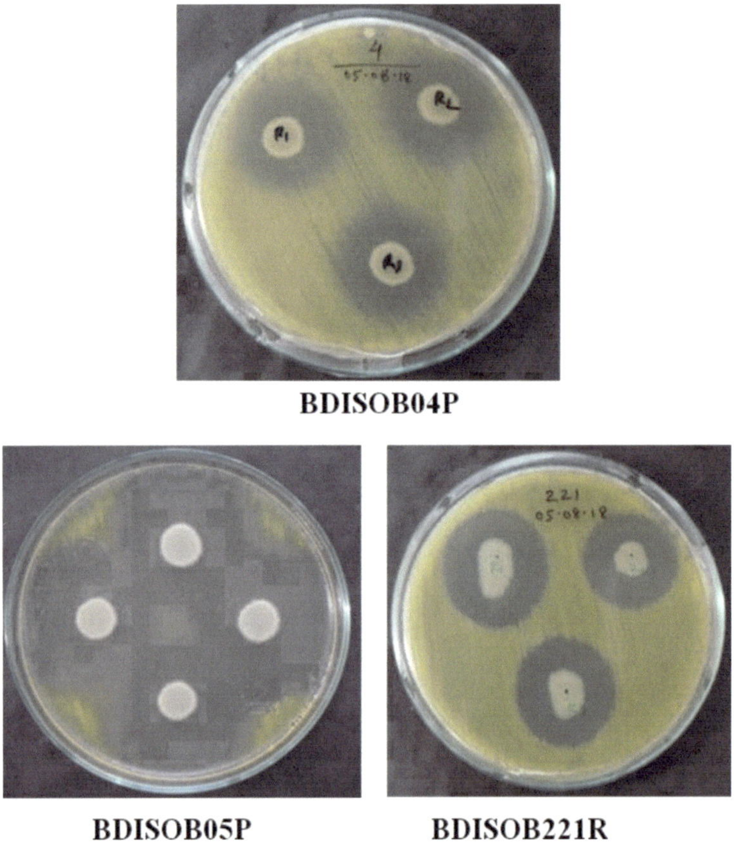

Figure 2.
Representative photographs of in vitro *growth inhibition of* X. oryzae *pv.* oryzae *by different potential bacterial isolates. BDISOB04P: an isolate from Cox's Bazar, BDISOB05P: an isolate from Mymensingh and BDISOB221R: an isolate from Chattagram.*

by 50.83–61.545% (**Table 4**). The maximum (61.54%%) growth inhibition of *X. oryzae* pv. *oryzae in vitro* was recorded by BDISOB54R while the minimum (50.93%) growth inhibition was exhibited by BDISOB12R. These antagonistic bacterial isolates were identified by sequencing of 16SrDNA gene (**Figure 3D**). The bacterial species were BDISOB70R [*Serratia marcescens*], BDISOB54R [*B. gladioli*], BDISOB08R [*Serratia marcescens*], BDISOB31R [*Serratia marcescens*], BDISOB06R [*Serratia marcescens*], BDISOB171R [*Alcaligenes faecalis*], BDISOB46R [*Serratia marcescens*], BDISOB09R [*Serratia marcescens*], BDISOB33R [[*Serratia marcescens*], BDISOB11R [*Serratia marcescens*], BDISOB36R [*Serratia marcescens*], BDISOB07R [*Serratia nematodiphila*], BDISOB172R [*B. aerophilus*] and BDISOB12R [*Serratia marcescens*] by sequencing of bacterial 16SrDNA.

3.2 Assessment of plant growth promoting determinants

Three plant growth promoting determinants viz. siderophore and IAA production as well as phosphate solubilization capability were assessed. The results revealed that the development of yellow-orange halo zone around the bacterial growth on chrome azurol S agar plates was considered as positive (+) for siderophore production, formation of pink color by the culture supernatant of the bacterial isolates in presence of Salkowski's reagent confirmed IAA production which was indicated by '+" sign and observation of clear halo zone in National Botanical Research Institute's phosphate (NBRIP) agar plates indicated the bacterial

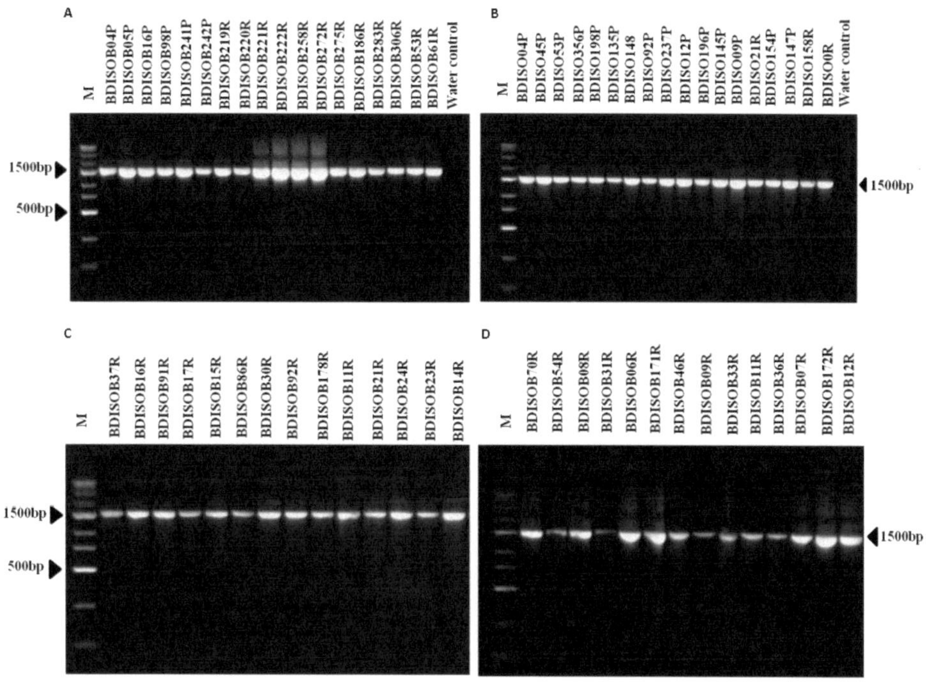

Figure 3.
PCR confirmation of the antagonistic bacterial isolates by amplification of 16S rDNA using primers 27F and 1518R obtained from plant samples collected in irrigated and rainfed seasons. These PCR products were then used for sequencing. Bacterial isolates obtained from (A) irrigated: BDISOB04P, BDISOB05P, BDISOB16P, BDISOB98P, BDISOB241P, BDISOB242P, BDISOB219R, BDISOB220R, BDISOB221R, BDISOB222R, BDISOB258R, BDISOB272R, BDISOB275R, BDISOB186R, BDISOB283R, BDISOB306R, BDISOB53R and BDISOB61R. (B) Rainfed: BDISO04P, BDISO45P, BDISO53P, BDISO356P, BDISO198P, BDISO135P, BDISO148P, BDISO92P, BDISO237P, BDISO12P, BDISO196P, BDISO145P, BDISO09P, BDISO21R, BDISO154P, BDISO147P, BDISO158R, BDISO0R, (C) irrigated: BDISOB37R, BDISOB16R, BDISOB91R, BDISOB17R, BDISOB15R, BDISOB86R, BDISOB30R, BDISOB92R, BDISOB178R, BDISOB11R, BDISOB21R, BDISOB24R, BDISOB23R and BDISOB14R: a rhizosphere isolate, Cumilla and (D) rainfed: BDISOB70R, BDISOB54R, BDISOB08R, BDISOB31R, BDISOB06R, BDISOB171R, BDISOB46R, BDISOB09R, BDISOB33R, BDISOB11R, BDISOB36R, BDISOB07R, BDISOB172R and BDISOB12R.

isolates are capable of phosphate solubilization which was denoted by "+" sign (**Figure 3**). Out of these bacterial species, Out of these bacterial species, 48 bacterial species were found as positive for IAA production, all 63 bacterial species were found positive for siderophore production and 48 were found capable to solubilize insoluble phosphate. In case of Indole Acetic Acid (IAA), BDISOB92FarR (*Pseudomonas fluorescens*), BDISOB172ThaR (*B. aerophilus*), BDISOB45PanP (*Bacillus paramycoides*), BDISOB01MymR (*Bacillus amyloliquefacience*) showed highest IAA production. Whereas, BDISOB186KusR (*Bacillus paramycoides*) showed lowest IAA production. BDISOB54KhuR (*B. gladioli*) *and* BDISOB21ChaR (*S. maltophilia*) indicataed moderate IAA production. BDISOB198HabP (*Seratiaplymuthica*), BDISOB148JoyP (*Seratia marcescens*), BDISOB145JoyP (*Seratia marcescens*), BDISOB07FarR (*Seratianematodiphilia*), BDISOB12FarR (*Seratia marcescens*), BDISOB31MagR (*Seratia marcescens*), BDISOB46GopR (*Seratia marcescens*) and BDISOB70KusR (*Seratia marcescens*) were statistically similar. The bacterial isolatesBDISOB222GaiR (*P. plecoglossicida*), BDISOB45PanP (*Bacillus paramycoides*) BDISOB01MymR (*B. amyloliquefaciens*) BDISOB04KhaP (*P. putida*), BDISOB05MymP (*P. putida*), BDISOB221GaiR (*Pseudomonas sp.*) showed highest siderophore production. Whereas, BDISOB135SerP (*Bacillus sp.*), BDISOB145JoyP (*Seratia marcescens*) *and* BDISOB21ChaR (*Stenotrophomonas maltophilia*) showed

Isolate ID	Closest relatives	Accession no.	Alignment	Homology	Growth inhibition of *X. oryzae* pv. *oryzae* (%)
BDISO04P	*P. putida*	FR749878.1	827/1080	96	46.37
BDISO45P	*Bacillus paramycoides*	MK467557.1	1027/1133	91	50.00
BDISO356P	*P. hibiscicola*	KJ396817.1	1125/1148	98	46.83
BDISO198P	*Serratia plymuthica*	KU821695.1	472/530	89	50.00
BDISO135P	*Bacillus sp.*	KU146461.1	189/237	80	38.33
BDISO148P	*Serratia marcescens*	MN691926.1	929/990	94	54.26
BDISO92P	*Serratia marcescens*	MG996733.1	568/616	92	44.18
BDISO237P	*Alcaligenes faecalis*	KR827435.1	1048/1102	95	57.19
BDISO12P	*Alcaligenes faecalis*	MN513225.1	927/1094	85	57.44
BDISO196P	*Alcaligenes faecalis*	MN513225.1	901/1111	81	46.18
BDISO145P	*Serratia marcescens*	MF360051.1	545/630	87	40.00
BDISO09P	*Serratia marcescens*	MN252007.1	171/185	92	44.47
BDISO21R	*Serratia marcescens*	MG557818.1	194/200	97	54.60
BDISO154P	*P. taiwanensis*	MN416314.1	161/178	90	47.22
BDISO147P	*Serratia marcescens*	MF716688.1	1086/1130	96	60.66
BDISO158R	*Serratia marcescens*	MK346258.1	866/953	91	47.27
BDISO0R	*B. amyloliquefaciens*	KC888017.1	1151/1153	99	50.00

Table 2.
List of antagonistic bacterial isolates identified by homology search ofsequences of 16SrDNA by BLAST program obtained from plant samples collected in aman season 2018.

lowest siderophore production. The Sierophore production found in BDISOB219GaiR (*P. taiwanensis*), BDISOB186KusR (*Pseudomonas sp.*), BDISOB283KisR (*Pseudomonas fluorescens*), BDISOB198HabP (*Seratiaplymuthic*), BDISOB54KhuR (*B. gladioli*) and BDISOB21ChaR (*S. maltophilia*) BDISOB198HabP (*Seratia plymuthica*), BDISOB148JoyP (*Seratia marcescens*), BDISOB158ChaR (*Seratia marcescens*) BDISOB148JoyP (*Seratia marcescens*), BDISOB145JoyP (*Seratia marcescens*), BDISOB07FarR (*Seratia nematodiphilia*), BDISOB12FarR (*Seratia marcescens*), BDISOB31MagR (*Seratia marcescens*), BDISOB46GopR (*Seratia marcescens*) and BDISOB70KusR (*Seratia marcescens*) were statistically similar. The bacterial isolates BDISOB05MymP (*P. putida*), BDISOB45PanP (*Bacillus paramycoides*) and BDISOB01MymR (*B. amyloliquefaciens*) showed highest siderophore production. Whereas, BDISOB186KusR (*Pseudomonas sp.*), BDISOB258GaiR (*P. putida) and* BDISOB70KusR (*Seratia marcescens*) showed lowest phosphate solubilization activity. The others bacteria in case of phosphate solubilization were statistically similar.

3.2.1 IAA production

In case of Indole Acetic Acid (IAA), four isolates those were BDISOB92FarR (*Pseudomonas fluorescens*), BDISOB172ThaR (*B. aerophilus*), BDISOB45PanP (*Bacillus paramycoides*), BDISOB01MymR (*Bacillus amyloliquefacience*) revealed highest IAA production. Conversely, only one BDISOB186KusR (*Bacillus paramycoides*) depicted lowest IAA production. Around, twelve isolates exhibited upper-moderate IAA production, besides, seven showed lower and lower-moderate IAA production. BDISOB198HabP (*Seratia plymuthica*), BDISOB148JoyP (*Seratia marcescens*),

Isolate ID	Closest relatives	Accession no.	Alignment	Homology	Growth inhibition of *X. oryzae* pv. *oryzae* (%)
BDISOB37R	*Pseudochrobactrum asaccharolyticum*	KC456599.1	275/298	92%	60.87
BDISOB16R	*Pseudochrobactrum asaccharolyticum*	KC456599.1	275/298	92	57.09
BDISOB91R	*Pseudochrobactrumsaccharolyticum*	KC456543.1	748/841	89	56.55
BDISOB17R	*Limnolyngbyacircumcreta*	KR697754.1	86/105	82	43.42
BDISOB15R	*Pseudochrobactrum asaccharolyticum*	KM921740.1	399/535	75	49.94
BDISOB86R	*Enterobacteraerogenes*	KM503142.1	444/483	92	45.75
BDISOB30R	*Pseudochrobactrum asaccharolyticum*	MK100767.1	166/177	94	47.73
BDISOB92R	*Pseudomonas fluorescens*	KJ027533.1	29/29	100	45.44
BDISOB178R	*Serratia marcescens*	MN691653.1	635/679	94	45.91
BDISOB11R	*Pseudochrobactrumsaccharolyticum*	MK377096.1	770/827	93	40.00
BDISOB21R	*Stenotrophomonas maltophilia*	MN173472.1	994/1084	92	38.42
BDISOB24R	*Pseudochrobactrumsaccharolyticum*	FJ950551.1	994/1084	92	36.55
BDISOB23R	*Pseudochrobactrum asaccharolyticum*	KC456600.1	1082/1122	96	32.46
BDISOB14R	*Pseudochrobactrum asaccharolyticum*	KC456600.1	535/541	99	20.83

Table 3.
List of antagonistic bacterial isolates identified by homology search ofsequences of 16SrDNA by BLAST program obtained from plant samples collected in boro season 2019.

Isolate ID	Closest relatives	Accession no.	Alignment	Homology	Growth inhibition of *X. oryzae* pv. *oryzae* (%)
BDISOB70R	*Serratia marcescens*	MG571677.1	239/300	80	52.38
BDISOB54R	*B. gladioli*	MH748601.1	1050/1108	95	61.54
BDISOB08R	*Serratia marcescens*	KU963569.1	100/114	88	59.31
BDISOB31R	*Serratia marcescens*	MN691926.1	929/990	94	59.17
BDISOB06R	*Serratia marcescens*	MG571677.1	111/127	87	59.26
BDISOB171R	*Alcaligenes faecalis*	MN513225.1	927/1094	85	57.37
BDISOB46R	*Serratia marcescens*	MF360051.1	545/630	87	55.53
BDISOB09R	*Serratia marcescens*	MN252007.1	171/185	92	55.92
BDISOB33R	*Serratia marcescens*	KJ535346.1	127/143	89	52.27
BDISOB11R	*Serratia marcescens*	MK806681.1	88/98	90	53.57
BDISOB36R	*Serratia marcescens*	MK961214.1	787/910	86	58.33
BDISOB07R	*Serratia nematodiphila*	MN691930.1	572/639	90	52.00
BDISOB172R	*B. aerophilus*	KY307912.1	874/1043	84	51.19
BDISOB12R	*Serratia marcescens*	MH074778.1	780/841	93	50.93

Table 4.
List of antagonistic bacterial isolates identified by homology search ofsequences of 16SrDNA by BLAST program obtained from plant samples collected in aman season 2019.

BDISOB145JoyP (*Seratia marcescens*), BDISOB07FarR (*Seratia nematodiphilia*), BDISOB12FarR (*Seratia marcescens*), BDISOB31MagR (*Seratia marcescens*), BDISOB46GopR (*Seratia marcescens*) and BDISOB70KusR (*Seratia marcescens*) were statistically similar as well as BDISOB172ThaR, BDISO1MymR, BDISO45PanP and BDISOB92FarR were statistically similar, apart from these all were under the group of statistically dissimilar (**Table 5** and **Figure 3**).

3.2.2 Siderophore production

Six bacterial isolates BDISOB222GaiR (*P. plecoglossicida*), BDISOB45PanP (*Bacillus paramycoides*), BDISOB01MymR (*B. amyloliquefaciens*), BDISOB04KhaP (*P. putida*), BDISOB05MymP (*P. putida*), BDISOB221GaiR (*Pseudomonas sp.*) exposed highest siderophore production. On the opposite, three of them which were BDISOB135SerP (*Bacillus sp.*), BDISO04DinP (*P. putida*) *and* BDISOB21ChaR (*S. maltophilia*) in the list of lowest siderophore production. Nine of them produced upper-moderate level of siderophore as well as thirteen isolates were released lower-moderate level of siderophore. Sixteen isolates those who produced siderophore including BDISOB219GaiR (*P. taiwanensis*), BDISOB186KusR (*Pseudomonas sp.*), BDISOB283KisR (*Pseudomonas fluorescens*), BDISOB198HabP (*Seratia plymuthic*), BDISOB54KhuR (*B. gladioli*) *and* BDISOB21ChaR (*S. maltophilia*) BDISOB198HabP (*Seratia plymuthica*), BDISOB148JoyP (*Seratia marcescens*), BDISOB158ChaR (*Seratia marcescens*), BDISOB148JoyP (*Seratia marcescens*), BDISOB145JoyP (*Seratia marcescens*), BDISOB07FarR (*Seratia nematodiphilia*), BDISOB12FarR (*Seratia marcescens*), BDISOB31MagR (*Seratia marcescens*), BDISOB46GopR (*Seratia marcescens*) and BDISOB70KusR (*Seratia marcescens*) were statistically similar and rest of them were statistically dissimilar (**Table 5** and **Figure 3**).

Treatments/ bacterial isolates	Name of bacteria	Indole acetic acid (IAA) (ng/ml)	Siderophore production (orange color halo zone) (mm)	Phosphate solubilization (clear halo zone) (mm)
Control	—	0.00 o	0.00 h	0.00f
BDISOB04KhaP	*P. putida*	44.88kl	28.67a	8.17c-e
BDISOB05MymP	*P. putida*	44.54 l	29.00a	14.33a
BDISOB219GaiR	*P. taiwanensis*	70.98c-g	20.13b	7.83 c-e
BDISOB221GaiR	*Pseudomonas sp.*	42.93 lm	28.00a	8.67c
BDISOB222GaiR	*P. plecoglossicida*	41.46 m	29.83a	11.67b
BDISOB258GaiR	*P. putida*	49.27j	14.50d-f	6.83de
BDISOB186KusR	*Pseudomonas sp.*	36.83n	19.50bc	6.50e
BDISOB283KisR	*Pseudomonas fluorescens*	43.90 l	18.33bc	8.33 cd
BDISO04DinP	*P. putida*	46.59 k	13.00 fg	8.17 c-e
BDISO45PanP	*Bacillus paramycoides*	81.46a	28.17a	14.33a
BDISO198HabP	*S. plymuthica*	71.22c-f	20.00b	7.50 c-e
BDISO135SerP	*Bacillus sp.*	67.80 h	10.83 g	8.33 cd
BDISO148JoyP	*S. marcescens*	71.22c-f	20.00b	7.50 c-e
BDISO1MymR	*B. amyloliquefaciens*	81.46a	29.83a	14.17a
BDISO145JoyP	*S. marcescens*	71.71c-e	13.17 fg	6.83de
BDISO158ChaR	*S. marcescens*	69.60e-h	20.00b	7.50 c-e
BDISOB37KhaR	*Pseudochrobactrum asaccharolyticum*	69.93d-g	14.33d-f	8.33 cd
BDISOB16CumR	*Pseudochrobactrum asaccharolyticum*	61.46i	16.50c-e	8.17 c-e
BDISOB92FarR	*Pseudomonas fluorescens*	82.68a	0.00 h	7.50 c-e
BDISOB21ChaR	*S. maltophilia*	78.78b	11.17 g	7.00 c-e
BDISOB17CumR	*Limnolyngbya circumcreta*	68.93gh	18.33bc	7.67 c-e
BDISOB15CumR	*Pseudochrobactrum asaccharolyticum*	70.27c-g	18.06bc	8.17 c-e
BDISOB86FarR	*E. aerogenes*	68.93 h	18.33bc	7.33 c-e
BDISOB30ChaR	*Pseudochrobactrum asaccharolyticum*	69.27f-h	18.06bc	8.17 c-e
BDISOB07FarR	*S. nematodiphila*	71.22c-f	13.50e-g	7.50 c-e
BDISOB12FarR	*S. marcescens*	72.22c	20.17b	6.83de
BDISOB31MagR	*S. marcescens*	70.89c-g	17.50b-d	7.50 c-e
BDISOB36MagR	*S. marcescens*	71.55c-e	20.00b	7.33 c-e
BDISOB46GopR	*S. marcescens*	71.89 cd	20.00b	7.17 c-e
BDISOB54KhuR	*B. gladioli*	77.56b	18.33bc	7.33 c-e
BDISOB70KusR	*S. marcescens*	71.22c-f	20.00b	6.83de
BDISOB172ThaR	*B. aerophilus*	81.71a	20.17b	8.00 c-e
Level of significance		*	*	*
LSD	—	1.839	3.101	1.702
CV (%)	—	1.78	10.34	12.88

** indicated 5% level of significance.*
Values in columns followed by the same letters indicate no significant differences.

Table 5.
Growth promoting determinants produced by different bacterial isolates antagonistic to X. oryzae *pv.* oryzae.

3.2.3 Phosphate solubilization

Among all bacterial isolates three of them those were BDISOB05MymP (*P. putida*), BDISOB45PanP (*Bacillus paramycoides*) and BDISOB01MymR (*B. amyloliquefaciens*) manifestedsupreme amount of phosphate solubilization activity. Whereas, another three of them which wereBDISOB186KusR (*Pseudomonas sp.*), BDISOB258GaiR (*P. putida) and* BDISOB70KusR (*Seratia marcescens*) showed lowest phosphate solubilization activity. Except highest and lowest phosphate solubilization producing isolates rest of them were showed moderate type activity. In this case, a noticeable differences were observed that except two isolates BDISOB221GaiR and BDISOB222GaiR all other isolates are statistically similar. The others bacteria in case of phosphate solubilization were statistically similar (**Table 5** and **Figure 4**).

3.3 Plant growth promotion by bacterial isolates antagonistic to *X. oryzae* pv. *oryzae*

Based on the growth inhibition of *X. oryzae* pv. *oryzae* by these antagonistic bacterial species, 32 bacterial isolates were selected for plant growth promotion assay and for subsequent assessment of their net house and field performances. Different plant growth promoting bacterial antagonists enhanced the root length, shoot length and vigor index at 14, 21 and 28 DAS (**Table 6**). Among 32 bacterial

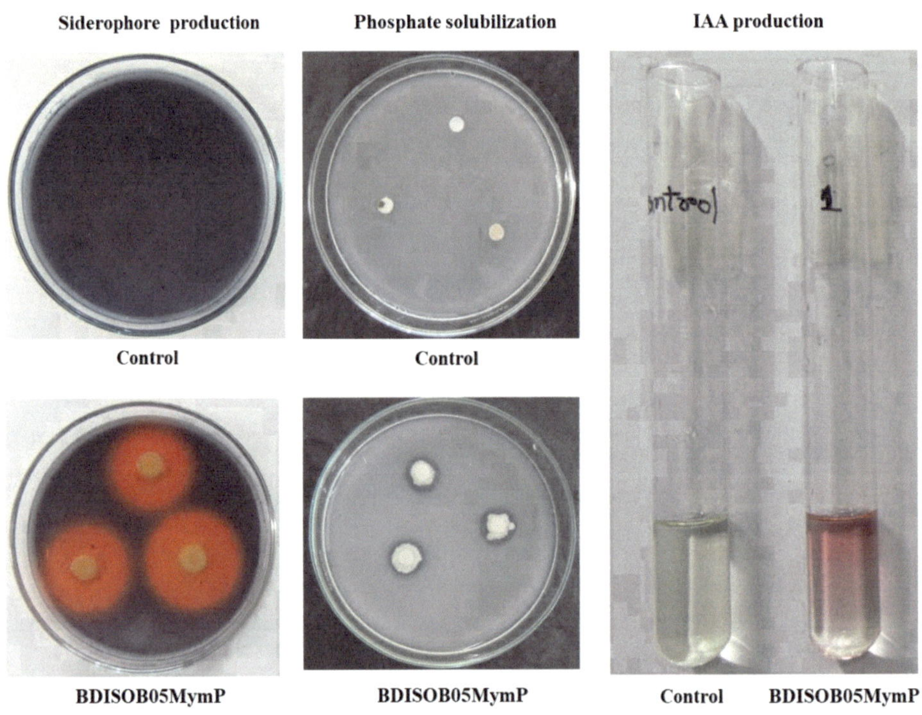

Figure 4.
Representative photographs showing the assessment of different plant growth promoting determinants. Siderophore production: antagonistic bacterial isolates showed positive siderophore production activity as indicated by orange halo zone around bacterial colony on CAS agar plates, phosphate solubilization: antagonistic bacterial isolates showed positive phosphate solubilizing activity by producing clear halo zone around the bacterial colony on National Botanical Research Institute's Phosphate (NBRIP) agar plates and indole acetic acid (IAA) production: IAA activity by different antagonistic bacterial isolatesindicated by the presence of pink color when bacterial culture supernatant mixed with Salkowskis reagent. BDISOB05P: isolate from Mymensingh.

Treatments		Root length (cm)			% Increase of vigor index over control			Shoot length (cm)			% Increase of root length over control			Vigor index			% Increase of shoot length over control		
		Days after sowing (DAS)																	
		14	21	28	14	21	28	14	21	28	14	21	28	14	21	28	14.00	21.00	28.00
Control		6.76	9.20	11.28	0	0.00	0.00	10.72	11.97	17.23	0.00	0.00	0.00	1316.32	2046.56	2449.34	0.00	0.00	0.00
BDISOB04P	*P. putida*	9.12	12.31	13.20	34.93	33.77	17.05	12.37	16.77	23.07	15.40	40.11	33.89	1697.18	2306.48	2877.95	28.93	12.70	17.50
BDISOB05P	*P. putida*	8.23	12.22	12.84	21.85	32.83	13.80	12.37	16.53	18.32	15.40	38.16	6.29	1634.27	2549.46	2658.42	24.15	24.57	8.54
BDISOB219R	*P. taiwanensis*	8.69	12.22	12.58	28.56	32.83	11.55	12.40	16.53	18.88	15.71	38.16	9.57	1869.68	2549.46	2790.04	42.04	24.57	13.91
BDISOB221R	*Pseudomonas sp.*	8.43	11.13	11.30	24.81	21.01	0.18	11.90	15.65	19.53	11.04	30.78	13.35	1647.00	2169.45	2497.50	25.12	6.00	1.97
BDISOB222R	*P. plecoglossicida*	10.63	14.95	16.23	57.38	62.50	43.91	15.12	21.15	27.85	41.06	76.74	61.61	2360.42	3309.17	4040.97	79.32	61.69	64.98
BDISOB258R	*P. putida*	9.12	13.04	13.37	34.93	41.78	18.56	12.37	17.60	23.42	15.40	47.08	35.88	1697.18	2420.82	2906.41	28.93	18.29	18.66
BDISOB186R	*Pseudomonas sp.*	8.12	11.75	13.50	20.13	27.75	19.68	12.00	17.38	22.32	11.98	45.26	29.52	1595.92	2311.51	2841.72	21.24	12.95	16.02
BDISOB283R	*Pseudomonas fluorescens*	10.90	14.87	16.11	61.32	61.59	42.79	14.68	21.22	29.65	37.01	77.30	72.05	2285.44	3223.44	4087.60	73.62	57.51	66.89
BDISOB04P	*P. putida*	7.72	12.42	12.84	14.21	35.00	13.80	11.88	17.37	18.32	10.89	45.13	6.29	1672.53	2541.80	2658.42	27.06	24.20	8.54
BDISOB45P	*Bacillus paramycoides*	10.32	14.25	15.63	52.69	54.89	38.59	14.18	21.73	30.33	32.35	81.62	76.02	2237.67	3286.48	4198.29	69.99	60.59	71.41
BDISOB198P	*S. plymuthica*	8.65	11.38	12.33	28.02	23.73	9.34	11.43	13.35	20.07	6.69	11.56	16.44	1687.00	2127.07	2689.20	28.16	3.93	9.79
BDISOB135P	*Bacillus sp.*	7.82	11.45	12.05	15.69	24.46	6.83	12.90	15.53	20.05	20.37	29.81	16.34	1788.54	2329.56	2418.03	35.87	13.83	−1.28
BDISOB148P	*S. marcescens*	8.65	11.38	13.36	28.02	23.73	18.41	11.43	13.35	20.57	6.69	11.56	19.38	1687.00	2127.07	2567.37	28.16	3.93	4.82
BDISOB01R	*B. amyloliquefaciens*	8.33	11.38	13.42	23.33	23.73	18.94	12.72	13.35	23.65	18.66	11.56	37.23	1810.30	2127.07	3187.73	37.53	3.93	30.15
BDISOB145P	*S. marcescens*	8.65	11.38	13.36	28.02	23.73	18.41	11.43	13.35	20.57	6.69	11.56	19.38	1687.00	2127.07	2567.37	28.16	3.93	4.82
BDISOB158R	*S. marcescens*	8.65	11.38	13.36	28.02	23.73	18.41	11.43	13.35	20.57	6.69	11.56	19.38	1687.00	2127.07	2567.37	28.16	3.93	4.82
BDISOB37R	*P. asaccharolyticum*	8.13	12.66	12.33	20.37	37.61	9.34	12.18	16.52	20.07	13.69	38.02	16.44	1632.11	2324.41	2689.20	23.99	13.58	9.79

Treatments		Root length (cm)			% Increase of vigor index over control			Shoot length (cm)			% Increase of root length over control			Vigor index			% Increase of shoot length over control		
		Days after sowing (DAS)																	
BDISOB16R	*Pseudochractrum asaccharolyticum*	8.34	11.95	12.12	23.38	29.89	7.42	11.57	18.52	24.45	7.93	54.74	41.88	1585.63	2528.73	3071.60	20.46	23.56	25.41
BDISOB92R	*Pseudomonas fluorescens*	7.10	13.06	12.38	5.08	41.92	9.78	12.02	15.87	20.28	12.19	32.59	17.70	1587.24	2429.56	2613.33	20.58	18.71	6.70
BDISOB21R	*S. marcescens*	8.65	11.62	13.52	28.02	26.30	19.86	11.43	12.50	19.43	6.69	4.46	12.77	1687.00	1792.92	2449.53	28.16	−12.39	0.01
BDISOB17R	*Limnolyngbya circumcreta*	7.10	11.45	13.36	5.08	24.46	18.41	12.02	15.53	20.57	12.19	29.81	19.38	1587.24	2329.56	2567.37	20.58	13.83	4.82
BDISOB15R	*P. asaccharolyticum*	8.13	12.66	12.33	20.37	37.61	9.34	12.18	16.52	20.07	13.69	38.02	16.44	1632.11	2324.41	2689.20	23.99	13.58	9.79
BDISOB86R	*E. aerogenes*	8.13	12.66	12.33	20.37	37.61	9.34	12.18	16.52	20.07	13.69	38.02	16.44	1632.11	2324.41	2689.20	23.99	13.58	9.79
BDISOB30R	*P. asaccharolyticum*	8.13	12.66	12.33	20.37	37.61	9.34	12.18	16.52	20.07	13.69	38.02	16.44	1632.11	2324.41	2689.20	23.99	13.58	9.79
BDISOB07R	*S. nematodiphila*	8.65	11.38	13.36	28.02	23.73	18.41	11.43	13.35	20.57	6.69	11.56	19.38	1687.00	2127.07	2567.37	28.16	3.93	4.82
BDISOB12R	*S. marcescens*	8.65	11.38	13.36	28.02	23.73	18.41	11.43	13.35	20.57	6.69	11.56	19.38	1687.00	2127.07	2567.37	28.16	3.93	4.82
BDISOB31R	*S. marcescens*	8.49	11.38	13.36	25.60	23.73	18.41	12.72	13.35	20.57	18.66	11.56	19.38	1604.39	2127.07	2567.37	21.88	3.93	4.82
BDISOB36R	*S. marcescens*	8.65	11.38	13.36	28.02	23.73	18.41	11.43	13.35	20.57	6.69	11.56	19.38	1687.00	2127.07	2567.37	28.16	3.93	4.82
BDISOB46R	*S. marcescens*	8.65	11.38	13.36	28.02	23.73	18.41	11.43	13.35	20.57	6.69	11.56	19.38	1687.00	2127.07	2567.37	28.16	3.93	4.82
BDISOB54R	*B. gladioli*	7.87	11.62	13.52	16.43	26.30	19.86	11.77	12.50	19.43	9.80	4.46	12.77	1459.41	1792.92	2449.53	10.87	−12.39	0.01
BDISOB70R	*S. marcescens*	8.65	11.38	13.36	28.02	23.73	18.41	11.43	13.35	20.57	6.69	11.56	19.38	1687.00	2127.07	2567.37	28.16	3.93	4.82
BDISOB172R	*B. aerophilus*	8.40	12.35	12.84	24.32	34.24	13.80	13.00	16.92	22.92	21.31	41.36	32.98	1719.13	2351.09	2872.18	30.60	14.88	17.26

Table 6.
Effect of different antagonistic bacteria on plant growth promotion of rice (cv. IR24).

isolates, the maximum vigor index (4198.29) was recorded in seedlings raised from seeds treated with BDISOB45PanR (*Bacillus paramycoides*) followed by BDISOB283R (*Pseudomonas fluorescens*) (4087.60), BDISOB222R (*P. plecoglossicida*) (4040.97) while the minimum (2418.03) vigor index was obtained in BDISOB135SheR (*Bacillus* sp.) followed by BDISOBP (*S. marcescens*) (2449.53) and BDISOB54R (*B. gladioli*) (2449.53) at 30 DAS. However, all the antagonistic bacterial isolates exhibited the increase of vigor index ranged by 0.01 to 71.41. This result implies that some of the selected antagonistic bacterial isolates have the potentiality in enhancing plant growth.

3.4 Plant growth promotion by different bacterial isolates antagonistic to *Xanthomonasoryzae* pv. *oryzae*

3.4.1 Fresh shoot weight

At 28 days after sowing the highest shoot weight (2260 mg) was recorded in plants raised from the seed treated with the bacterial isolate BDISOB01MymR followed by the bacterial isolates BDISOB05MymP (2250 mg), BDISOB45PanP (2173 mg), BDISOB04DinP (2033 mg), BDISOB86FarR (2033 mg), BDISOB07FarR, (2033 mg) BDISOB283KisR (1950 mg). But the lowest shoot weight was observed in control (untreated seed) (933 mg) Rest of the isolates were showed moderate fresh shoot weight. Among all bacterial isolates seventeen were statistically similar and others denoted statistically dissimilar (**Table** 7).

3.4.2 Dry shoot weight

At 28 days after sowing the highest shoot weight (546 mg) was recorded in plants raised from the seed treated with the bacterial isolate BDISOB01Mym) followed by the bacterial isolates BDISOB04DinP (473mgmg), BDISOB04KhaP (470 mg), BDISOB92Far (466 mg), BDISOB222GaiR (443 mg) were statistically similar Whereas, the lowest (260 mg) was reorded in the plants raised from untreated seed followed by the plants sprayed with [Bactroban (inducer) + SICOGREEN® (nutrient and hormonal solution) + Hemoxy (Copper fungicide)] (313 mg), BDISOB172ThaR (266 mg), BDISOB07FarR (270 mg), BDISOB86FarR (273 mg), BDISOB70KusR (276 mg), BDISOB54KhuRwere statistically similar. On the otherhand, the plants raised from the seed treated with the bacterial isolatesBDISOB21ChaR (376 mg), BDISOB186KusR (330 mg), BDISOB219GaiR (373 mg), BDISOB21ChaR (376 mg) were statistically similar (**Table** 7).

3.4.3 Fresh root weight

At 28 days after sowing the highest rootweight (1350 mg) was recorded in plants raised from the seed treated with the bacterial isolate BDISOB45PanPfollowed by the bacterial isolates BDISOB05MymP (1316 mg), BDISOB21ChaR (1306 mg) BDISOB15CumR (1256 mg), BDISOB01MymR (1253 mg), BDISOB92Far (1246 mg), BDISOB16CumR (1213 mg) were statistically similar Whereas, the lowest (830 mg) was recorded in the plants raised from untreated seed followed by thebacterial isolate BDISOB219GaiR (983 mg), plants sprayed with [Bactroban (inducer) + SICOGREEN® (nutrient and hormonal solution) + Hemoxy (Copper fungicide)] (1016 mg), BDISOB30ChaR (1080 mg). Other bacterial isolates were statistically similar (**Table** 7).

Treament	Isolate ID		Fresh shoot weight (mg)	Dry shoot weight (mg)	Fresh root weight (mg)	Dry root weight (mg)
To	Control	—	933.33 k	333.33d-g	830.00 g	170.00 l
T1	Positive control	—	1300.00j	360.00 cd	1016.67ef	220.00jk
T2	BDISOB04KhaP	*P. putida*	1693.33f-i	470.00a	1166.67a-f	403.33b
T3	BDISOB05MymP	*P. putida*	2250.00ab	450.00a	1316.67ab	416.67b
T4	BDISOB219GaiR	*P. taiwanensis*	1816.67d-i	410.00b	983.33 fg	246.67 hi
T5	BDISOB221GaiR	*Pseudomonas sp.*	1533.33ij	293.33 h	1113.33c-f	240.00ij
T6	BDISOB222GaiR	*P. plecoglossicida*	1883.33c-h	443.33a	1116.67c-f	440.00a
T7	BDISOB258GaiR	*P. putida*	1666.67f-i	323.33e-h	1166.67a-f	220.00jk
T8	BDISOB186KusR	*Pseudomonas sp.*	1633.33f-i	330.00d-g	1133.33b-f	233.33i-k
T9	BDISOB283KisR	*Pseudomonas fluorescens*	1950.00a-f	320.00e-h	1116.67c-f	266.67 h
T10	BDISO04DinP	*P. putida*	2033.33a-e	473.33a	1120.00b-f	246.67 hi
T11	BDISO45PanP	*Bacillus paramycoides*	2173.33a-c	326.67e-g	1350.00a	343.33d-f
T12	BDISO198HabP	*S. plymuthica*	1660.00f-i	350.00c-f	1093.33d-f	326.67 fg
T13	BDISO135SerP	*Bacillus sp.*	1766.67d-i	336.67d-g	1133.33b-f	323.33 fg
T14	BDISO148JoyP	*S. marcescens*	1693.33f-i	320.00e-h	1100.00b-f	313.33 g
T15	BDISO1MymR	*B. amyloliquefaciens*	2260.00a	346.67c-g	1253.33a-d	450.00a
T16	BDISO145JoyP	*S. marcescens*	1950.00a-f	313.33 gh	1136.67b-f	240.00ij
T17	BDISO158ChaR	*S. marcescens*	1763.33d-i	293.33 h	1180.00a-f	246.67 hi
T18	BDISOB37KhaR	*P. asaccharolyticum*	1686.67f-i	363.33 cd	1190.00a-e	226.67i-k
T19	BDISOB16CumR	*P. asaccharolyticum*	1730.00e-i	406.67b	1213.33a-e	230.00i-k
T20	BDISOB92FarR	*Pseudomonas fluorescens*	1933.33b-g	466.67a	1246.67a-d	326.67 fg
T21	BDISOB21ChaR	*S. maltophilia*	1800.00d-i	376.67c	1306.67a-c	336.67ef

Treament	Isolate ID		Fresh shoot weight (mg)	Dry shoot weight (mg)	Fresh root weight (mg)	Dry root weight (mg)
T22	BDISOB17CumR	*Limnolyngbya circumcreta*	2066.67a-d	363.33 cd	1220.00a-d	310.00 g
T23	BDISOB15CumR	*P. asaccharolyticum*	1866.67c-h	346.67c-g	1256.67a-d	363.33 cd
T24	BDISOB86FarR	*E. aerogenes*	2033.33a-e	326.67e-g	1170.00a-f	353.33c-e
T25	BDISOB30ChaR	*P. asaccharolyticum*	1733.33e-i	363.33 cd	1080.00d-f	266.67 h
T26	BDISOB07FarR	*S. nematodiphila*	2033.33a-e	316.67f-h	1146.67b-f	373.33c
T27	BDISOB12FarR	*S. marcescens*	1816.67d-i	320.00e-h	1113.33c-f	310.00 g
T28	BDISOB31MagR	*S. marcescens*	1580.00 h-j	323.33e-h	1116.67c-f	236.67i-k
T29	BDISOB36MagR	*S. marcescens*	1613.33 g-i	376.67c	1120.00b-f	230.00i-k
T30	BDISOB46GopR	*S. marcescens*	1700.00f-i	353.33c-e	1123.33b-f	246.67 hi
T31	BDISOB54KhuR	*B. gladioli*	1513.33ij	353.33c-e	1126.67b-f	213.33 k
T32	BDISOB70KusR	*S. marcescens*	1566.67 h-j	363.33 cd	1113.33c-f	233.33i-k
T33	BDISOB172ThaR	*B. aerophilus*	1510.00ij	360.00 cd	1160.00a-f	246.67 hi
Level of significance	—	—	*	*	*	*
LSD	—		270.7	27.85	161.9	20.58
CV	—		9.39	4.73	8.65	4.36

** indicated 5% level of significance.*
Values in columns followed by the same letters indicate no significant differences.

Table 7.
Plant growth promotion by different bacterial isolates antagonistic to X. oryzae *pv.* oryzae.

Isolate ID	Name of bacteria	Lesion length* (mm)	Reduction of lesion length (%)
Control	—	23.67a	0
Positive control	—	6.33b-d	73.31
BDISOB04P	*P. putida*	1.50ij	92.61
BDISOB05P	*P. putida*	1.00j	95.71
BDISOB219R	*P. taiwanensis*	5.67c-f	76.04
BDISOB221R	*Pseudomonas sp.*	5.00d-g	78.85
BDISOB222R	*P. plecoglossicida*	0.83j	96.56
BDISOB258R	*P. putida*	1.50ij	93.61
BDISOB186R	*Pseudomonas sp.*	5.33c-g	77.38
BDISOB283R	*Pseudomonas fluorescens*	1.33ij	94.38
BDISOB04P	*P. putida*	5.83c-e	75.25
BDISOB45R	*Bacillus paramycoides*	2.00ij	91.55
BDISOB198P	*S. plymuthica*	5.83c-e	52.36
BDISOB135R	*Bacillus sp.*	2.83hi	88.08
BDISOB148P	*Serratia marcescens*	5.83c-e	75.69
BDISOB1R	*B. amyloliquefaciens*	2.33ij	90.16
BDISOB145P	*S. marcescens*	6.83bc	71.12
BDISOB158R	*S. marcescens*	6.83bc	50.14
BDISOB37R	*P. asaccharolyticum*	5.33c-g	77.44
BDISOB16R	*P. asaccharolyticum*	5.17d-g	78.01
BDISOB92R	*Pseudomonas fluorescens*	4.50e-g	80.85
BDISOB21R	*S. marcescens*	2.17ij	93.80
BDISOB17R	*Limnolyngbyacir cumcreta*	4.00 gh	83.33
BDISOB15R	*P. asaccharolyticum*	5.33c-g	54.03
BDISOB86R	*E. aerogenes*	4.00 gh	83.33
BDISOB30R	*P. asaccharolyticum*	4.33e-h	81.64
BDISOB07R	*S. nematodiphila*	4.00 gh	83.33
BDISOB12R	*S. marcescens*	4.00 gh	83.06
BDISOB31R	*S. marcescens*	5.00d-g	78.97
BDISOB36R	*S. marcescens*	5.83c-e	75.49
BDISOB46R	*S. marcescens*	4.17f-h	82.28
BDISOB54R	*B. gladioli*	4.17f-h	82.41
BDISOB70R	*S. marcescens*	2.83hi	87.96
BDISOB172R	*B. aerophilus*	7.50b	68.21
Level of significance		*	
CV (%)		16.80	

** indicated 5% level of significance.*
Values in columns followed by the same letters indicate no significant differences.

Table 8.
Effect of some selected antagonistic bacterial isolates on the reduction of lesion length in susceptible check cultivar (IR24) caused by X. oryzae *pv.* oryzae.

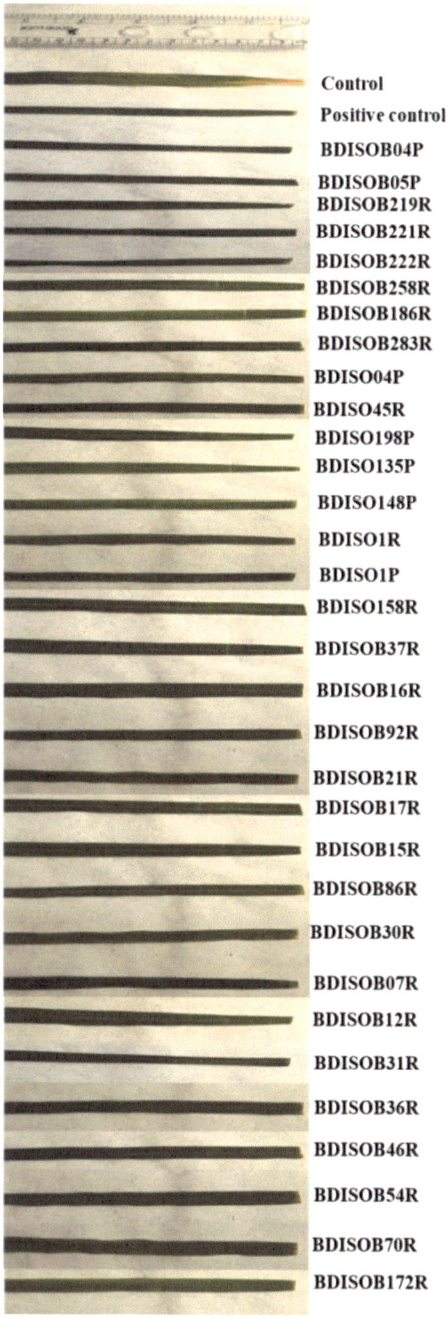

Figure 5.
Reduction of lesion length by some selected antagonistic bacterial in susceptible check cultivar (IR24). Photographs were taken at 14 days after inoculation.. BDISOB04P (P. putida*), BDISOB05P (*P. putida*), BDISOB219R (*P. taiwanensis), *BDISOB221R* (Pseudomonas *sp.)], BDISOB222R (*P. plecoglossicida), *BDISOB258R* (P. putida), *BDISOB186R* (Pseudomonas *sp.*), *BDISOB283R* (Pseudomonas fluorescens), *BDISO04P (*P. putida*), BDISO45R (*Bacillus paramycoides*), BDISO198P (*S. plymuthica*), BDISO135R (*Bacillus sp.*), BDISO148P (*S. marcescens*), BDISOB01R (*B. amyloliquefaciens*), BDISO145P (*S. marcescens*), BDISO158R (*S. marcescens*), BDISOB37R (*P. asaccharolyticum*), BDISOB16R (*P. asaccharolyticum*), BDISOB92R* (Pseudomonas fluorescens), *BDISOB21R (*S. maltophilia*), BDISOB17R (*Limnolyngbya circumcreta*), BDISOB15R (*P. asaccharolyticum*), BDISOB86R (*E. aerogenes*), BDISOB30R* (P. asaccharolyticum), *BDISOB07R (*Serratia nematodiphila*), BDISOB12R (*Serratia marcescens*), BDISOB31R (*Serratia marcescens), *BDISOB36R (*Serratia marcescens*), BDISOB46R (*Serratia marcescens*), BDISOB54R (*B. gladioli*), BDISOB70R (*Serratia marcescens*) and BDISOB172R (*B. aerophilus).

3.4.4 Dry root weight

At 28 days after sowing the highest dry root weight (450 mg) was recorded in plants raised from the seed treated with the bacterial isolateBDISOB01MymR, BDISOB222GaiR (440 mg) followed by the bacterial isolates BDISOB05MymP (413 mg), BDISOB04KhaP (403 mg). Whereas, the lowest (170 mg) was reorded in the plants raised from untreated seed followed by the plants sprayed with [Bactroban (inducer) + SICOGREEN® (nutrient and hormonal solution) + Hemoxy (Copper fungicide)] (220 mg), BDISOB54KhuR (213 mg). Other bacterial isolates were statistically similar (**Table 7**).

3.5 Effect of some selected antagonistic bacterial isolates on the reduction of lesion length in susceptible check cultivar (IR24) caused by *X. oryzae* pv. *oryzae*

To evaluate the mechanisms of BB severity reductionby plant growth promoting antagonistic bacteria, susceptible check variety IR24 was used. The results of plant inoculation showed a significant reduction of lesion length in plants sprayed with formulated bacterial bioagents as compared with untreated control.

(**Table 8**). 96.56% reduction of lesion length was marked as highest spraying with BDISOB222R followed by BDISOB05P (95.71%), BDISOB283R (94.38%), BDISOB21R (93.80%), BDISOB258R (93.61%), BDISOB04P (92.61%), BDISO45P (91.55%) and BDISO1R (90.16%). The minimum (50.145%) reduction of lesion length were observed in plants sprayed with BDISO158R followed by BDISO198P (52.36%) and BDISOB15R (54.03%). Ten bacterial isolates were revealed upper-moderate level of lesion length reduction and eleven isolates were marked their place at lower-moderate level of lesion length reduction. However, all other bacterial isolates reduced lesion length significantly as compared with the untreated plants (**Table 8** and **Figure 5**).

4. Discussion

Antagonistic bacterial isolates were identified mostly as different species of *Pseudomonas*, *Bacillus*, *Serratia* and *Delftia*. In a previous study, frequency of antagonistic bacteria on LB medium was low [62], but another study revealed that using different growth media such as King's B, and Gould's S1 and Nutrient Agar were effective for the isolation of higher number of antagonistic bacteria [63]. It was reported that some antagonistic bacteria such as *B. subtilis*, *B. amyloliquefaciens*, *B. valismortis*, *Streptomyces* sp., *P. chlororaphis* and *Acinetobacter baumannii* were identified based on 16S rRNA sequence analysis [64]. A number of bacteria from species *Alcaligens*, *Arthobacter*, *Burkholderia*, *Alcaligens*, *Arthobacter*, *Burkholderia*, *Bacillus*, *Azospirillum*, *Azotobacter*, *Klebsiella*, *Enterobacter* and *Serratia* have been observed to develop plant growth. However, as bio control agents, isolates of *fluorescens, Pseudomonas,* and *Bacillus* have been the most exploited and studied [65–68]. Nowadays, antagonistic bacteria were also used for plant roots as a biological control infecting by numerous plant pathogens [26, 69]. Out of 300 bacterial isolates sixteen isolates of several species were evaluated *in vitro* and they exposed antagonistic activity to *X. oryzae* pv. *oryzae*. It was observed that 54.03% to 96.56% of lesion length was diminished when treating with antagonistic bacteria. These findings were identical to the reported by Monteiro *et al.* [63] because they also showed that BB pathogen was suppressed by antagonistic bacteria. According to Ranjbariyan *et al.* [70] who also experimented that three bacterial isolates significantly acted higher growth inhibition of *X. oryzae* pv. *oryzae*. Antibiotics, enzymes like chitinases, glucanases, proteases, and siderophore produce directly or indirect mechanisms in which the antagonistic bacteria compete with the pathogen for a niche or nutrient sites [34].

Out of the 63 bacterial isolates, 48 bacterial species were found as positive for IAA (Indole Acetic Acid) production, all 63 bacterial species were found positive for siderophore production and 48 were found capable to solubilize insoluble phosphate. IAA also has been speculated to fasten the overall fitness of plant-microbe associations [57]. It was proved that numerous plant-associated bacteria have the ability to produce IAA by stimulating plant roots development and improving absorption of water and nutrients from soil [71, 72]. The IAA producing bacteria encouraged adventitious root formation, produced the greatest roots and shoots weight [73]. All 63 bacterial isolates were found to produce siderophore. When iron availability is in stress microorganism those who produce siderophore supplied Fe nutrition to enhance plant growth [74]. Siderophore also assists when it comes to the growth condition of shoots, roots as well as nutrition in plants [75]. Siderophore plays a crucial role in selecting a potential bioagent [76], besides, it has been considered as an alternative to ruinous pesticides effects [77]. The biological control mechanism depended on the role of siderophore as competitors for Fe in order to reduce Fe availability for the phytopathogen [78]. Siderophores produced by numerous bacteria had a significant role in the biocontrol and negatively affected the growth of several pathogens [78, 79]. Forty eight bacterial isolates showed the capability of phosphate solubilization. It has been also experimented that phosphate solubilizing bacteria (PSB) can also triggered plant growth promotion [80]. This PSB inoculants have been exploited as a possible alternative for phosphate fertilizers which is inorganic [81] and it also influences phosphate uptake and plant growth [82, 83]. It has also been documented that the application rates of phosphate fertilizers reduced to 50% by inoculating phosphate solubilizing microbes (PSM) added phosphate fertilizers reduced the disease incidence up to 50% [84].

Among the bacterial isolates, 32 were selected based on their antagonistic capability and growth promoting determinants. PGPB have significant impact in surging root length, vigor index and shoot length. Sakthivel *et al.* [85] and Mishra and Sinha [86] reported to enhance growth of rice seedling with bioagent application. Van Peer and Schippers [87] stated that shoot, root and fresh weight was raised for cucumber, lettuce, tomatoand potato as a result of bacterization with *Pseudomonas* strains. The results of the present study depicts that the effect of plant growth promoting bacterial isolates on growth and vigor of rice plants was significantly higher than control. It has been reported that *P. fluorescens* and other plant growth promoting rhizobacteria can show antagonisms to potentially harmful bacterial pathogens and eventually those bacteria contribute to enhance plant growth [88]. Biological agents like plant growth promoting bacteria (PGPB) can be used as bio-fertilizer [89].

Forty eight bacterial species were found positive for phosphate solubilization out of 63 antagonistic bacterial species identified in this study. It has been reported that phosphate solubilizing bacteria (PSB) induced plant growth promotion [80]. Plant roots-associated PSB have been considered as one of the possible alternatives for inorganic phosphate fertilizers for promoting plant growth and yield [81]. Plant growth and phosphate uptake have increased in many crop species due to the results of PSB inoculants [82, 83]. It has also been documented that the application rates of phosphate fertilizers reduced to 50% by inoculating phosphate solubilizing microbes (PSM) in crops without significantly reducing crop yield [84]. In sustainable agriculture, specific plant pathogens can be supressed by biological agents such as plant growth promoting bacteria (PGPB) which can also be used as bio-fertilizer [89]. There are a plenty of PGPB strains that reported to suppress numerous of plant pathogen, reduced disease incidence, triggered plant growth factor and provides nutrition for the growth of the plant [63, 90]. Thus, it has been considerable research interest in the potential use of antagonistic bacteria as PGPB [91, 92]. To evaluate plant-interaction with bacteria, such as endophytes, biocontrol agents, phytopathogens, and symbionts needs to be re-infection and development of those experimental strains in or on field grown plants [93]. Effective root colonization by fluorescent *Pseudomonas* spp. has been manifested to take an inevitable part in controlling plant pathogens as a biocontrol agent [94]. The significance ofthis study is that functionally characterized all antagonistic bacteria may be used for biocontrol of BB along with enhanced rice growth. Even though, *Pseudomonas* spp. are indigenous and involve in various rhizomicrobiomes but few of them have the ability to grow above 37°C and become opportunistic pathogens, thuspredictable biosafety regulations are needed to implement this technology practically for field application [95]. In a nutshell, based on all results achieved from during this study, bacterial strains may be an effective bio-inoculant for controlling BB of rice by ensuring its biosafety aspects.

5. Conclusion

Thirty two potential bacterial isolates were identified belong to the genera mostly *Pseudomonas*, *Bacillus* and *Serratia* from rice phylloplane and rhizosphere among sixty three that inhibited the growth of *X. oryzae* pv. *oryzae* in *in vitro* significantly and were found positive for enhancing plant growth promotion by the production of plant growth promoting determinants viz. IAA, siderophore and phosphate solubilization. Formulated bacterial isolates can be viable in talcum powder for at least three months post formulation. Reduction of lesion length caused by *X. oryzae* pv. *oryzae* on susceptible cultivar IR24 by the formulated bacterial isolates primarily indicates their potentiality in controlling BB of rice. Patenting, registration, large scale formulation and commercilization of these PGP bacteria would be the next step of this work.

Acknowledgements

This research work was carried out with the financial support from National Agricultural Technology (NATP), Phase-2, under Program Based Research Grant (PBRG), Bangladesh Agricultural Research Council (BARC), Farmgate, Dhaka, Bangladesh to Dr. Md. Rashidul Isalm (Grant ID No.: 091).

Conflict of interest

There is no conflict of interest among the authors.

Author details

Md. Mahfujur Rahman[1], Md. Mostafa Masud[1], Muhammad Iqbal Hossain[1], Nur-E-Tajkia Islam[1], Md. Zahangir Alam[1], Md. Mamunur Rashid[3], Mohammad Ashik Iqbal Khan[2], Md. Abdul Latif[2], Krishna Pada Halder[4] and Md. Rashidul Islam[1*]

1 Department of Plant Pathology, Bangladesh Agricultural University, Mymensingh, Bangladesh

2 Plant Pathology Division, Bangladesh Rice Research Institute (BRRI), Joydebpur, Gazipur, Bangladesh

3 Plant Pathology Division, BRRI Regional Station, Cumilla, Bangladesh

4 Director Research, Bangladesh Rice Research Institute (BRRI), Joydebpur, Gazipur, Bangladesh

*Address all correspondence to: rashidul.islam@bau.edu.bd

IntechOpen

© 2021 The Author(s). Licensee IntechOpen. This chapter is distributed under the terms of the Creative Commons Attribution License (http://creativecommons.org/licenses/by/3.0), which permits unrestricted use, distribution, and reproduction in any medium, provided the original work is properly cited. CC BY

References

[1] Latif MA, Kabir MS, Sharma NR and Hossain MA. 2007: Integrated management of five major diseases of rice. (Dhaner pachti prodhan roger somonnito babostapona- in Bangla). 1st edition, published by Bangladesh Rice Research Institute (BRRI), Gazipur 1701, Bangladesh. BRRI Press.

[2] Latif MA, Rafii MY, Rahman MM, Talukdar MRB. 2011: Microsatellite and minisatellite markers based DNA fingerprinting and Genetic diversity of blast and ufra resistant genotypes. Comptes Rendus Biologies. 334: 282-289.

[3] Miah SA. 1973: Recent research results on rice diseases in Bangladesh. A paper presented at the Annual International Rice Conference held at IRRI, Los Baños, Philippines, during April 1973.

[4] Miah SA, Shahjahan AKM, Hossain MA and NR Sharma 1985: A survey of rice diseases in Bangladesh. Tropical Pest Management 31(3): 208-213.

[5] Shahjahan AKM. 1993: Practical approaches to crop pest and disease management in Bangladesh. Bangladesh Agriculture Research Council. P.168.

[6] Sharma NR, Hossain MA, Haque MA, Mondal AH and Rahman MM. 1991: Effect of rate and split application of nitrogen on bacterial leaf blight incidence in rice.Bangladesh J.Plant Pathology.7 (1 & 2): 17-20.

[7] Nino-Liu, D., Ronald, P., and Bogdanove, A. (2006). Xanthomonas oryzae pathovars: model pathogens of a model crop. Mol. Plant Pathol. 7, 303–324. doi: 10.1111/j.1364-3703.2006. 00344.x.

[8] Chattopadhyay SB and Mukherjee N. 1973. Progress of two pathogenic diseases on rice crop grown under high nitrogen fertilized and intensive rice cropping condition. Int. *Rice Comm. Newsl*. 22(2): 43-48.

[9] Devadath S, Dath AP and Jain RK. 1987. Effect of shade, nitrogen fertilization, waterlogging and inoculum concentration on the incidence of bacterial leaf blight of rice. Indian Phytopathology. 40: 529-530.

[10] Kim CH and Cho YS. 1970: Effect of NPK fertilizer levels and growth condition on the development of bacterial blight in rice plants. Kor. J. Plant Prot 9(1): 7-13.

[11] Kauffman HE and Rao PS. 1972: Resistance to bacterial leaf blight-India. *In RiceBreeding*. IRRI, Los Banos, Laguna, Philippines. 283-287.

[12] Mohanty SK, Reddy PR and Sridhar R. 1983: Effect of major nutrients on the susceptibility of rice pants to bacterial leaf blight. *Z. Pflanzenkr. Pflanzensch*. J. Plant Dis. Prot. 90: 50-54.

[13] Rashid, M. M; Nihad, S. A. I.; Khan, M. A. I.; Haque, A.l; Ara, A.; Ferdous, T.; Hasan, M.A.; Latif, M. A. (2021). Pathotype profiling, distribution and virulence analysis of *Xanthomonas oryzae* pv. oryzae causing bacterial blight disease of rice in Bangladesh, J. Phytopathol. 169 (7-8); 438-446.

[14] Alam, S, Islam, R., Hossain, I., Bhuiyan, M. R. (2016). Pathotypic variation of *X. oryzae* pv. oryzae in Bangladesh. *Archives of Phytopathology and Plant Protection*, 323, 1477-2906

[15] Islam, M. R., Alam, M. S., Khan, A. I., Hossain, I., Adam, L. R., Daayf, F. (2016). Analyses of genetic diversity of bacterial blight pathogen, *X. oryzae* pv. oryzae using IS1112 in Bangladesh. *Comptes Rendus Biologies*, 339(9-10), 399-407

[16] Brar, D.S., Khush, GS. (1997). Alien introgression in rice. *Plant Molecular Biology*, 35, 35-47.

[17] Jalaluddin M and Kashem MA. 1999: Pathogenic variability in *Xanthominasoryzae* pv *oryzae* in Bangladesh. Indian J. Agril. Sci. 69: 25-27.

[18] Noda T, Yamamoto T, Ogawa T and Kaku H. 1996. Pathogenic races of Xanthomonas oryzae pv. oryzae in South and East Asia. JIRCAS J. 3: 9-15.

[19] Webster RK and Gunnel PS. 1992: Compendium of Rice Disease. p. 62.

[20] Lee KS, Rasabandith S, Angeles ER and Khush GS. 2003: Inheritance of resistance to bacterial blight in 21 cultivars of rice. Phytopathology. 93: 147-152.

[21] Ou 1973: A Handbook of Rice Diseases in the Tropics. Los Baños: International Rice Research Institute.

[22] Gnanamanickam, S. S. 2009. An overview of progress in biological control. In: Biological control of rice diseases: Progress in biological control Series, ed. by S. S. Gnanamanickam, pp. 43-51. Springer, Dordrecht, Netherlands.

[23] MacManus, P. S., Stockwell, V. O., Sundin, G. W., and Jones, A. L. (2002). Antibiotic use in plant agriculture. Ann. Rev. Phytopathol. 40, 443–465. doi: 10.1146/annurev.phyto.40. 120301.093927

[24] Shanti ML, Shenoy VV, Lalitha Devi G, Mohan Kumar V, Premalatha P, Naveen Kumar G, Shashidhar HE, Zehr UB and Freeman WH. 2010: Marker-assisted breeding for resistance to bacterial blight in popular cultivar and parental lines of hybrid rice. J Plant Pathol. 92: 495-501.

[25] Ali B, Sabri AN, Hasnain S. 2010. Rhizobacterial potential to alter auxin content and growth of *V. radiata* (L.) World Journal of Microbiology and Biotechnology. 26: 1379-1384.

[26] Kloepper, J.W., Tuzun, S., Liu, L., Wei, G., 1993: Plant growth promoting rhizobacteria as inducers of systemic disease resistance. In: Lumsden, R.D., Waughn, J.L. (Eds.), Pest Management: Biologically Based Technologies. American Chemical Society Books, Washington, DC, pp. 156-165.

[27] Van Loon, L.C., 1997: Induced resistance in plants and the role of pathogenesis related proteins. Eur. J. Plant Pathol. 103, 753-765.

[28] Viswanathan, R., 1999: Induction of systemic resistance against red rot disease in sugarcane by plant growth promoting rhizobacteria.Ph.D. Thesis, TNAU, Coimbatore, India, 175 p.

[29] Wei G, Kloepper JW, Tuzun S, 1991. Induction of systemic resistance of cucumber to Colletotrichum orbiculare by select strains of plant growth-promoting rhizobacteria. Phytopathology 81, 1508–1512.

[30] Viswanathan, R., Samiyappan, R., 1999: Induction of systemic resistance by plant growth promoting rhizobacteria against red rot disease caused by Collectotrichum falcatum went in sugarcane.Proceedings of Sugar Technology Association of India, 61 24-39.

[31] Ji, G.-H., Lan-Fang, W., Yue-Qiu, H., Ya-Peng, W., and Xue-Hui, B. (2008). Biological control of rice bacterial blight by Lysobacter antibioticus strain 13-1. Biol. Control 45, 288–296. doi: 10.1016/j. biocontrol.2008.01.004.

[32] Chithrashree, Udayashankar, A. C., Chandra, N. S., Reddy, M. S., and Srinivas, C. (2011). Plant growth-promoting rhizobacteria mediate induced systemic resistance in rice against bacterial leaf blight caused by *X. oryzae* pv.

[33] Li, Q., Jiang, Y., Ning, P., Zheng, L., Huang, J., Li, G., et al. (2011). Suppression of Magnaporthe oryzae by culture filtrates of Streptomyces globisporus JK-1. Biol. Control 58, 139–148. doi: 10.1016/j.biocontrol.2011.04.013

[34] Mousivand, M., Jouzani, G. S., Monazah, M., and Kowsari, M. (2012). Characterization andantagonistic potential of some native biofilm forming and surfactant producing Bacillus subtilis strains against six pathotypes of Rhizoctonia solani. J. Plant Pathol. 94, 171–180. doi: 10.4454/jpp.v94i1.017

[35] Boukaew, S., Chanasirin, K., and Poonsuk, P. (2013). Potential for the integration ofbiological and chemical control of sheath blight disease caused by Rhizoctonia solani on rice. World J. Microbiol. Biotechnol. 10, 1885–1893. doi: 10.1007/s11274-013-1353-x

[36] Spence, C. A., Raman, V., Donofrio, N. M., and Bais, H. P. (2014). Global gene expression in rice blast pathogen Magnaporthe oryzae treated with a natural rice soil isolate. Planta 239, 171–185. doi: 10.1007/s00425-013-1974-1

[37] Wei G, Kloepper JW, Tuzun S. 1996. Induced systemic resistance to cucumber diseases and increased plant growth by plant growth promoting rhizobacteria under field condition. Phytopathology 86:221224

[38] P. Vidhyasekaran and M. Muthamilan (1999) Evaluation of a Powder Formulation of Pseudomonas fluorescens Pf1 for Control of Rice Sheath Blight, Biocontrol Science and Technology, 9:1, 67-74, DOI: 10.1080/09583159929910

[39] Bardin M, Ajouz S, Comby M, Lopez-Ferber M, Graillot B, Siegwart M, Nicot PC. 2015. Is the efficacy of biological control against plant diseases likely to be more durable than that of chemical pesticides? Frontiers in Plant Science 6. doi: 10.3389/fpls.2015.00566.

[40] Montano, P. F., Alias-Villegas, C., Bellogin, R. A., del-Cerro, P., Espuny, M. R., Jimenez-Guerrero, I., et al. (2014). Plant growth promotion in cereal and leguminous agricultural important plants: from microorganism capacities to crop production. Microbiol. Res. 169, 325–336. doi: 10.1016/j.micres.2013.09.011

[41] Van Loon LC, Bakker PAHM and Pieterse CMJ. 1998: Systemic resistance induced by rhizosphere bacteria. Annu Rev Phytopathol 36 453-483.

[42] Kloepper JW, Ryu C-M and Zhang SA 2004: Induced systemic resistance and promotion of plant growth by Bacillus spp.Phytopathology 94: 1259-1266.

[43] Becker, G.J., Conrath, U., 2007. Priming for stress resistance; from the lab to field. Current Opinion in Plant Biology 10, 425–431

[44] Conrath, U. et al. (2002) Priming in plant–pathogen interactions. Trends Plant Sci. 7, 210–216

[45] Chen S, Xu CG, Lin XH, Zhang Q. 2001. Improving bacterial blight resistance of '6078', an elite restorer line of hybrid rice, by molecular marker-assisted selection. Plant Breeding. 120: 133-137.

[46] Ramamoorthy, V., Raghuchander, T., Samiyappan, R., 2002. Induction of defence- related proteins in tomato roots treated with Pseudomonas fluorescens Pf1 and Fusarium oxysporum f. sp. lycopersici. Plant and Soil 239, 55–68.

[47] Schneider, S., Ullrich, W.R., 1994. Differential induction of resistance and enhanced enzyme activities in cucumber and tobacoo caused by treatment with abiotic and biotic

inducers. Physiological and Molecular Plant Pathology 45, 291–304.

[48] Babul RM, Sajeena A, Samundeeswari AM, Sreedhar A, Vidhyasekaran P, Seetharaman K, Reddy, M. S., 2003: Induction of systemic resistance to *Xanthomona oryzae* pv. *oryzae* by salicylic acid in *O. sativa* (L.). Journal of Plant Diseases and Protection110 (5), 419–431.

[49] Choong-Min, R., Murphy, J.F., Reddy, M.S., Kloepper, J.W., 2007. A two-strain mixture of rhizobacteria elicits induction of systemic resistance against Pseudomonas syringae and Cucumber mosaic virus coupled to promotion of plant growth on Arabidopsis thaliana. World Journal of Microbiology andBiotechnology 17, 280–286.

[50] Jetiyanon, K., Kloepper, J.W., 2002. Mixtures of plant growth promoting rhizobacteria for induction of systemic resistance against multiple plant diseases. Biological Control 24, 285–291

[51] Murphy, J.F., Reddy, M.S., Choong-Min, R., Kloepper, J.W., Li, R., 2003. Rhizobacteria mediated growth promotion of tomato leads to protection against cucumber mosaic virus. Phytopathology 93, 1301–1307.

[52] Niranjan-Raj, S., Chaluvaraju, G., Amruthesh, K.N., Shetty, H.S., Reddy, M.S., Kloepper, J.W., 2003a. Induction of growth promotion and resistance against downy mildew on pearl millet (Pennisetum glaucum) by rhizobacteria. Plant Disease 87, 380–384.

[53] Niranjan-Raj, S., Deepak, S.A., Basavaraju, P., Shetty, H.S., Reddy, M. S., Kloepper, J.W., 2003b. Comparative performance of formulations of plant growth promoting rhizobacteria in growth promotion and suppression of downy mildew in pearl millet. Crop Protection 22, 579–588

[54] Raupach, G.S., Kloepper, J.W., 2000. Biocontrol of cucumber diseases in the field by plant growth-promoting rhizobacteria with and without methyl bromide fumigation. Plant Disease 84, 1073–1075

[55] Udayashankar, A.C., Nayaka, C.S., Niranjan-Raj, S., Kumar, B.H., Reddy, M.S., Niranjana, S.R., Prakash, H.S., 2009. Rhizobacteria mediated resistance against Bean common mosaic virus strain blackeye cowpea mosaic in cowpea. Pest Management Science 65, 1059–1106

[56] Panhwar QA, Radziah O, Zaharah AR, Sariah M, Mohd Razi I (2012) Isolation and characterization of phosphorus solubilizing bacteria from aerobic rice. *African J Biotech* 11 11: 2711–2719.

[57] Gupta G, Parihar SS, Ahirwar NK, Snehi SK, Singh V. 2015. Plant Growth Promoting Rhizobacteria (PGPR): Current and Future Prospects for Development of Sustainable Agriculture. J Microb Biochem Technol 7:096-102.

[58] Gordon SA, Weber RP. 1951. Colorimetric estimation of indoleacetic acid. Plant Physiology 26:192-195. doi: 10.1104/pp.26.1.192

[59] Azman NA, Sijam k, Hata EM, Othman R and Saud HM. 2017. Screening of Bacteria as Antagonist against *Xanthomonas oryzae* pv. *oryzae*, the Causal Agent of Bacterial Leaf Blight of Paddy and as Plant Growth Promoter Journal of Experimental Agriculture International 16(4): 1-15.

[60] Nautiyal, C.S. (1999) An efficient microbiological growth medium for screening phosphate solubilizing microorganisms. FEMS Microbiol. Lett. 170, 265-270

[61] Giovannoni JJ, DellaPenna D, Bennett A and Fischer R (1991).

Polygalacturonase and tomato fruit ripening. Hort. Rev.13: 67-103

[62] Wang W, Chen LN, Wu H, Zang H, Gao S, Yang Y, Xie S and Gao X. 2013: Comparative Proteomic Analysis of Rice Seedlings in Response to Inoculation with Bacillus cereus. Lettersin Applied Microbiology. 56: 208-215.

[63] Gomez LCC, Schiliro E, Valverde CA, Mercado BJ. The biocontrol endophytic bacterium Pseudomonas fluorescens PICF7 induces systemic defense responses in aerial tissues upon colonization of olive roots. Front Microbiol. 2014; 5: 427. doi: 10.3389/fmicb.2014.00427 PMID: 25250017

[64] Monteiro L, Mariano RdLR and Souto-Maior AM. 2005: "Antagonism of Bacillus spp. Against *X. campestris* pv. *campestris*." Brazilian Archives of Biology and Technology 48: 23-29.

[65] Alexander DB and Zuberer DA. 1991. Use of chrome azurol S reagents to evaluate siderophore production by rhizosphere bacteria. Biology and Fertilization of Soil. 12: 39-45.

[66] Patten CL, Glick BR. 2002. Role of Pseudomonas putida indoleacetic acid in development of the host plant root system. Applied and Environmental Microbiology 68:3795–3801. doi: 10.1128/aem.68.8.3795-3801.2002.

[67] Yasmin S, Zaka A, Imran A, Zahid MA, Yousaf S and Rasul G. 2016: Plant Growth Promotion and Suppression of Bacterial Leaf Blight in Rice by Inoculated Bacteria. PLoS ONE 11(8): e0160688.

[68] Yasmin S, Hafeez FY, Mirza MS, Rasul M, Arshad HMI, Zubair M and Iqbal M. 2017. Biocontrol of Bacterial Leaf Blight of Rice and Profiling of Secondary Metabolites Produced by Rhizospheric Pseudomonas aeruginosa BRp3. Front. Microbiol. 8: 1895.

[69] Van Peer R, Schippers B. 1991. Plant growth responses to bacterization with selected Pseudomonas spp. strains and rhizosphere microbial development in hydroponic cultures. Canadian Journal of Microbiology 35:456-463. doi: 10.1139/m89-070.

[70] Ranjbariyan AR, Shams-Ghahfarokhi M, Kalantari S, Razzaghi-Abyaneh M. 2011. Molecular identification of antagonistic bacteria from Tehran soils and evaluation of their inhibitory activi- ties toward pathogenic fungi. Iranian journal of microbiology 3: 140–146.

[71] Fuhrmann, J., Wollum, A.G. Nodulation competition among *Bradyrhizobium japonicum* strains as influenced by rhizosphere bacteria and iron availability. *Biol Fert Soils* 7, 108–112 (1989). https://doi.org/10.1007/BF00292567

[72] Glick, B. R. 1995. The enhancement of plant growth by free-living bacteria. Can. J. Microbiol. 41:109-117.

[73] Kloepper JW, Lifshitz R, Zablotowicz RM. 1989. Freeliving bacterial inocula for enhancing crop productivity. Trends in Biotechnology 7: 39–44. doi: 10.1016/0167-7799(89) 90057-7.

[74] Okon Y, Labandera-Gonzalez CA. 1994. Improving Plant Productivity with Rhizosphere Bacteria. Commonwealth Scientific and Industrial Research Organization, Adelaide, Australia. Agronomic Applications of Azospirill.

[75] Rahman MA, Kadir J, Mahmud TMM, Rahman RA, Begum MM. 2007. Screening of antagonistic bacteria for biocontrol activities on Colletotrichum gloeosporioides in papaya. Asian Journal

of Plant Sciences 6:12–20. doi: 10.3923/ajps.2007.12.20.

[76] Aslantaş R, Çakmakçi R, Şahin F. 2007. Effect of plant growth promoting rhizobacteria on young apple tree growth and fruit yield under orchard conditions. Scientia Horticulturae 111:371–377. doi: 10.1016/j.scienta.2006.12.016

[77] Wu SC, Cao ZH, Li ZG, Cheung KC, Wong MH. 2005. Effects of biofertilizer containing N-fixer, P and K solubilizers and AM fungi on maize growth: a greenhouse trial. Geoderma 125:155–166. doi: 10.1016/j.geoderma.2004.07.003.

[78] Cakmakci R, Dönmez MF, Erdoğan Ü. 2007. The effect of plant growth promoting rhizobacteria on barley seedling growth, nutrient uptake, some soil properties, and bacterial counts. Turkish Journal of Agriculture and Forestry 31:189–199.

[79] Crowley DE (2006) Microbial siderophores in the plant rhizos- pheric. In: Barton LL and Abadía J (eds) Iron Nutrition in Plants and Rhizospheric Microorganisms. Springer, Dor- drecht, pp 169-198.

[80] Verma VC, Singh SK, Prakash S 2011: Bio-controland plant growth promotion potential of siderophore producing endophytic Streptomyces from Azadirachta indica Juss. J Basic Microbiol 51: 550–555.

[81] Manninen M, Mattila-Sandholm T. 1994. Methods for the detection of Pseudomonas siderophores. Journal of Microbiological Methods 19:223–234. doi: 10.1016/0167-7012(94)90073-6.

[82] Schenk PM, Carvalhais LC, Kazan K 2012: Unraveling plant–microbe interactions: can multi- species transcriptomics help? Trends Biotechnol 30: 177–184.

[83] Beneduzi A, Ambrosini A, Passaglia LM. 2012. Plant growth-promoting rhizobacteria (PGPR): their potential as antagonists and biocontrol agents. Genetics and Molecular Biology 35:1044–1051. doi: 10.1590/s1415-47572012000600020.

[84] Yu X, Ai C, Xin L, Zhou G. 2011. The siderophoreproducing bacterium, bacillus subtilis CAS15, has a biocontrol effect on fusarium wilt and promotes the growth of pepper. European Journal of Soil Biology 47:138–145. doi: 10.1016/j.ejsobi.2010.11.001.

[85] Oteino N, Lally RD, Kiwanuka S, Lloyd A, Ryan D, Germaine KJ, Dowling DN. 2015. Plant growth promotion induced by phosphate solubilizing endophytic Pseudomonas isolates. Frontiers in Mi- crobiology 6: 745. doi: 10.3389/fmicb.2015.00745.

[86] Thakuria D, Talukdar N, Goswami C, Hazarika S, Boro R, Khan M. 2004. Characterization and screening of bacteria from rhizosphere of rice grown in acidic soils of Assam. Current Science :978–985.

[87] Fankem H, TChakounte G, Nkot L, Mafokoua H, Dondjou D, Simo C, Nwaga D, Etoa FX, et al. 2015. Common bean (Phaseolus vulgaris L.) and soya bean (Glycine max) growth and nodulation as influenced by rock phosphate solubil- ising bacteria under pot grown conditions. In- ternational Journal of Agricultural Policy and Research 5:242–250.

[88] Gusain YS, Kamal R, Mehta CM, Singh US, Sharma AK. 2015. Phosphate solubilizing and indole3-acetic acid producing bacteria from the soil of Garhwal Himalaya aimed to improve the growth of rice. Journal of environmental biology 36:301–307.

[89] Yazdani M, Bahmanyar MA, Pirdashti H, Esmaili MA. 2009. Effect of phosphate solubilization microorganisms (PSM) and plant growth promoting rhizobacteria (PGPR) on yield and yield

components of corn (Zea mays L.). World Academy of Science, Engineering and Technology 49:90–92.

[90] Sakthivel N, Sivamani E, Unnamalai N, Gnanamanickam SS. 1986. Plant growth-promoting rhizobacteria in enhancing plant growth and suppressing plant pathogens. Current Science :22– 25.

[91] Mishra DS, Sinha AP. 1998. Plant growth promotion by some biocontrol agents. In: Fiftieth Annual Meeting of Indian Phytopathological Society and National Symposium on "Present Scenario in Diseases of Oilseeds and Pulses", Maharastra, India.

[92] Van Peer R, Schippers B. 1989. Plant growth responses to bacterization with selected Pseudomonas spp. strains and rhizosphere microbial development in hydroponic cultures. Canadian Journal of Microbiology 35:456–463. doi: 10.1139/m89-070.

[93] Kloepper JW, Leong J, Teintze M, Schiroth MN 1980: Enhanced plant growth by siderophores produced by plant growth promoting Rhizobacteria. Nature 286: 885–886.

[94] Bonaldi, M., Chen, X., Kunova, A., Pizzatti, C., Saracchi, M., and Cortesi, P. (2015). Colonization of lettuce rhizosphere and roots by tagged Streptomyces. Front. Microbiol. 6:25. doi: 10.3389/fmicb.2015.00025

[95] Shanthi AT, Vittal RR. 2013. Biocontrol potentials of plant growth promoting rhizobacteria against Fusarium wilt disease of cucurbit. ESci Journal of Plant Pathology 2:156–161.

Section 3

Crop Quality Control and Food Processing

Chapter 12

Near-Infrared Spectroscopy and Machine Learning: Analysis and Classification Methods of Rice

Pedro S. Sampaio and Carla M. Brites

Abstract

Nowadays, the conventional biochemical methods used to differentiate and characterize rice types, biochemical properties, authentication, and contamination issues are difficult to implement due to the high cost of reagents, time requirement and environmental issues. Actually, the success of agri-food technology is directly related to the quality of analysis of experimental data acquired by sensors or techniques such as the infrared-spectroscopy. To overcome these technical limitations, a rapid and non-destructive methodology for discrimination and classification of rice has been investigated. Near-infrared spectroscopy is considered as fast, clean, and non-destructive analytical tools and its spectra present significant biomolecular information that must be analysed by sophisticated methodologies. Machine learning plays an important role in the analysis of the spectral data being used several methods such as Partial Least Squares, Principal Component Analysis, Partial Least Squares-Discriminant Analysis, Support Vector Machine, Artificial Neuronal Network, among others which can successfully be applied for food classification and discrimination as well as in terms of authentication and contamination issues. The quality control of rice is extremely important at every stage of production, beginning with estimation of raw agricultural materials and monitoring their quality during storage, estimating food quality during the production process and of the final products as well as the determination of their authenticity and the detection of adulterants.

Keywords: Authentication, Classification, Machine Learning, Near-Infrared, Rice, Spectroscopy

1. Introduction

1.1 Rice (*Oryza sativa* L.): biochemical and physical characteristics

Rice (*Oryza sativa* L.), considered as the principal staple food for half of the world's population, is consumed from ancient times being considered one of the most important sources of dietary proteins, carbohydrates, vitamins, minerals and fiber [1]. Rice belongs to the family of cereal grasses, along with wheat, corn, millet, oats, barley, rye, and numerous others. Rice is a plant that normally grows for only one year, consisting of rounded, hollow, and articulated stalks (stems), has flat-looking leaves and a terminal panicle. Rice is considered the only cereal adapted to grow in either flooded or non-flooded soil. Rice is cultivated in different

IntechOpen

climatic and geographic conditions and is the basis of food for a significant part of the world population. The diversity of rice grains and their quality are important factors for producers and consumers and depend on genetic characteristics and growing conditions. The grain is the seed of rice which, when the egg is fertilized, contains an embryo that has an ability to germinate and give rise to a new plant. It consists of the mature ovary, the lemma and palea (shell), the rachilla, the sterile lemmas and the wing (not always present). The embryo, present on the ventral side of the spikelet, close to the spikelet, has an embryonic root. The rest of the grain structure consists mainly of the endosperm (the edible portion), which contains starch, proteins, carbohydrates, fat, crude fiber and inorganic substance. The rough rice kernel includes the husks or hulls and pedicel, as well as the caryopsis (**Figure 1**). The weight distribution of rice caryopsis throughout the maturation phase is defined as follows: pericarp (1–2%), tegument and aleurone (5%), starchy endosperm (89–91%) and embryo (2–3%) [3]. A rice caryopsis (rice seed or whole rice grain) tends to accumulate rapidly during the developmental phase, over 5 to 15 days after fertilization under ideal conditions for development. Starch is accumulated in higher concentration in the starchy endosperm. Small amounts of starch are found in the subaleurone layer and very small amounts are present in the embryo and aleurone layer [4]. Functional proteins are present in different tissues of the embryo during development; the proteins considered storage are found accumulated in these tissues [5]. Storage proteins are found in high amounts in the starchy endosperm, however, the protein concentration is higher in the aleurone layer compared to the subaleurone layer and in the starchy endosperm [3]. Lipids, in the

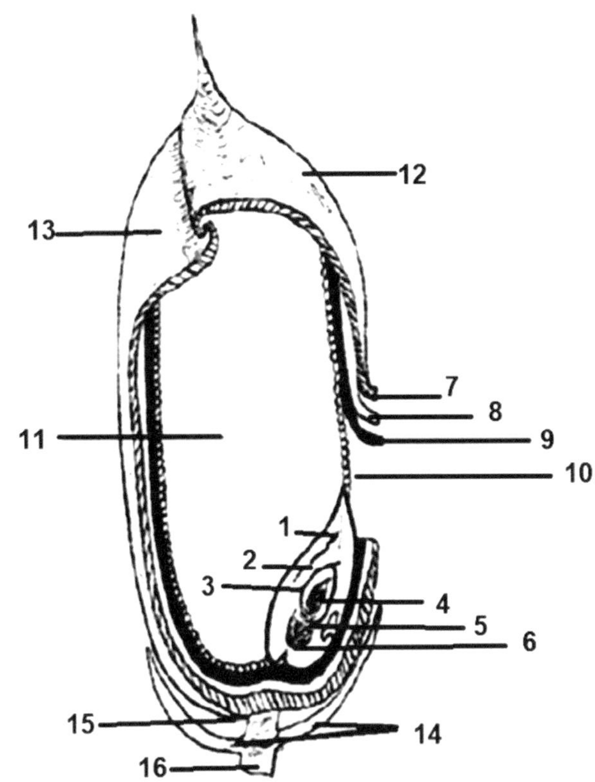

Figure 1.
Parts of rough rice grain. 1-Scutelium (Cotyledon); 2-Coleoptile; 3-Epicotyl (Plumule); 4-Apical meristem; 5-Radicle; 6-Coleorhiza; 7-Pericarp; 8-Tegmen (Seed coat); 9-Aleurone layer; 10-Subaleurone layer; 11-Starchy endosperm; 12-Lemma; 13-Palea; 14-Sterile lemmas; 15-Rachilla; 16-Part of pedicel. Adapted from: [2].

form of lipid bodies, begin to accumulate about five days after anthesis and increase in content in conjunction with starch and protein it can be accumulated for a longer period [6]. The biological activity of the pericarp and seed coat during development is important for cereals, including rice, but the synthetic activity of the seed covering the maternal tissue begins to decline before the endosperm and embryo maturity [7].

Many characteristics of grain quality, such as milling behaviour, appearance, nutritional properties, and cooking qualities, have been routinely evaluated [8]. The evaluation methods of rice varieties are based on their chemical composition, namely (protein, moisture, fat, and ash), apparent amylose concentration, gelatinization temperature, gel consistency and dough viscosity. These procedures are based on standardized methods, which are often considered to be slow and expensive [8]. The classification and characterization of different types of rice depends on several physicochemical parameters, namely, biometric data and protein, fat, ash, moisture, starch, amylose, among other.

Starch is one of main components in rice grain, being the essential carbohydrate reserve in the grain, and so its impact in the evaluated physico-chemical parameters. Starch is a complex polysaccharide of α-D-glucose units exclusively, which are joined by a sequence of α-D-(1,4)-glucosidic linkages thus giving rise to a linear or helical chain, being composed by two classes of glucose polymers: amylopectin and amylose. Amylose is a linear polymer of D-glucose units, and amylopectin is a highly branched polymer of glucose. These are referred to as amylose (20–30%). The much less frequent α-(1,6)-glucosidic linkages form the branch points between the chains thereby creating highly branched domains, denominated amylopectin (70–80%) [9]. Amylose is considered the most important determinant of the eating quality of rice and based on their contents, rice varieties can be classified as: waxy (0–2%); very low (3–12%); low (13–20%); intermediate (21–25%) and high (>26%) [10]. The classical and still commonly used method for the amylose and amylopectin determination is the iodine reaction coupled with potentiometric or amperometric titration. There are also other methods such as: differential scanning calorimetry [11], potentiometric [12], spectrophotometric [13], and chromatographic [14, 15] that can be used for classification and a detailed analysis. The fine structure of amylose, both molecular size and chain-length distribution, are also significant factors of the hardness of cooked rice [16]. Amylose content is correlated with the retrogradation behavior, influencing the textural properties of cooked rice and the viscoelasticity dynamic of rice starch gel [17]. The elongation of grains, volume expansion as well as water absorption characteristics are accounted for cooked rice quality [18].

Proteins and lipid content are also characteristics currently accepted to define rice quality [19]. After starch, the protein is the second main component of rice, being found by four fractions: albumin (soluble in water), globulin (soluble in salt), glutelin (soluble in alkali), which represents the dominant protein in brown rice and white rice, and prolamine (soluble alcohol), a secondary protein in all rice mill fractions [20, 21]. Lipids are the third major component of brown rice, next to carbohydrates and protein, playing a major role in the quality of rice during processing and storage. Fats or lipids are mainly concentrated in the outer bran layer of brown rice, up to 20% by mass; therefore, the lipids content of brown rice is greater than that of milled rice [19, 22].

Appearance quality is how the rice appears after milling and it is associated with grain length, width, length-width ratio (shape) and translucency/chalkiness of the endosperm. Generally, most markets prefer translucent rice as opposed to chalky ones. Appearance quality has a direct influence on marketability and success of commercial varieties. The physical properties of rice grain include all of its external

Figure 2.
Rice grains aspects.

or integral characteristics, such as its appearance (size, shape, smoothness, colour), weight, hardness, volume, flow properties and so on (**Figure 2**).

Rice classification and consequent analysis is a comprehensive quality indicator not only in terms of the appearance but also for its cooking and processing qualities. Physical properties of rice are fundamental in all activities related to the production, preservation and utilisation of rice [23]. The parameters such as dimensions, density, hardness, friction and mechanical properties are affected by the moisture content of the grain and its degree of milling, and also to a small extent by temperature. Cereal research, as well as grading and evaluation of food products, have encouraged the development of non-destructive, rapid and accurate analytical techniques to evaluate grain quality and safety being characterized by a huge amount of experimental data that must be accurately analysed [24]. Different types of rice vary in terms of size, shape, color and constitution, which cannot be accurately identified by human visualization. Often, rice seed cultivars, characterized by high quality, can be faked using low quality cultivars or confused with other cultivars, which complicates rice quality, yield and value. For this reason, the identification of rice seed cultivars is extremely important.

Grain appearance is characterized by biometric parameters (length, width, length/width ratio), total whiteness, vitreous whiteness, and chalkiness, being considered as crucial factor that affects its market acceptability. Grain shape can be described by biometric parameters, which are closely associated with grain weight [25, 26]. The ratio of the length and the width is used internationally to describe the shape and class of the variety. Grain weight provides information about the size and density of the grain. Grains of different density mill differently, and are likely to retain moisture differently and cook differently. Uniform grain weight is important for consistent grain quality [27]. Chalkiness, an opaque white discoloration of the endosperm, reduces the value of head rice kernels and decreases the ratio of head to broken rice produced during the milling process [28]. Viscosity is a characteristic that indicates some of the cooking properties of rice, being evaluated by Rapid Visco Analysis (RVA), which mimics the process of cooking and monitors the changes to a slurry of rice flour and water, during the test. Starch viscosity curves are useful for breeding because the shape of the curve is unique to each class of rice [29]. The primary RVA parameters include peak viscosity, PV (first peak viscosity after gelatinization); trough or hot paste viscosity, HPV (paste viscosity at the end of the 95 °C holding period) and final or cool paste viscosity, CPV (paste viscosity at the end of the test) [30]. The breakdown (BD = PV – HPV); setback (SB = CPV – PV); consistency (CS = CPV – HPV); set back ratio (SBR = CPV/HPV) and stability

(ST = HPV/PV) are considered as secondary parameters, once are derived from primary ones [30–32]. Other factors include peak time (time required to reach peak viscosity), and pasting temperature (temperature of initial viscosity increase) [33].

Industrial processing parameters such as the milling yield husked, milling yield milled, and milling industrial can influence positive and negatively the acceptability of rice by the industrials, can also affect the commercial value of rice. Rice yield and milling quality determine the economic value of rice from the field to the mill and in the industrial market. The rice commercial quality depends on several parameters that are evaluated separately or are involved several time-consuming experimental procedures. The evaluation of some parameters are related to biochemical or biological properties that allow more esasily its determination or prediction. Milling quality aspects affected by temperature during rice ripening include chalkiness, immature kernels, kernel dimensions, fissuring, protein content, amylose content, and amylopectin chain length [10]. Rice milling process can be subjected to dehusking of paddy which results in brown rice, and removing the bran from the kernel by polishing the brown rice to yield white rice. The milling quality of rice determines the yield and appearance of the rice after the milling process.

1.2 Near-infrared spectroscopy

Beer's law is generally applied in analytical spectroscopy to correlate the concentrations of standard samples with corresponding analyte absorbances to develop the calibration curve that is later used to evaluate the concentration of analyte of unknown samples, typically at lambda (λ_{max}). Variation in other wavelengths/wavenumber regions is often not considered but contains significant information that may be selected to represent analyte absorption fingerprint signatures and spectral profiles for ultimate pattern recognition and/or quantification of analytes in unknown samples.

Analytical infrared spectra are focus on the absorption or reflection of the electromagnetic radiation can be divided in three regions of IR: near IR (NIR) in the 12.000–4000 cm^{-1} region, mid IR (MIR) in the 4000–400 cm^{-1} region, and far IR (FIR) beyond 400 cm^{-1} (**Figure 3**). The MIR region (4000–400 cm^{-1}) is a well-recognized and reliable method through which different compounds can be identified and quantified, being used for biological applications, which includes the so-called fingerprint regions representative for lipids, proteins, amide I/II, carbohydrates, and nucleic acids (**Figure 3**). FIR spectroscopy (400–20 cm^{-1}) provides information on the highly ordered structures such as fibrillar formation and protein dynamics [35] since it is more sensitive to the vibrations from the peptide skeletons and hydrogen bonds than MIR [36]. NIR, known also "far-visible spectroscopy" or "overtone vibrational spectroscopy", can measure the chemical composition of biological materials using the diffuse reflectance or transmittance of the sample at several wavelengths [37]. The NIR spectrum, from 12.000 to 4000 cm^{-1} lies between the visible and mid-infrared regions of the electromagnetic spectrum, is characterized by a number of absorption bands that vary in intensity due to energy absorption by specific functional groups in a sample [38].

NIR is a spectroscopic technique used to study of hydrogen bonding because it evaluates the overtones and combinations of the molecule's vibrational modes, principally those involving hydrogen. NIR spectroscopy can measure the concentration of components, characterized by different molecular composition such as protein, water, or starch [39]. The chemical bonds present in food and crop components such as fats, water, and carbohydrates are easily detected by NIR spectroscopy due to the specificity of the radiation, in terms of the groups of interest such as N-H, C-H, and O-H bonds. Due to the macromolecular complexity of the rice sample, it is normal for these bands to overlap one another.

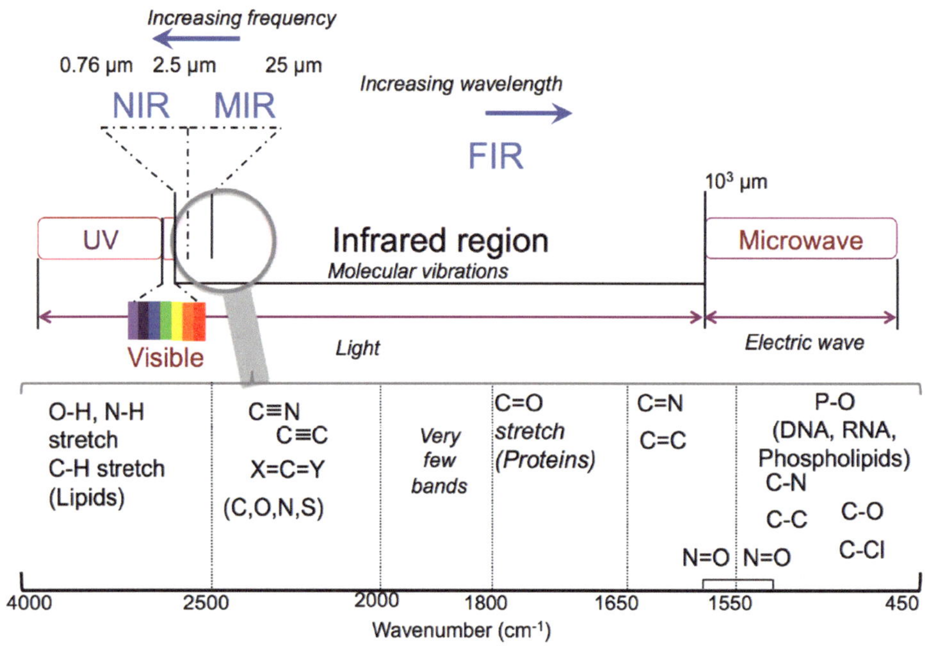

Figure 3.
Infrared spectral region (adapted by Balan et al. [34]).

The transmission and reflection are defined as the two major modes of NIR spectroscopy, that are used based on physical state of the sample. Transmission modes are more suitable for liquids, thin solids, and thick solids when inspecting a food item for its ripeness, or whether it contains pests or defects. In another side, reflectance mode is applied for measuring content in whole grains such as lipids, starch, amylose, protein, moisture, and oil content. Low reflectivity indicates that energy diffuses readily beneath the surface of most samples, including visually opaque samples. Low absorptivity represents that NIR light energy easily penetrates the samples without fast attenuation [40]. This technique is extensively used in breeding procedures for quality improvement of any cereals, and crop management, receivable testing, and on-line process control [41, 42].

The NIR methodology presents some advantages such as no sample preparation or pre-treatment process, no need for dangerous reagents or solvents, and no disposal problem, either. These advantages can eliminate sampling errors caused by manual sample handling and reagent contamination. The samples also can be used in additional studies, being carried out by technically untrained personnel. On the other hand, through NIR analysis, it is possible to obtain a set of spectra, simultaneously, in a certain range of wavelengths, which may serve as a basis for the development of specific calibration curves for each analyte. In the calibration process are transformed during modelling using, for this purpose, chemometric techniques that use a representative set of training to use the program to discriminate slight differences that exist in the specific spectra of the sample [43]. A single spectrum can be subjected to many different calibration models, to measure any number of constituents.

Different techniques such as machine vision and Visible/Near-Infrared spectroscopy have been developed and applied to determine and characterize rice varieties and evaluate the biochemical characteristics. Traditional techniques used for rice variety evaluation such as High-pressure Liquid Chromatography (HPLC) or Gas chromatography-mass spectrometry (GC-MS) are time-consuming and hard to apply [44]. NIR spectroscopy, compared to the traditional analysis methods, is characterized by many advantages, such as is easy-to-use, real-time analysis, fast and accurate, highly reproducible results, non-destructive sampling, no sample preparation, multiple components analysis with a single measurement, high precision and non-destructive detection, being widely used in the measurement of agricultural and food products [45, 46].

1.3 Spectral pre-processing techniques

Over the years, several multivariate regression analysis methods have been developed in order to provide significant information from spectral data, due in part to the limitations of univariate spectral analysis. The processing of spectral data for chemical analysis usually uses the field of statistics and advanced mathematics for an analysis in terms of multivariate regression of spectral data. Simultaneous investigation of several wavenumbers or wavenumbers for biochemical analysis can be carried out through multivariate regression techniques, as these allow the analysis of different sample components without the need for spectral resolution and spectral deconvolutions. Pre-processing methods allowed eliminating noise caused by spectral data, which allow to remove the non-informative variability present in the spectra. Data pre-processing techniques such as normal variable transformation (SNV), multiplicative dispersion correction (MSC) and smoothing derivative are required for raw NIR spectra for proper qualitative classification and development of quantitative calibration models. MSC is used to compensate for particle size effects as it rotates the spectra to remove part of that effect, adjusting as close to the average spectrum as possible [47]. The first and second derivatives are calculated according to the Savitzky–Golay approach using a 19 point window and a 2nd or 3rd order polynomial, which allows to remove noise such as baseline drift, large, reverse and so on [48–50] (**Figure 4**).

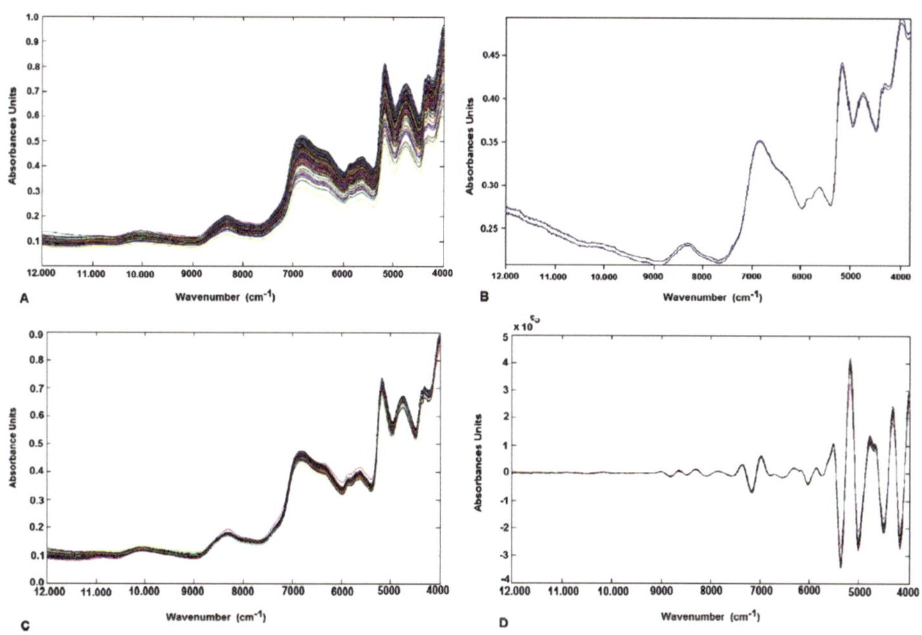

Figure 4.
Rice NIR spectra data without treatment (a); and after pre-processing procedure: baseline correction; (b, c) and first derivative process. (Adapted from Sampaio et al. [51]).

1.4 Machine learning methods

Machine learning is one of the most promising technologies in the field of artificial intelligence, that involve the use of algorithms that allow machines to learn by imitating the way humans learn step. Machine learning based on experimental data allows to optimize grouping or classification, developing models that allow to predict the behavior or properties of systems. There are two main types of machine learning: the supervised and the unsupervised process. Supervised machine learning uses algorithms that "learn" from the labeled data entered by a person without an algorithm. The algorithm generates expected output data as long as the input has been labelled and prior primary. There are two types of data that can be used in the development of the algorithm: (a) classification, which classifies an object into different classes, for example, it allows determining the type of rice according to its physical characteristics; (b) Regression, predicts a numerical value such as the concentration of any biochemical parameters such as the protein, lipids, or carbohydrates, etc. Supervised learning consists of learning a function from training examples, based on their attributes (inputs) and labels (outputs). In the unsupervised machine learning, unlike the previous case, there is no human intervention, and the algorithms learn process is based on the data with unlabeled elements, looking for patterns between them without human intervention. In this case two types of algorithms have been developed: (a) clustering, classifies the output data into groups according to its similarity; (b) association, the algorithm discovers rules within the data set. In semi-supervised learning, both labeled and unlabeled data is used for training, with usually only a small amount of labeled data, but a large amount of unlabeled data. Instead, the learning system receives some sort of a reward after each action, and the goal is to maximize the cumulative reward for the whole process. The much recognized machine learning methods are: Principal Component Analysis (PCA), the most basic feature extraction unsupervised techniques, based on the analysis of the variance of features within the full spectrum; the clustering unsupervised methods, used to identify biological subtypes within a sample, such as Hierarchical Cluster Analysis (HCA), k-Nearest Neighbors (KNN), Artificial Neural Networks (ANN), discriminant analysis (DA), Partial Least-Squares-Discriminant Analysis (PLS-DA), Partial Least-Squares (PLS), and Support Vector Machines (SVM).

1.4.1 Principal Component Analysis

Principal Component Analysis (PCA) is an unsupervised technique that allows the dimensionality reduction of the multivariate data to *n* principal components that preserves the variance of initial data as possible in the lower dimensionality output data [52]. The huge number of data are transformed into a reduced number of uncorrelated variables called principal components (PC) where each component represents a linear combination of the original data and the number of PCs is equal to the original variables. Early PCs explain most of the sample data, which allows for the reduction of data size. A PCA can reveal as variables that determine some inherent structure of the data, which can be interpreted in chemical or physicochemical terms. The scatter plot of PC1 and PC2 scores represent the most expressive variability among themselves, which account for most of the variability between samples and contain information from the entire spectrum. The PCA has been coupled with Mahalanobis distances to reduce dimensionality before carrying out the discriminant analysis [53]. Plots of PCs *versus* each other represents how the variables that they account for are related. To monitor the cluster together is important to determine a set of scaling coefficients, the scores. The scores for each factor can be evaluate for every spectrum in the training set. The original spectra are constructed when the scores are multiplied by the load vectors and the results summed. In this way, knowing the set of charge vectors, how scores represent the spectra with the precision of the original responses at all wavelengths. PCA avoids the problem of overfitting by selecting too many wavelengths. This pattern recognition method was used to determine the Mahalanobis distances that are determined in units of standard deviations from the center (mean) of the training set cluster. Cross-validation is one method that is employed for evaluating the suitable number of factors. For performing this evaluation, each sample present in the calibration set is eliminated one by one and the remaining samples are used to build a Mahalanobis matrix for one, two, three factors, and so on. Then, the excluded sample is predicted, using the models developed for Mahalanobis grouping. The excluded sample is then put back to the calibration set, and a new sample is removed. The process continues until all changes have been removed from the calibration and prediction set. This represents an advantage of cross-validation compared to other methods, since the favors are not the same in relation to those used to define the model.

1.4.2 Discriminant Analysis

A Discriminant Analysis is a strategy that has been used successfully for a qualitative analysis, being called pattern recognition. This methodology aims to classify groups as groups into well-defined groups according to the similarities of a "training set" despite limited knowledge of the composition of those belonging to the group. Johnson and Wichern [54] concluded that the use of discriminant analysis uses several variables and analyzed how to solve the grouping together. The development of calibration models in discriminant analysis is based on two methods: Mahalanobis distances, considered the unit distance vector in multidimensional space, and PCA coupled with Mahalanobis distances [54, 55]. The Mahalanobis distance can be defined by an ellipsoid in a multidimensional space that circumscribes the data. This method is based on a matrix that represents the inverse of the matrix formed by combining the covariance matrices within the group of all groups, which is generated by combining information from all different materials of interest in a single matrix. Studies developed by and Williams considered the Mahalanobis distance as the mathematical number that defines the position, size and shape of the ellipsoid for all clusters [38]. According to of statistical perspective, the Mahalanobis distance considers the sample variability to be valid, while the Euclidean distance method does not consider the variability of values in all dimensions to be valid. The Mahalanobis distances look at not only variation between the responses at the same wavelengths, but also at the inter-wavelength variations. Instead of treating all values equally when calculating the distance from the mean point, it weights the differences by the range of variability in the direction of the sample point. The place of each cluster in multidimensional space is defined by the mean value of the absorbances (the group mean) at each wavelength. Dunmire and Williams indicated that the sample can be classified clearly if it falls within three times the Mahalanobis distance from the respective centroid and at least six times the Mahalanobis distance from the ellipses of other groups [38]. Meanwhile, the Mahalanobis distance represents a multidimensional distance *D* defined by the matrix equation as follows (Eq. (1)) [55]:

$$D^2 = (x - x')M(x - x') \tag{1}$$

where x represents a vector related to optical readings at several wavelengths which describes the position in multidimensional space corresponding to the spectrum of a given sample, x' is a vector that represents the position of a reference point in space, while M is the pooled inverse covariance matrix describing distance measures in the multidimensional space.

1.4.3 Partial Least Squares-Discriminant Analysis

Partial Least Squares-Discriminant Analysis (PLS-DA) is defined as a linear classification method that permits to estimate the predictive models based on partial least squares regression algorithm that follows for latent variables with maximum covariance, representing the significative sources of data variability with linear combinations of the original variables is considered an example of machine learning tool applied to conduct a global cellular analysis of bioprocess as an exploratory technique, gaining increasing attention as a useful feature selector and classifier [56–60]. Multivariate classification methods aimed at finding mathematical models able to recognize the membership of each sample to its appropriate class, by a set of measurements. PLS-DA have shown promising results in the detection of food adulteration without identifying specific compounds [61]. PLS-DA is a discriminant classifier, being particularly suitable for handling correlated features (e.g., spectroscopic variables). The predicted value is a number, but not a dummy integer. Thus, a cut off value needs to be set to determine which class the sample belongs to. PLS-DA is computed based to full cross validation methods. More specifically, a predictor block is used to estimate (by PLS) a binary response called dummy Y (a binary response matrix encoding the class-belonging). Mathematically, the regression relation between the data matrix X and the dummy vector y for a two-class case is represented by the model represented in Eq. (2)

$$y = \hat{y} + e = X_b + e \tag{2}$$

where $\hat{y}$, b, and e represents, respectively, the vectors of predicted responses, regression coefficients, and residuals. When new samples (test set) need to be classified, their predicted responses, $ynew$, are calculated based on the measurements, X_{new}, and the regression coefficients, b, estimated on the training set, and the classification rule is then applied to assign each individual to one of the categories under study.

1.4.4 Support Vector Machine

Support Vector Machine (SVM) is a widely used supervised statistical learning algorithm, considered as a nonlinear classification technique, which works with supervised learning models that analyze data used for classification and regression analysis, producing linear boundaries between objects groups in a transformed space of the x-variables [62–64]. SVM was previously used to detect and quantify milk adulteration by mid-infrared spectrometry [64] and to identify rice seed cultivars [65]. SVM reveals advantages in dealing with small sample, non-linear and high dimensional data. The model performance depends of the selection of kernel function in SVM models, and the commonly used Radial Bias Function (RBF) is used as kernel function. The regularization parameter c, controls trade-off between the minimum training error and minimum model complexity, along with the kernel parameter g of the kernel function. The parameter c reflects the degree of generalization, represents the width of the kernel function and reflects the degree of generalization are determined by a grid-search procedure in SVM.

1.4.5 Partial Least Squares

Partial Least Squares (PLS) regression and principal component regression (PCR) are examples of quantitative regression algorithms that are currently used for linear data, being considered as factor-based models. PLS and PCR use information from all wavelengths in the entire NIR spectrum to predict sample composition, instead of using a few selected wavelengths. PLS is similar to PCR but more sensitive in terms of variations in sample concentration. Studies performed by Wehling described that PLS and PCR, based on data reduction approaches, allowed to decrease a huge number of variables to a much smaller number of new variables that account for most of the variability in the samples [66]. The amount of a constituent in samples can then be predicted by these new variables. PLS is the most widely used supervised multivariate data analysis method that estimates and quantify components in a specific sample. Each training example is defined as a pair $(x, f(x))$, where x represents the input, and $f(x)$ is the output of the underlying unknown function. The objective of supervised learning is given a set of examples of f, return a function h that best approximates f. Osborne et al. indicated that PLS tends to generate solutions that need fewer factors than calibrations of comparable performance produced by PCR [53]. PLS is defined as a regression algorithm that uses concentration data during the decomposition process and involves information as much as possible into the first few loading vectors [67]. It performs, simultaneously, a decomposition on the spectral and concentration data. A small number of factors are developed as specific data linear and regression on the scores of the factors used to derive a prediction equation. To remove irrelevant spectral variables and to improve model performance, several methods have been studied to select the optimal variables for multivariate calibration. The multivariate calibration allowed builds a predictive model, relating variables (wavenumbers) to properties of interest (concentration data). To address this common problem, a variety of linear regression methods based on latent variables (LVs) have been developed, such as partial least squares (PLS), but due to several drawbacks such as the noise in spectral data, the calibration and prediction errors are high, and the model can be affected [68]. Regardless of the regression method, the initial stage of this process is related to a typical development, optimization and refinement. The main objective of any multivariate regression is to predict unknown the samples' with a degree of certainty and great accuracy using a process known in multivariate analysis as "validation". The established regression models must be sufficiently validated, usually with independent validation samples of known concentrations. Root-mean-square-error-of-prediction (RMSEP) and root-mean-square-percent-relative error (RMSRE) are utilized to calculate the reliability and performance of the regression model for accurate determination of analyte concentrations of validation or future samples.

The matrices containing the data provided by the NIR spectra, denominated by X and the vector Y containing the parameters that it will be determine are employed to build the regression model. The performance of the final PLS model is evaluated according to the RMSEP and the correlation coefficient (R). RMSEP was defined as:

$$RMSEP = \sqrt{\frac{\sum_{i=1}^{n}\left(y_i - \hat{y}_i\right)^2}{n}} \quad (3)$$

$$R = \sqrt{1 - \frac{\sum_{i=1}^{n}\left(\hat{y}_i - y_i\right)^2}{\sum_{i=1}^{n}\left(y_i - \bar{y}\right)^2}} \quad (4)$$

where n represents the number of samples in test set validation, y_i is the reference measurement result for the test set sample i and $\hat{y}_i$ is the estimated result of the model for the test sample i. (Eq. (3)). Correlation coefficient (R) relatively to the predicted and the quantified value are determined for both the calibration and the test set which is determined based on the (Eq. (4)), where $\bar{y}$ represents the mean of the reference measurement for all samples in the calibration and test set. The best combination of spectral regions and the pre-processing techniques were selected by picking the PLS model with a small RMSEP, a high R and a low number of latent variables covering enough data variance. The model construction was based on test set validation composed by randomly chosen samples from the entire dataset, not used for model calibration. Based on PLS models, there are some procedures that depends on specific algorithm, spectral region selection, can considerably improve the performance of the full-spectrum calibration techniques, avoiding non-modeled interferences and building a well-fitted model [69–71]. Studies then performed showed that it is fundamental to conduct a spectra region selection responsible for the property of interest to increase the prediction performance [72, 73]. These procedures can be categorized into two classes: single wavelength selection and wavelength interval selection. Different strategies have been suggested for selection of optimal set of spectral regions such an interval PLS (iPLS), synergy PLS (siPLS), and moving window PLS (mwPLS) [69, 74, 75]. The principle of iPLS involves of splitting the spectra into equal-width intervals, and developing sub-PLS models for each one. The sub-intervals with the lowest value of the root mean squared error of prediction (RMSEP) must be chosen as the best. Several methods based on iPLS were developed to optimize the combination of the selected intervals, such as synergy iPLS (siPLS) [74]. These methods present a significant advantage because it uses a graphical presentation to focus on a selection of better sub-intervals and perform comparison among the prediction execution of local models and the full-spectrum model. Instead of just testing a series of adjacent but non overlapping intervals, which would miss some more informative ones, mwPLS was proposed to overcome this drawback. This strategy develops a series in a window that moves through the complete spectra and then selects the informative intervals with low model complexity and low value of the sum of residuals. Because it considers all the possible continuous intervals, it can select all the possible informative intervals but not the optimized ones [76].

1.4.6 Soft Independent Modeling of Class Analogy

Soft Independent Modeling of Class Analogy (SIMCA) is a supervised discriminant analysis method based on PCA [77]. This methodology is a class-modeling approach, meaning that, in defining the class boundaries, the method focuses on the similarities among samples from the same category [61, 78]. For each class, a PCA model is created and consequently the residual variance of the modeled class with the residual variance of the unknown sample is compared to determine which category the sample belongs to. The number of PCs used in each class should be selected to achieve the best classification results. SIMCA results are presented in terms of "sensitivity" and "specificity", where the former specifies the percentage of samples truly belonging to the category correctly accepted by the class model, while the latter expresses the percentage of the objects from other classes which have been correctly rejected. SIMCA starts from a principal component analysis (PCA) of only the training objects belonging to the category to be modeled, to "capture" the regular variability due to the similarities among samples of the same class [79, 80]. Once the PCA is calculated, objects are accepted or rejected by the class-model based to their reduced distance from the class space, referred as d. For a generic i^{th} sample, the d value is calculated by Eq. (5),

$$d_i = \sqrt{\left(\frac{T_i^2}{T_{0.95}^2}\right)^2 + \left(\frac{Q_i}{Q_{0.95}}\right)^2} = \sqrt{\left(T_{i,red}^2\right)^2 + Q_{i,red}^2} \qquad (5)$$

where T^2 is the Mahalanobis distance of the sample from the center of the class space and Q is its orthogonal distance from the PC subspace. These values are divided by $T^2_{0.95}$ and $Q_{0.95}$, which are the 95th percentiles of the T^2 and $Q_{0.95}$ distributions, obtaining the reduced T^2 (T^2_{red}) and the reduced Q (Q_{red}), respectively [79]. Due to the normalization, T^2 and Q limit values are equal to 1; a sample will then be accepted by the class model if $d <$ sqrt(2), otherwise it is rejected.

1.4.7 k-Nearest Neighbor

k-Nearest Neighbor (k-NN) is methodology used for a classification step based on the closest training examples in the feature space. If most an unknown sample's *k*-Nearest Neighbors in training set belong to a specific class, then this unidentified sample is classified as this class. The parameter *k* affects the performance of k-NN model. The Euclidean distance is the most common algorithm used in k-NN [81].

1.4.8 Random Forest

Random Forest (RF) is a novel machine learning algorithm that presents many decision trees, and each tree is grown from a bootstrap sample of the response variable. The optimal split is chosen from a random subset of variables at each node of the tree, and then extends the tree to the maximum extent without cutting. Prediction procedure can be performed from new data by combining the outputs of all trees. RF is suitable and fast to deal with a large amount of data, showing the advantages to reduce variance and achieve comparable classification accuracy [82, 83].

1.4.9 Artificial Neural Networks

Artificial Neural Networks (ANNs) is defined a non-parametric regression models that capture any phenomena, to any degree of accuracy (depending on the adequacy of the data and the power of the predictors), without prior knowledge of the phenomena. ANNs are applied for classification and function mapping difficulties which are tolerant of some inaccuracy and have lots of training data available, but to which hard and fast rules cannot easily be applied [84]. In the ANN the input layer is linked to an output layer, either directly or through one or numerous hidden layers of interconnected neurons. The amount of hidden layers defines the depth of a ANN, and the width depends on the amount of neurons of each layer. Rapid optimization algorithms are used to iteratively develop forward and backward passes for minimization of a loss function and to learn the weights and biases of the layer. The activation functions are applied to the present values of the weights at each layer in the forward pass. The final result of a forward pass is new predicted outputs. The backward pass computes the error derivatives among the expected outputs and the real outputs. These errors are then disseminated backwards updating the weights and calculating new error terms for each layer. Iterative repetitions of this process is designated as back-propagation [85]. A neural network is an adaptable system that learns relationships from the input and output data sets and then can predict a previously unseen data set of similar characteristics to the input set [86, 87]. Multilayer perceptron (MLP) and radial basis function (RBF) are widely used neural network architecture in literature for regression problems [88–90]. MLPs are usually used for prediction and classification using suitable training algorithms for the network weights. The MLP trained with the use of back propagation learning algorithm. **Figure 5a** represents a three-layer structure (MLP) the most basic ANN and its minimum configuration that consists of three layers of nodes (1) input layer, (2) hidden layer, and (3) output layer. The input layer accepts the data and the hidden layer processes them and finally the output layer displays the resultant outputs of the model [91, 92]. Each node, with the exception of the input, is a neuron that is based on a non-linear activation function. The MLP can be regarded as a hierarchical mathematical function planning some set of input values to output values via many simpler functions. Normally, the nodes are fully linked between layers and therefore the quantity of parameters quickly increases to huge numbers with a considerable risk of overfitting [93]. The RBF is considered the most broadly used structural design in ANN and simpler than MLP neural network (**Figure 5b**). The RBF has also an input, hidden and output layer. There are different types of radial basis functions, but the most widely used type is the Gaussian function.

1.4.10 Multiple Linear Regression

Multiple Linear Regression (MLR) is a commonly used machine learning algorithm that allows to determine a mathematical relationship among a number of random variables, analyzing how multiple independent variables are related to one dependent variable. Since each of the independent factors has been determined to predict the dependent variable, information about the multiple variables is used to develop an accurate prediction about the level of effect they have on the outcome variable. The model generates a relationship in the form of a straight line (linear) that best approximates all the individual data points. The most important advantage of MLR is it helps us to understand the relationships among variables present in the dataset. This will further help in understanding the correlation between dependent

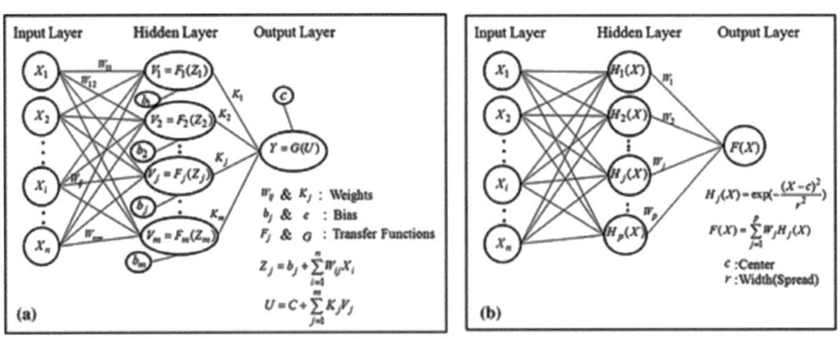

Figure 5.
A comparative study of artificial neural network (MLP, RBF) models for rice biochemical parameters prediction. Simple configuration of (a) MLP and; (b) RBF neural networks [86].

and independent variables. MLR is one of the oldest regression methods, being used to establish linear relationships between several independent variables (X_i) and the dependent variable (sample property) (Y) that depends by them. The developed model can be represented in the following the Eq. (6):

$$y_i = b_0 + \sum_{i=1}^{N} b_i x_i + e_{i,j} \tag{6}$$

where y_i represents the sample property, b_i represents the computed coefficient for each variable x_i, while $e_{i,j}$ is the error. Each independent variable is analyzed and correlated with the specific property y_j. Regression coefficients b_i represent the effects of each determined term. After the MLR model has been developed the accuracy in prediction of the dependent variable is evaluated by computation of the correlation coefficient, which is calculated when true values are compared to predicted ones. Coefficient of determination R^2 is not reserved for MLR, as it is one of the most frequently used statistic parameters for assessment of validity of the developed model regardless of the model type (Eq. (7)).

$$R^2 = 1 - \frac{\Sigma\left(y_i - \hat{y}_i\right)}{\Sigma\left(y_i - \bar{y}\right)} \tag{7}$$

2. Practical applications of NIR spectroscopy and chemometrics

2.1 NIR spectroscopy in rice analysis: identification and classification

There are several studies that discribe the quantitative analysis by NIR spectroscopy in different types of food, providing an exceptional method for the evaluation of chemical composition (*i.e.* protein, starch, lipid, amylose, and moisture contents) in raw pork and beef [94], in cheese or other dairy products [95]. However, it is most widely used in the field of grains and cereal products. In some cases, such measurements are important to achieve the end-used objectives of a plant breeding program. The use of NIR spectroscopy for the quality assessment of processed foods has generated a lot of interest during the review period. Access to food with high quality is essential to human health. Thus, the accurate collection of agricultural food quality data in real-time is utmost importance, such as grains and flours. NIR spectroscopy has proven beneficial for the analysis of various cereals, grains, flours, and baked goods, including specific quality parameters, which influence classification, safety, grading, and price. By analyzing numerous factors and properties of crops during different steps in their development, crop quality can be expected early on. To maximize efficiency and lessen waste of produce, it is important that these data collection methods be non-invasive, non-destructive, and economical. Gas chromatography (GC), high-pressure liquid chromatography (HPLC), or mass spectrometry (MS) represent some quantitative instrumental techniques used for quality assessment of foods. However, these techniques are not applicable for real-time measurements. Spectroscopic instrumentation have recently utilized in agricultural industries for quality analysis. NIR spectroscopy allows a detailed food analysts to examine the quality, composition, and the authenticity of agricultural and food products quickly and accurately, based on physicochemical properties of crops. Machine learning methodologies have been coupled with NIR spectroscopy for the prediction of rice quality factors [96] and the quantitative determination of amylose values [51]. There have been numerous applications of portable NIR instruments in recent years for specific analyses such as determining adulteration in rice and other food quality parameters. NIR spectroscopy is highly useful in analyzing shelf life and maturity of agricultural products like rice. However, the data collection and modeling are still time consuming for portable spectrometers to be efficient in some applications. This can be potentially overcome by combining NIR spectroscopy with other analytical methods. Studies developed by [97] allowed to develop a tandem approach of monitoring rice germ shelf life during storage using NIR and a portable *e*-nose. Le et al. proposed a study that combines deep learning with NIR to provide a much faster method of cereal analysis compratively to traditional NIR models [98]. The deep learning algorithm removes interference of spectral signal developing modeling significantly efficient. Jiang et al. developed a portable NIR spectrometer system to dynamically evaluate the fatty acid content of rice during storage [99]. Another challenge in NIR spectroscopy is determining authenticity and the geographical location of certain agricultural products like grains. Studies carried out by Sampaio et al. developed a strong and accurate classification model based on machine learning methods and NIR spectroscopy, allowing to sorting two genotypes of rice with high accuracy based on these characteristics [100]. Barnaby et al. correlated the grain chalk of rice to the genomic regions of NIR spectra. These spectral regions can be applied in the automation of grain chalk quantification or for other grain products as well [101].

There are several studies based on NIR to predict viscosity properties of rice. Delwiche et al. developed calibration models on whole-grain milled rice using PLS regression to predict viscosity properties of a flour-water paste as recorded by the RVA, that determine the cooking and processing characteristics of rice [102]. Meadows and Barton later used NIR to predict RVA data in rice flour [103]. A PLS regression of NIR spectra *vs*. RVA viscosity showed a highest correlation ($R = 0.961$–0.903) to NIR was at 212–228 sec, which is between the initial pasting time and peak viscosity. Furthermore, the pasting parameters of setback and break down, and gelatinization peak temperature of rice flour were predicted successfully using NIR [104]. Texture of cooked rice was also predicted by NIR analysis of whole grain rice [105]. Five of seven sensory texture attributes were predicted by NIR using PLS analysis, whose calibration models were developed based onf second derivative spectra. RVA peak viscosity and breakdown were also successfuly predicted based on NIR spectra and PLS regression models. Calibrations were developed using PLS and ANN analyses. The results showed limited precision of this method. However, it can be used as a rough screening method for starch amylose content. Xie et al. later reported that NIR spectra correlated strongly with differential scanning calorimetry (DSC) for measuring amylopectin retrogradation in bread staling [106]. Nowadays, requirements of quality control in grain milling and food processing increasingly call for on-line analyses [41]. Studies developed by Sampaio et al. based on NIR spectroscopy associated to PCA, PLS-DA, and SVM for discrimination and classification of rice varieties (Indica and Japonica) were explored after different spectra processing steps such as MSC, first derivative and second derivative [100]. The PCA allowed revealing the pattern and relationship of each variety and chemical similarities that were effectively distinguished by PLS-DA and SVM, according to their specific properties. The SVM model, showed a significant fitting accuracy (97%), cross-validation (93%), and prediction (91%). These data support the strength of the model for efficient rice types classification. The principal differences between both rice types were present at range 7476–7095 cm^{-1}, 7046 cm^{-1} and 4264–4153 cm^{-1}, which can be used for its discrimination, being possible to develop a robust classification model for rice samples based on their specific physicochemical properties. The classification models developed using SVM tools were very robust compared to PLS-DA models, allowing to classify with high confidence both rice varieties. The machine learning tools can facilitate the process of classification and identification of different types of grains being possible, in the next future, to discriminate their origin, harvest season, state of conservation as well as the presence of contaminants and adulteration issues based on robust classification method, allowing to create a rice database and making *in situ*, real-time in classifying the types and origins of rice.

Studies developed by Osborne et al. using near infrared transmission spectroscopy allowed to discriminate between Basmati and other long-grain rice samples. A discriminant rule was derived using the Fisher linear discriminant function calculated from the first few principal component scores of the NIR spectra [107]. The discriminant rule was assessed by cross-validation. Based on this study, nine Basmati varieties and 53 other rice samples were classified correctly from NIR spectra, but 8% of the Basmatis and 14% of the others were misclassified on the basis of spectra of individual grains. NIR spectroscopy technique also offers effective quantitative capability for moisture, fat, protein and gluten content in rice cookies [108].

According to studies performed by Chen et al., the NIR diffuse reflectance spectroscopy of multi-grain seeds, a spectral discriminant analysis method for the variety identification of multi-grain rice seed was developed using the PLS-DA [109]. Due to the slight differences of seeds spectra in various varieties, it's necessary to propose the novel and valid methods. In this study, the SNV pretreatment combined with wavelength-screening methods improved the accuracy of the discriminant models. The selected optimal wavelength model was the combination of 54 discrete wavelengths within NIR region. NIR spectral discrimination total recognition accuracy rates reached 94.3% for a study that involves the identification of one type of differentiation (negative and excellent hybrid variety) and several interference groups (positive, four pure groups and four mixed groups).

The Hyperspectral Imaging (HSI) technique coupled with visible (vis) and/or NIR spectroscopy is generally used to identify or inspect different substances of seed by recognizing the molecular bonds in the sample, being considered the most feasible methods for rapidly and non-destructively detecting the substances of agricultural products, combining the technologies of spectroscopy and digital imaging. Studies developed by He et al. used the system NIR-HSI combined with multiple data preprocessing methods [110]. This approach allowed simultaneously to obtain spectral and spatial information from testing samples in the form of a hypercube constituted by two spatial dimensions and one spectral dimension. The HSI technique has the ability to collect hyperspectral information from samples of different sizes and shapes based on the spatial data. The detection speed of HSI is faster than that of point-based techniques, as many samples can be scanned and analyzed at the same time by using an HSI camera [111]. The classification models was developed to identify the vitality of rice seeds, presenting a great potential for identifying vitality and vigor of rice seeds. When detecting the seed vitality of the three different years, the extreme learning machine model with Savitzky–Golay preprocessing reached a significant classification accuracy of 93.67% by spectral data. In terms of the non-viable seeds identification from viable seeds of different years, the least squares support vector machine model coupled with raw data and selected wavelengths achieved a significant classification achievement (94.38% accuracy), and can be adopted as an optimal combination to identify non-viable seeds from viable seeds. In another study, carried out by Barnaby et al., NIR hyperspectral image consists of numerous bands with small spectrum gaps (every 4 nm in our study) and can assess grain traits such as fat, starch, protein, moisture, color, and many other physico-chemical compounds at once [101]. Genome wide association study allowed to confirm known genes and to identify new genes that can affect grain quality traits based on hyperspectral imaging technique. The PLS-DA models of hyperspectral data identify spectral ranges that distinguished genetic and production environment differences, and this data can support to resolve the genetics of complex traits such as rice grain quality.

The nitrogen content is an important chemical indicator used for monitoring and management of plant due to its role in photosynthesis, productivity as well as its effect on carbon and oxygen cycle. The nitrogen content can be measured by laboratory analysis, meanwhile, its spectral reflectance of NIR (700–1075 nm) in the field was measured using hand held spectroradiometer. Studies performed by Afandia et al. evaluated nitrogen content in rice crop based on NIR reflectance using ANN [111]. The reported study allowed to conclude that the organic molecules (nitrogen, water, etc) present a specific absorption pattern in the NIR region and the comparison between measured and model estimation of nitrogen content presented a RMSE of 0.32.

A study developed by Lin et al., based on the imaging method, a system constituted by a NIR camera, filters, an automatically exchange filters device, and the imaging processing techniques allowed to detect the rice protein content based on the spectrum absorption. The NIR data allowed to establish the calibration model based on MLR, PLS, and ANN analysis models. In the MLR model, the NIR imaging system used the calibration model that take in account 5 wavelengths (880 nm, 910 nm, 920 nm, 1000 nm, and 1014 nm) to predict the rice protein content, and had R^2 validation (0.782) and standard error of predicition (SEP) 0.274%, and respectively. The NIR imaging system used 15 filters ranging from 870 to 1014 nm in the PLS model, the predictive results expressed a significant performance (R^2_{val} = 0.782, and SEP = 0.274%) comparatively tothe MLR model. The ANN model, the net input using the 5 spectrum wavelengths selected by the MLR, simplified the model, and the predicting results (R^2_{val} = 0.806, and SEP = 0.266%) were similar to those of the PLS. The prediction results indicated that the developed NIR imaging system has the advantages of simple, convenient operation, and high detection accuracy as well as it presents commercial potential in non-destructive high accurate predicting capability detection of rice protein content [112].

NIR spectroscopy was used to develop a new discrimination method of varieties of rice. The several variables compressed by PCA were used as inputs of multiple discriminant analysis (MDA). The study showed that the combinantion of spectroscopy and computer data processing technology based on PCA and MDA for the identification of rice from different areas allowed to identify correctly about 98% for the calibration process, and 100% for the prediction process. These results showed that the proposed alternative method is a feasible way for the identification of the specific production areas of rice [113].

2.2 NIR spectroscopy in rice authentication

NIR spectroscopy has been widely used in the evaluation of agricultural products due to its many advantages, such as being easy-to-use, non-destructive, fast and accurate, providing highly reproducible results, requiring minimum or, often, no sample preparation, and allowing the analysis of several constituents based on a single measurement. As consequence of the importance of rice at global level, in the literature it is possible to find several studies aimed at their analysis and characterization. Due to environmental reasons and the rice the market, non-destructive approaches are generally preferred. NIR spectroscopy has emerged as an important tool to determine fraud, adulteration, contamination in grains and flours. A substantial instrumental improvements (e.g., hyperspectral imaging, FT-NIR) and advances in data analysis (e.g., deep learning) have allowed for the development of screening methods for detecting the presence of pests (e.g., rice weevil) across a range of stored grains [114–116].

Direct spectroscopic measurements have been widely applied for several foods and commodities, especially in the grain, cereal products, such for classification of rice [117–121]. Furthermore, in the structure of the evaluation of rice quality, NIR spectroscopy has been used for the discrimination of rice [122, 123]; varieties classificationand transgenic rice detection [124]; the physico-chemical properties quantification (such as moisture content, sound whole kernel, whiteness, translucency, color, and amylogram characteristics) [125]; cultivars classification [126], protein and amylose content prediction [127, 128]; wax rice detection [129]; and eating quality prediction [130]. Barnaby et al. correlated the grain chalk of rice to the genomic regions of NIR spectra [101]. These spectral regions can be applied in the automation of grain chalk quantification and potentially for other grain products as well [131].

Rapid and nondestructive detection of rice authenticity and quality were performed based on hand-held NIR spectrometer coupled with the appropriate chemometrics. The selection of different preprocessing methods with PCA and modeling with KNN and SVM multivariate calibration model showed that MSC + PCA plus KNN showed superiority in this study with more than 90% classification rate for all categories of rice samples studied. Based on these results, the hand-held spectrometer associated to an appropriate multivariate calibration model could be used for quick and non-destructive detection of rice quality and authenticity [132].

Food fraud remains a significant problem for food regulators, importers, merchants, law enforcement personnel, and the consumer. A key feature of food fraud is the use of a lower value ingredient to imitate an authentic product. NIR analysis technology, PLS-DA, and SVM have been used to detect whether high-quality rice was mixed with other varieties of rice. NIR spectral data analyzed using PLS-DA and a SVM algorithm, was shown to be a feasible method (5% detection limit) for the rapid identification of fraudulent rice varieties blended with authentic Wuchang rice samples [133].

Studies performed by Liu et al. showed that those techniques represent a significant support to qualitative discrimination [133]. PLS was used to establish the quantitative analysis model to support in the recognition of the degree of fraud. As consequence of the direct correlation between the results of NIR analysis and the homogeneity of the samples, four groups of samples with different physical forms (full granules, 40 mesh, 70 mesh, and 100 mesh) were prepared. Regarding qualitative analysis, the performance of the model has no obvious relationship with the physical state of the sample, the qualitative model of PLS-DA and SVM can detect the fraudulent rice with a 5% detection limit. The determination coefficient and root mean square errors of the optimal prediction result were 0.96 and 2.93, respectively. Based on this study, NIR analysis technology can be considered as a reliable and fast strategy to determine if the premium high-quality rice is adultered with inferior categories of rice.

Different preprocessing approache were used for NIR signals pretreatment. Besides considering raw data, the first derivative (Savitzky–Golay approach, 15 points window, 2nd order polynomial), second derivative (Savitzky–Golay approach, 15 points window, 3rd order polynomial), and standard normal variate (SNV) were also evaluated (**Figure 6**). NIR data were further mean-centered prior to the creation of any calibration model. The most suitable preprocessing approach, together with the optimal complexity (number of LVs or PCs to be extracted) of any classification model, were defined based on a cross-validation procedure. PLS-DA

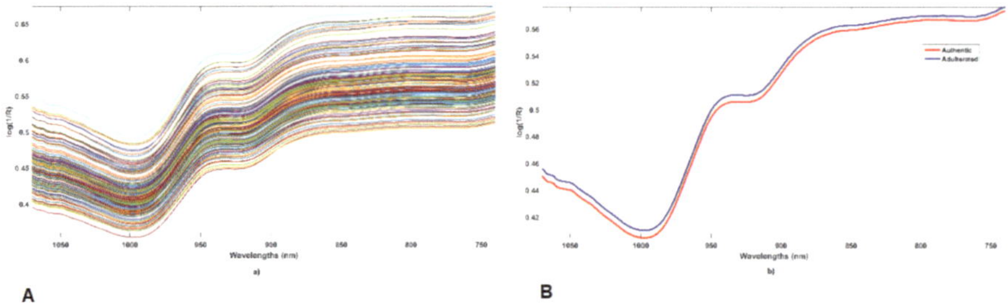

Figure 6.
NIR spectra (a) raw spectra of samples, (b) mean spectra of authentic (red line) and adulterated samples (blue line). (Adapted from [134]).

selection, specifically, was based on the combination of pre-processing and model complexity leading to the lowest mean classification error, whereas for SIMCA the maximum efficiency was sought. A study developed by Duy Le Nguyen Doan investigate the possibility of combination NIR spectroscopy and chemometric classifiers with the aim of detecting adulterated rice samples [134]. Two different strategies were exploited: discriminant classifier (PLS-DA), and class-modelling technique (SIMCA). Both strategies provided different results; in particular, SIMCA appeared unable to solve the investigated problem. On the other hand, PLS-DA analysis showed to be a suitable approach. These results indicate that the high within-class variability can have an impact on the possibility of detecting low levels of adulteration; simultaneously, was also suggested that the proposed approach could be useful for detecting samples adulterated. Then, this study demonstrates that the combination of NIR spectroscopy and PLS-DA can represent an effective, rapid and non-destructive tool for the determination of adulteration in jasmine rice [134].

2.3 NIR Spectroscopy in Rice Contamination

Fast determination of heavy metals is necessary and important to ensure the safety of crops. The potential of NIR spectroscopy coupled with chemometric technology for quantitative analysis of cadmium in rice was investigated. The spectrum was pre-processed using first derivation to reduce the baseline shift and several chemometric techniques, such as iPLS, mwPLS, siPLS, and biPLS were proposed to extract and optimize spectral interval from full-spectrum data. The PLS models based on four chemometric algorithms outperformed the full-spectrum PLS model then developed. Among the techniques, biPLS performed better with the optimal subinterval selection [135].

Heavy metals are spectrally featureless so that spectral responses could not be directly used for the assessment of heavy metals in rice. With a close combination of protein, crude fiber, and other ingredients, heavy metals present significant correlation with protein in rice [136]. The detection of heavy metal concentration in grain is mostly realized by physical and chemical direct methods that can exactly obtain the residual levels of heavy metal; however, it is time consuming, cumbersome, and inefficient. On the basis of the hypothesis that heavy metal concentration could be spectrally estimated through the correlation between heavy metal concentration and protein contents, the objectives of this study are to: (1) build quantitative model for the quick prediction of both heavy metal and protein content, and (2) to evaluate the feasibility of near-infrared spectroscopy in assessing heavy metal concentration in coarse rice.

Protecting people from heavy metal contamination is an important public-health concern and a major national environmental issue. The NIR spectral technique is used to identify heavy metal concentration such as lead (Pb) and copper (Cu) in rice. The NIR spectral data were treated by some methods, including, logarithm, baseline correction, standard normal variate, multiple scatter correction, first derivates, and continuum removal. The lead (Pb) was accumulated in rice at a high level (17.05) compared with the others heavy metals. MSC-PLSR models were developed, respectively, for Pb (R^2 = 0.49, RMSE = 2.01 mg/kg) and Cu (R^2 = 0.29, RMSE = 0.75 mg/kg). It is achievable to identify Pb and Cu content in rice by using NIR spectral technique. However, further studies should be performed on the application of spectral technique in discriminating the other heavy metals in rice due to the limitations of few samples and particles size interference.

3. Conclusions

Based on the reported studies, it was possible to develop a robust classification, authentication or fraud detection model for rice samples considering their specific physicochemical properties and using machine learning tools such as PLS-DA, KNN, ANN, and SVM among other methodologies applied to NIR spectroscopy data, revealing the pattern and relationship of each variety and chemical similarities, according to their specific properties. The classification models developed using several models allow to classify with high confidence rice varieties using the spectral data. The results show that the use of these chemometric tools, combined with spectroscopy capabilities, can facilitate the process of classification and identification of different rice types. The rice discrimination by their origin, harvest season, state of conservation as well as the presence of contaminants and adulteration issues based on robust classification methods can facilitate the creation of a data base, a useful tool for rice authenticity that can increase the confidence and producer-consumer engagement in rice-based foods.

Acknowledgements

Acknowledge of funding: The study was supported by project TRACE-RICE -Tracing rice and valorising side streams along Mediterranean blockchain, grant n° 1934, (call 2019, Section 1 Agrofood) of the PRIMA Programme supported under Horizon 2020, the European Union's Framework Programme for Research and Innovation, and Research Unit, UIDB/04551/2020 (GREEN-IT, Bioresources for Sustainability).

Conflict of interest

The authors declare no conflict of interest.

Author details

Pedro S. Sampaio[1,2,3]* and Carla M. Brites[1,3]

1 Instituto Nacional de Investigação Agrária e Veterinária (INIAV), Quinta do Marquês, Oeiras, Portugal

2 DREAMS-Centre for Interdisciplinary Development and Research on Environment, Applied Management and Space, Faculty of Engineering, Lusófona University (ULHT), Lisbon, Portugal

3 GREEN-IT Bioresources for Sustainability, ITQB NOVA, Oeiras, Portugal

*Address all correspondence to: pedro.sampaio@ulusofona.pt

IntechOpen

© 2021 The Author(s). Licensee IntechOpen. This chapter is distributed under the terms of the Creative Commons Attribution License (http://creativecommons.org/licenses/by/3.0), which permits unrestricted use, distribution, and reproduction in any medium, provided the original work is properly cited. (cc) BY

References

[1] "FOOD AND AGRICULTURE ORGANIZATION OF THE UNITED NATIONS Statistics.," 2016. [Online]. Available: http://faostat.fao.org/.

[2] Morphology of Rice Seed Development and Its Influence on Grain Quality. Paul A. Counce and Karen A. K. Moldenhauer; Nese Sreenivasulu (Ed.), Rice Grain Quality: Methods and Protocols, Methods in Molecular Biology, vol. 1892)

[3] L. Shin. B. S. Luh, " Properties of the rice caryopsis," in *Luh BS (ed) Rice.* , New York, Springer, 1991, p. 389-419.

[4] E.T. Champagne, D.F. Wood, B.O. Juliano, D.B. Bechtel, "The rice grain and its gross composition," in *In: Champagne ET (ed) Rice chemistry and technology.* , St. Paul, MN., AACC International, , 2004, p. 77-100.

[5] H. Yamagata, T. Sugimoto, K. Tanaka, Z. Kasai, "Biosynthesis of storage proteins in developing rice seeds.," *Plant Physiol,* vol. 70, no. 4, p. 1094-1100, 1982.

[6] K. Ichihara, N. Kobayashi, K. Saito, "Lipid synthesis and acyl-CoA synthetase in developing rice seeds," *Lipids,* vol. 38, no. 8, p. 881-884, 2003.

[7] N. Sreenivasulu, V. Radchuk, M. Strickert, O. Miersch, W. Weschke, U. Wobus, "Gene expression patterns reveal tissue-specific signaling networks controlling programmed cell death and ABA-regulated maturation in developing barley seeds," *Plant J,* vol. 47, no. 2, p. 310, 2006.

[8] X. Kong, P. Zhu, Z. Sui, J. Bao, "Physicochemical properties of starches from diverse rice cultivars varying in apparent amylose content and gelatinization temperature combinations," *Food Chem* , vol. 172, pp. 433-440, 2015.

[9] M.K. Pandey, N S. Rani, M. S. Madhav, R. M. Sundaram, G. S. Varaprasad, A.K.P. Sivaranjani, A. Bohra, G. R. Kumar, A. Kumar, "Different isoforms of starch-synthesizing enzymes controlling amylose and amylopectin content in rice (*Oryza sativa* L.)," *Biotechnology Advances,* vol. 30, p. 1697-1706, 2012.

[10] B. O. Juliano, "A simplified assay for milled-rice amylose," *Cereal Science Today,* vol. 60, no. 16, pp. 334-340, 1971.

[11] D. Sievert, J.H. Holm, "Determination of Amylose by Differential Scanning Calorimetry," *Starch/Stärke,* vol. 45, pp. 136-139, 1993.

[12] W. Banks, C.T. Greenwood, D.D. Muir, "Studies on Starches of High Amylose-Content. Part 14. The Fractionation of Amylomaize Starch by Aqueous Leaching.," *Starch*, vol. 23, 1971.

[13] W.R. Morrison, B. Laignelet., " An improved colorimetric procedure for determining apparent and total amylose content in cereals and other starches.," *Journal Cereal Science,* vol. 1, pp. 9-20, 1983.

[14] N. K. Matheson, L. A. Welsh, "Estimation and fractionation of the essentially unbranched amylose and branched amylopectin component of starches with concanavalin A.," *Carbohydrate Research,* vol. 180, pp. 301-313, 1988.

[15] Y.H. Yun, H.D. Li, L.R. Wood, W. Fan, J.J. Wang, D.S. Cao, Q.S. Xu, Y.Z. Liang, "An efficient method of wavelength interval selection based on random frog for multivariate spectral calibration.," *Spectrochimica Acta Part A: Molecular and Biomolecular Spectroscopy,* vol. 111, pp. 31-36, 2013.

[16] H. Li, S. Prakash, T.M. Nicholson, M.A. Fitzgerald, R.G. Gilbert, "The importance of amylose and amylopectin fine structure for textural properties of cooked rice grains.," *Food Chemistry,* vol. 196, pp. 702-711, 2016.

[17] Z.-H. Lu, T. Sasaki, Y.-Y. Li, T. Yoshihashi, L-T. Li, K. Kohyama, "Effect of Amylose Content and Rice Type on Dynamic Viscoelasticity of a Composite Rice Starch Gel.," *Food Hydrocolloids,* vol. 23, no. 7, pp. 1712-1719, 2009.

[18] X. J. Ge, Y.Z. Xing, C.G. Xu, Y.Q. He, "QTL analysis of rice grain elongation, volume expansion and water absorption using are combinant inbred population.," *Plant Breeding,* vol. 124, pp. 121-126, 2005.

[19] Z. Zhou, K. Robards, S. Helliwell, C. Blanchard, "Composition and functional properties of rice," *International Journal of Food Science and Technology,* vol. 37, p. 849-868, 2002.

[20] L. Amagliani, J. O'Regan, A. L. Kelly, J. A. O'Mahony, "The composition, extraction, functionality and applications of rice proteins: A review.," *Trends in Food Science & Technology,* vol. 64, pp. 1-12, 2017.

[21] R.J. Bryant, A. K. Jackson, K.M. Yeater, W. G. Yan, A. M. McClung, R. G. Fjellstrom, "Genetic variation and association mapping of protein concentration in brown rice using a diverse rice germplasm collection," *Cereal Chemistry,* vol. 90, no. 5, p. 445-452, 2013.

[22] Y. T. Thomas, R. Bath R. Y. Kuang, "Composition of amino acids, fatty acids, minerals and dietary fiber in some of the local and import rice varieties of Malaysia.," *International Food Research Journal,* vol. 22, no. 3, p. 1148-1155, 2015.

[23] K. R. Bhattacharya, P. V. Subba Rao, "Processing Conditions and Milling Yield in Parboiling of Rice," *Journal of Agricultural and Food Chemistry,* vol. 14, no. 5, pp. 473-475, 1966 .

[24] K. Liakos, P. Busato, "Sensors. Machine Learning in Agriculture: A Review.," *Sensors,* vol. 18, p. 2674, 2018.

[25] T. Q. Zheng J. L. Xu Z. K. Li H. Q. Zhai J.M. Wan "Genomic regions associated with milling quality and grain shape identified in a set of random introgression lines of rice (*Oryza sativa* L.)," *Plant Breed,* vol. 126, p. 158-163, 2007.

[26] X. Qiu, Y. Pang, Z. Yuan, D. Xing, J. Xu, M. Dingkuhn "Genome-wide association study of grain appearance and milling quality in a worldwide collection of Indica rice germplasm.," *PLoS ONE,* vol. 10, p. e0145577, 2015.

[27] X. Wang, Y. Pang, C. Wang, K. Chen, Y. Zhu, C. Shen, J. Ali, J. Xu, Z. Li, "New Candidate Genes Affecting Rice Grain Appearance and Milling Quality Detected by Genome-Wide and Gene-Based Association Analyses.," *Front. Plant Sci.* , vol. 7, p. 1998, 2017.

[28] A. J. Lisle, M. Martin, M. A. Fitzgerald "Chalky and Translucent Rice Grains Differ in Starch Composition and Structure and Cooking Properties," *Cereal Chemistry,* vol. 77, no. 5, 2000.

[29] B. Juliano, "Rice quality screening with the rapid visco analyser," in *Applications of the rapid visco analyser.*, H. J. In Walker CE, Ed., Sydney, Newport Scientific:, 1996, p. 19-24.

[30] J. Bao, Y. Xia, "Genetic control of paste viscosity characteristics in indica rice (*Oryza sativa* L.).," *Theoretical and Applied Genetics,* vol. 98, no. 6, p. 1120-1124, 1999.

[31] L.S. Collado, H. Corke, "Properties of starch noodles as affected by sweet potato genotype.," *Cereal Chemistry Journal,* vol. 74, no. 2, p. 182-187, 1997.

[32] J. Bao, S. Shen, M. Sun, H. Corke, "Analysis of genotypic diversity in the starch physicochemical properties of nonwaxy rice: apparent amylose content, pasting viscosity and gel texture.," *Starch—Starke,* vol. 58, no. 6, p. 259-67., 2006.

[33] L.Q. Wang, W.J.Liu, Y. Xu, Y.Q. He, L.J. Luo, Y.Z. Xing, C.G. Xu, Q. Zhang, "Genetic basis of 17 traits and viscosity parameters characterizing the eating and cooking quality of rice grain.," *Theoretical and Applied Genetics,* vol. 115, no. 4, p. 463-76., 2007.

[34] V. Balan, C.T. Mihai, F.D. Cojocaru, C.M. Uritu, G. Dodi, D. Botezat, I. Gardikiotis, "Vibrational Spectroscopy Fingerprinting in Medicine: from Molecular to Clinical Practice.," *Materials,* vol. 12, p. 2884, 2019.

[35] J. Durig, "Far-IR Spectroscopy, Applications.," in *In Encyclopedia of Spectroscopy and Spectrometry*, Amsterdam, NY, USA, Elsevier, 1999, p. pp. 498-504.

[36] Y. Han, S. Ling, Z. Qi, Z. Shao, X. Chen, "Application of far-infrared spectroscopy to the structural identification of protein materials.," *Phys. Chem. Chem. Phys.* , vol. 20, p. 11643-11648, 2018.

[37] J. Workman, J. Shenk, "In: Near-Infrared Spectroscopy in Agriculture. Roberts C.A., Workman, J., Jr., and Reeves III, J.B., American Society of Agronomy, Inc., Crop Science society of America, Inc., and Soil Science Society of America, Inc.,," 2004, pp. 3-10.

[38] D. W. R. Dunmire, "Automated qualitative and quantitative NIR reflectance analyses," *Cereal Foods World,* vol. 35, pp. 913-918, 1990.

[39] W. Murray, "Chemical principle of near-infrared technology.," in *In: Near-infrared technology in the agricultural and food industries.* , P. W. a. K. N. (Eds.), Ed., American Association of Cereal Chemists. St. Paul, MN, 1990, pp. 17-34.

[40] B. M. Nicolai, T. Defraeye, B. De Ketelaere, E. Herremans, M. Hertog, W. Saeys, A. Torricelli, T. Vandendriessche, P. Verboven, "Nondestructive measurement of fruit and vegetable quality by means of NIR spectroscopy: A review.," *Journal of Postharvest Biology and Technology,* vol. 46, pp. 99-118, 2007.

[41] B. Osborne, "Review: Applications of near infrared spectroscopy in quality 36 sceening of early-generation material in cereal breeding programmes.," *J. Near Infrared Spectrosc.* , vol. 14, pp. 93-101, 2006.

[42] B. Osborne, "Flours and breads. Ch. 8.1.," in *In: Near-Infrared Spectroscopy in Food Science and Technology.*, M. W. a. C. A. e. Ozaki Y., Ed., John Wiley & Sons, Inc., NJ,, 2007, pp. 281-296 pp.

[43] J.K.Drennen, B.D.Gebhart, E.G. Kraemer, R.A. Lodder, "Near-infrared spectrometric determination of hydrogen ion, glucose, and human serum albumin in a simulated biological matrix," *Spectroscopy,* vol. 6, pp. 28-32, 1990.

[44] J. Qu, "Applications of NIR-Spectroscopy in Food Safety Evaluation and Control: A Review of Recent Research Advances.," *Crit Rev Food Science &Nut,* vol. 55, p. 1939-54, 2015.

[45] J. Majumdar, "Analysis of agriculture data using data mining techniques: application of big data.," *J Big Data,* vol. 4, p. 20, 2017.

[46] M. Calingacion, "Diversity of Global Rice Markets and the Science Required for Consumer-Targeted Rice Breeding," *PLoSONE,* vol. 9, p. 85106, 2014.

[47] S. R. Delwiche, M. M. Bean, R. E. Miller, B. D. Webb, P. C. Williams

"Apparent amylose content of milled rice by nearinfrared reflectance spectrophotometry," Cereal Chemistry, vol. 72, p. 182-187, 1995.

[48] A. Savitzky, M. Golay, "Smoothing and Differentiation of Data by Simplified Least Squares Procedures," *Anal. Chem.* , vol. 36, p. 1627-1639, 1964.

[49] A. G. M. Savitzky, " Smoothing and differentiation of data by simplified least squares procedures," *Analytical Chemistry,* vol. 36, p. 1627-1639, 1964.

[50] S. Xie, B. Xiang, L. Yu, H. Deng, "Tailoring noise frequency spectrum to improve NIR determinations," *Talanta,* vol. 80, p. 895-902, 2009.

[51] P. Sampaio, A. Soares, A. Castanho, A. Almeida, J. Oliveira, C. Brites, "Optimization of rice amylose determination by NIR-spectroscopy using PLS chemometrics algorithms,," *Food Chemistry,* vol. 242, pp. 196-204, 2018.

[52] I.T.Jolliffe, J. Cadima, "Principal component analysis: a review and recent developments," *Phil. Trans. R. Soc.* , vol. A.3742015020220150202, 2016.

[53] B.G. Osborne, T. Fearn, P.H. Hindle "Near infrared calibration II.," in *Ch.7 In: Practical NIR Spectroscopy with Applications in Food and Beverage Analysis, 2nd ed.* , Longman Scientific & Technical, UK, 1993, pp. 121-144.

[54] R. A. Johnson, D. W. Wichern "Discrimination and classification. Ch.11," in *In: Applied Multivariate Statistical Analysis*, 4th ed. ed., New Jersey, A Simon and Schuster Company, Eaglewood Cliffs, 1998, pp. 629-725.

[55] H. L. Mark, D. Tunnel, "Qualitative near-infrared reflectance analysis using Mahalanobis distances.," *Anal. Chem.,* vol. 57, pp. 1449-1456, 1985.

[56] M. Barker, W. Rayens, "Partial least squares for discrimination.," *J. Chemom,* vol. 17, p. 166-173, 2003.

[57] H. Nocairi, E. Mostafa Qannari, E. Vigneau, D. Bertrand, "Discrimination on latent components with respect to patterns. Application to multicollinear data.," *Comput. Stat. Data Anal.,* vol. 48, pp. 139-147, 2005.

[58] U. Indahl, H. Martens, T. Næs, " From dummy regression to prior probabilities in PLS-DA.," *J. Chemom.,* vol. 21, pp. 529-536, 2007.

[59] M. Sjöström, S. Wold, B. Söderström, "PLS discriminant plots," in *In Pattern Recognition in Practice*, E. K. L. E. Gelsema, Ed., Amsterdam, The Netherlands, Elsevier , 1986, p. 461-470.

[60] L. Ståhle and S. Wold, "Partial least squares analysis with cross-validation for the two-class problem: A Monte Carlo study," *J. Chemom.* , vol. 1, p. 85-196, 1987.

[61] S. Wold, "Pattern recognition by means of disjoint principal components models," *Pattern Recognit,* vol. 8, pp. 127-139, 1976.

[62] S. ML, Support vector machines, vol 1., New York: Springer, 2008.

[63] R. Balabin, R. Safieva and E. Lomakina, "Near-infrared (NIR) spectroscopy for motor oil classification: from discriminant analysis to support vector machines," *Microchem J,* vol. 98, pp. 121-128, 2011.

[64] P. Santos, E. Pereira-Filho, L. Rodriguez-Saona, "Rapid detection and quantification of milk adulteration using infrared microspectroscopy and chemometrics analysis," *Food Chem,* vol. 138, pp. 19-24, 2013.

[65] W. Kong, C. Zhang, F. Liu, P. Nie, Y. He, "Rice seed cultivar identification using near-infrared hyperspectral

imaging and multivariate data analysis.," *Sensors,* vol. 13, pp. 8916-8927, 2013.

[66] R. Wehling, "Infrared Spectroscopy. Ch.27," in *Food Analysis*, 2nd ed ed., S. (. Nielson, Ed., Gaithersburg, MD, Aspen Publishers, Inc., 1998, pp. 413-424.

[67] F. E. Dowell, J. Throne, J. E. Baker, "Automated Nondestructive Detection of Internal Insect Infestation of Wheat Kernels by Using Near-Infrared Reflectance Spectroscopy.," *J. Econ. Entomol,* vol. 91, pp. 899-904, 1998.

[68] S. Wold, M. Sjostrom, "PLS-regression: a basic tool of chemometrics," *Chemometrics and Intelligent Laboratory Systems,* vol. 58, pp. 109-130, 2001.

[69] M. Friedel, C.-D. Patz, H. Dietrich, "Comparison of different measurement techniques and variable selection methods for FT-MIR in wine analysis.," *Food Chemistry,* vol. 141, pp. 4200-4207, 2013.

[70] H.W. Lee, A. Bawn, S. Yoon, "Reproducibility, complementary measure of predictability for robustness improvement of multivariate calibration models via variable selections.," *Analytica Chimica Acta,* vol. 757, pp. 11-18, 2012.

[71] L. Nørgaard, A. Saudland, J. Wagner, J.P. Nielsen, L. Munck L, S.B. Engelsen, "Interval Partial least-squares regression (iPLS): A comparative chemometric study with an example from Near-Infrared spectroscopy.," Applied Spectroscopy, vol. 54, pp. 413-419, 2000.

[72] J. H. Kalivas, "Two data set for near infrared spectra," *Chemometrics Intelligent Laboratories Systems,* vol. 37, pp. 255-259, 1997.

[73] C. H. Spiegelman, M. J. McShane, M. J. Goetz, M. Motamedi, Q. L. Yue, G. L. Coté, "Theoretical justification of wavelength selection in PLS calibration: development of a new algorithm.," *Analytical Chemistry,* vol. 70, pp. 35-44, 1998.

[74] L. Leardi, J. Nørgaard, "Sequential application of backward interval partial least squares and genetic algorithms for the selection of relevant spectral regions.," *Chemometrics,* vol. 18, pp. 486-497, 2004.

[75] H.L. Ma, J.W. Wang, Y.J. Chen, J.L. Cheng, Z.T. Lai, "Rapid authentication of starch adulteration in ultrafine granular powder of Shanyao by near-infrared spectroscopy coupled with chemometric methods.," *Food Chemistry,* vol. 215, pp. 108-115, 2017.

[76] Y.-H. Yun, H.-D. Li, L. R. E. Wood, W.Fan, J.-J. Wang, D.-S. Cao, Q.-S. Xu, Y.-Z. Liang, "An efficient method of wavelength interval selection based on random frog for multivariate spectral calibration.," *Spectrochimica Acta Part A: Molecular and Biomolecular Spectroscopy,* vol. 111, pp. 31-36, 2013.

[77] M. Daszykowski, J. Orzel, M. Wrobel, H. Czarnik-Matusewicz, B. Walczak, "Improvement of classification using robust soft classification rules for near-infrared reflectance spectral data Improvement of classification using robust soft classification rules for near-infrared reflectance spectral data.," *Chemometr. Intell. Lab.,* vol. 109, p. 86-93, 2011.

[78] S. Wold, M. Sjöström, "SIMCA: A Method for Analyzing Chemical Data in Terms of Similarity and Analogy.," in *In Chemometrics: Theory and Application*, vol. 52, Washington, DC, American Chemical Society, 1977, pp. 243-282.

[79] H. Yue, S. Qin, "Reconstruction-Based Fault Identification Using a Combined Index.," *Ind. Eng. Chem. Res.,* vol. 40, p. 4403-4414, 2001.

[80] M. Cocchi, A. Biancolillo, F. Marini, "Chapter Ten-Chemometric Methods for Classification and Feature Selection.," in *In Comprehensive Analytical Chemistry;*, J. B. C. T. R. E. Jaumot, Ed., Amsterdam, The Netherlands, Elsevier, 2018.

[81] J. Ruiz, T. Parello, R. Gomez, "Comparative study of multivariate methods to identify paper finishes using infrared spectroscopy.," *IEEE Trans. Instrum. Meas.,* vol. 61, pp. 1029-1036, 2012.

[82] L. Breiman, "Random forests.," *Mach. Learn.,* vol. 45, pp. 5-32, 2001.

[83] L. Fraiwan, K. Lweesy, N. Khasawneh, H. Wenz, H. Dickhaus, "Automated sleep stage identification system based on time-frequency analysis of a single EEG channel and random forest classifier.," *Comput. Meth. Programs Biomed.,* vol. 108, pp. 10-19, 2012.

[84] M.N. Vrahatis, G. D. Magoulas, K. Parsopoulos, V. P. Plagianakos, "Introduction to artificial neural network training and applications Conference Paper," 2000.

[85] D.E. Rumelhart, G. E. Hinton, R. J. Williams, "Learning representations by back-propagating errors.," *Nature,* vol. 323, pp. 533-536, 1986.

[86] S. Haykin, Neural Networks: a comprehensive foundation., New York.: Macmillan, 1999.

[87] ASCE Task Committee on Application of Artificial Neural Networks in Hydrology, "Artificial neural networks in hydrology I: preliminary concepts.," *J Hydrol Eng 5:115-123,* 2000.

[88] S. Cohen, N. Intrator, "Automatic model selection in a hybrid perceptron/radial network.," *Inf Fusion Special Issue Mult Experts,* vol. 3, no. 4, pp. 259-266, 2002.

[89] J. Kenneth, S. Wernter, J. MacInyre, "Knowledge extraction from radial basis function networks and multilayer perceptrons.," *Int J Comput Intell Appl,* vol. 1, no. 3, pp. 369-382, 2001.

[90] W. Loh, L. Tim, "A comparison of prediction accuracy, complexity, and training time of thirty three old and new classification algorithm.," *Mach Learn,* vol. 40, no. 3, pp. 203-238, 2000.

[91] Mohammad Ali Ghorbani, Rahman Khatibi, Behrouz Hosseini, Mehmet Bilgili, "Relative importance of parameters affecting wind speed prediction using artificial neural networks.," *Theor Appl Climatol,* vol. 114, no. 1, pp. 107-114, 2013.

[92] F. Rosenblatt, Principles of Neurodynamics, New York: Spartan, 1962.

[93] S. Theodoridis, Machine Learning: A Bayesian and Optimization Perspective, London, UK: Academic Press, 2015.

[94] E. Lanza, "Determination of Moisture, Protein, Fat, and Calories in Raw Pork and Beef By Near Infrared Spectroscopy.," *Journal of Food Science,* vol. 48, no. 2, pp. 471-474, 1983.

[95] J. Rodriguez-Otero, M. Hermida, A. Cepeda, "Determination of fat, protein, and total solids in cheese by near-infrared reflectance spectroscopy.," *J AOAC Int.,* vol. 78, no. 3, pp. 802-806, 1995.

[96] P. Williams, "Application of chemometrics to prediction of some wheat quality factors by near-infrared spectroscopy," *Cereal Chem.,* vol. 97, pp. 958-966, 2020.

[97] C. Malegori, S. Buratti, S. Benedetti, *P. oliveri*, S. Ratti, C. Cappa, M. Lucisano, "A modified mid-level data fusion approach on electronic nose and FT-NIR data for evaluating the effect of

different storage conditions on rice germ shelf life.," *Talanta,* vol. 206, p. 120208, 2020.

[98] B. Le, "Application of deep learning and near infrared spectroscopy in cereal analysis.," *Vib. Spectrosc.,* vol. 106, p. 103009, 2020.

[99] H. Jiang, T. Liu, Q. Chen, "Dynamic monitoring of fatty acid value in rice storage based on a portable near-infrared spectroscopy system.," *Spectrochim. Acta Part A Mol. Biomol. Spectrosc.,* vol. 240, p. 118620., 2020.

[100] P. Sampaio, A. Castanho, A. Almeida, J. Oliveira, C. Brites, "Identification of rice flour types with near-infrared spectroscopy associated with PLS-DA and SVM methods.," *Eur. Food Res. Technol.,* vol. 246, pp. 527-537, 2019.

[101] J. Barnaby, T. Huggins, H. Lee, A. McClung, S. Pinson, M. Oh, G. Bauchan, L. Tarpley, K. Lee, M. Kim, "Vis/NIR hyperspectral imaging distinguishes sub-population, production environment, and physicochemical grain properties in rice.," *Sci. Rep.,* vol. 10, pp. 1-13, 2020.

[102] S. Delwiche, K. McKenzie, B.D. Webb, "Quality characteristics in rice by near-infrared reflectance analysis of whole-grain milled samples.," *Cereal Chem,* vol. 73, pp. 257-263, 1996.

[103] F. Meadows, F. Barton, "Determination of Rapid Visco Analyser Parameters in Rice by Near-Infrared Spectroscopy," *Cereal Chemistry,* vol. 79, pp. 563-566, 2002.

[104] J. Bao, Y. Cai, H. Corke, "Prediction of Rice Starch Quality Parameters by Near-Infrared Reflectance Spectroscopy," *Journal of Food Science,* vol. 66, pp. 936-939, 2001.

[105] J.-F. Meullenet, A. Mauromoustakos, T. Horner and B. Marks, "Prediction of Texture of Cooked White Rice by Near-Infrared Reflectance Analysis of Whole-Grain Milled Samples," *Cereal Chemistry,* vol. 79, pp. 52-57, 2002.

[106] F. Xie, F. Dowell and X. Sun, "Using visible and near-infrared reflectance spectroscopy and differential scanning calorimetry to study starch, protein and temperature effects on bread staling.," *Cereal Chem,* vol. 81, pp. 249-254, 2004.

[107] B. Osborne, B. Mertens, M. Thompson, T. Fearn, "The Authentication of Basmati Rice Using near Infrared Spectroscopy," *J. Near Infrared Spectrosc,* vol. 1, pp. 77-83, 1993.

[108] L. Wimonsiri, P. Ritthiruangdej, S. Kasemsumran, N. Therdthai, W. Chanput, Y. Ozaki, "Rapid analysis of chemical composition in intact and milled rice cookies using near infrared spectroscopy," *J. Near Infrared Spectrosc.,* vol. 25, pp. 330-337 , 2017.

[109] j. Chen, M. Li, T. Pan, L. Pang, L. Yao, J. Zhang, "Rapid and non-destructive analysis for the identification of multi-grain rice seeds with near-infrared spectroscopy," *Spectrochimica Acta Part A: Molecular and Biomolecular Spectroscopy,* vol. 219, pp. 179-185, 2019.

[110] X. He, X. Feng, D. Sun, F. Liu, Y. Bao, Y. He, "Rapid and Nondestructive Measurement of Rice Seed Vitality of Different Years Using Near-Infrared Hyperspectral Imaging," *Molecules,* vol. 24, no. 12, p. 2227, 2019.

[111] S.D. Afandi, Y.Herdiyeni, L. B. Prasetyo, W. Hasbi, K. Arai, H. Okumura, "Nitrogen Content Estimation of Rice Crop Based on Near Infrared (NIR) Reflectance Using Artificial Neural Network (ANN)," *Procedia Environmental Sciences,* vol. 33, pp. 63-69, 2016.

[112] L.-H. Lin, F.-M. Lu, Y.-C. Chang, "Prediction of protein content in rice using a near-infrared imaging system as diagnostic technique.," *Int J Agric & Biol Eng,* vol. 12, no. 2, pp. 195-200, 2019.

[113] Z. Zi-li, W. Chun-Feng, J. Di, H. Yong, L. Xiao-li and S. Yong-ni, "Discrimination of varieties of rice using near infrared spectral by PCA and MDA model," in *6th International Conference on Computer Science & Education (ICCSE)*, 2011.

[114] J. Johnson, "An overview of near-infrared spectroscopy (NIRS) for the detection of insect pests in stored grains.," *J. Stored Prod. Res.,* vol. 86, p. 101558, 2020.

[115] F. Kosmowski, T. Worku, "Evaluation of a miniaturized NIR spectrometer for cultivar identification: The case of barley, chickpea and sorghum in Ethiopia," *PLoS ONE,* vol. 13, p. e0193620/1–e0193620/17., 2018.

[116] C. Blanch-Perez-del-Notario, W. Saeys, A. Lambrechts, "Fast ingredient quantification in multigrain flour mixes using hyperspectral imaging.," *Food Control ,* vol. 118, p. 107366, 2020.

[117] C. Viejo, D. Torrico, F. Dunshea, S. Fuentes, "Emerging Technologies Based on Artificial Intelligence to Assess the Quality and Consumer Preference of Beverages.," *Beverages,* vol. 5, p. 62, 2019.

[118] A. Kaya, A. Keçeli, C. Catal and B. Tekinerdogan, "Sensor Failure Tolerable Machine Learning-Based Food Quality Prediction Model.," *Sensors,* vol. 20, p. 3173, 2020.

[119] K. Böhme, P. Calo-Mata, J. Barros-Velázquez, I. Ortea, "Review of Recent DNA-Based Methods for Main Food-Authentication Topics," *J. Agric. Food Chem,* vol. 67, pp. 3854-3864, 2019.

[120] W. Liu, X. Wang, J. Tao, B. Xi, M. Xue, W. Sun, "A Multiplex PCR Assay Mediated by Universal Primers for the Detection of Adulterated Meat in Mutton.," *J. Food Prot. ,* vol. 82, p. 325-330, 2019.

[121] R. Yin, Y. Sun, K. Wang, N. Feng, H. Zhang, M. Xiao, "Development of a PCR-based lateral flow strip assay for the simple, rapid, and accurate detection of pork in meat and meat products.," *Food Chem,* vol. 318, p. 126541, 2020.

[122] J. Spink, D. Moyer, "Understanding and combating food fraud.," *Food Technol,* vol. 67, pp. 30-35, 2013.

[123] M. Xiao, Y. Chen, H. Chu, R. Yin, "Development of a polymerase chain reaction—Nucleic acid sensor assay for the rapid detection of chicken adulteration.," *LWT,* vol. 131, p. 109679, 2020.

[124] A. Vinayaka, T. Ngo, K. Kant, P. Engelsmann, V. Dave, M.-A. Shahbazi, A. Wol, D. Bang, "Rapid detection of *Salmonella enterica* in food samples by a novel approach with combination of sample concentration and direct PCR.," *Biosens. Bioelectron,* vol. 129, pp. 224-230, 2019.

[125] B. Pakbin, A. Basti, A. Khanjari, L. Azimi, A. Karimi, "Differentiation of stx1A gene for detection of *Escherichia coli* serotype O157: H7 and Shigella dysenteriae type 1 in food samples using high resolution melting curve analysis.," *Food Sci. Nutr.,* vol. 8, p. 3665-3672, 2020.

[126] D.-D. Li, C.-B. Hao, Z.-M. Liu, S.-J. Wang, Y. Wang, Z. Chao, S.-Y. Gao, S. Chen, "Development of a novel dual priming oligonucleotide system-based PCR assay for specific detection of Salmonella from food samples.," *J. Food Saf ,* vol. 40, p. 12789, 2020.

[127] Y. Geng, G. Liu, L. Liu, Q. Deng, L. Zhao, X. Sun, J. Wang, B. Zhao, J. Wang, "Real-time recombinase polymerase

amplification assay for the rapid and sensitive detection of Campylobacter jejuni in food samples.," *J. Microbiol. Methods,* vol. 157, p. 31-36, 2019.

[128] N. Salihah, M. Hossain, M. Abdul Hamid, M. Ahmed, "A novel, rapid, and sensitive real-time PCR assay for cost-effective detection and quantification of *Staphylococcus aureus* in food samples with the ZEN double quenched probe chemistry," *Int. Food Res. J.* , vol. 26, p. 193-201, 2019.

[129] S. Rani, A. Pradhan, "Evaluation and meta-analysis of test accuracy of direct PCR and bioassay methods for detecting Toxoplasma gondii in meat samples.," *LWT,* vol. 131, p. 109666, 2020.

[130] R. Köppel, A. Sendic, H.-U. Waiblinger, "Two quantitative multiplex real-time PCR systems for the efficient GMO screening of food products.," *Eur. Food Res. Technol.* , vol. 239, p. 653-659, 2014.

[131] D. Bwambok, N. Siraj, S. Macchi, N. Larm, G. Baker, R-L. Pérez, "QCM Sensor Arrays, Electroanalytical Techniques and NIR Spectroscopy Coupled to Multivariate Analysis for Quality Assessment of Food Products, Raw Materials, Ingredients and Foodborne Pathogen Detection: Challenges and Breakthroughs," *Sensors,* vol. 20, no. 23, p. 6982, 2020.

[132] E. Teye, C.L.Y. Amuah, T. McGrath, C. Elliott, "Innovative and rapid analysis for rice authenticity using hand-held NIR spectrometry and chemometrics," *Spectrochimica Acta - Part A: Molecular and Biomolecular Spectroscopy,* vol. 217, pp. 147-154, 2019.

[133] Y. Liu, Y. Li, Y. Peng, Y. Yang, Q. Wang, "Detection of fraud in high-quality rice by near-infrared spectroscopy.," *J. Food Sci.,* vol. 85, p. 2773-2782, 2020.

[134] D. Le Nguyen Doan, Q. C. Nguyen, F. Marini, A. Biancolillo, "Authentication of Rice (*Oryza sativa* L.) Using Near Infrared Spectroscopy Combined with Different Chemometric Classification Strategies," *Appl. Sci.,* vol. 11, no. 1, p. 362, 2021.

[135] R. Hongyan, Z. Dafang, Y. Junxing, Y. Xinfang, "A Feasibility Study of NIR Spectra in Identifying Heavy Metal Contamination in Rice Around Abandoned Tailing Ponds: A Case Study in Guiyang County in South China Ren Hongyan.," *J Geophys Remote Sens,* vol. 2, p. 1, 2013.

[136] J. Yang, Y. Zha and H. Liu, "The distribution and chemical forms of Cd, Cu and Pb in polluted seeds.," *China Environmental Science,* vol. 19, pp. 500-504, 1999.

Chapter 13

Fermented Brown Rice as a Functional Food

Keiko Kataoka

Abstract

Brown rice, especially in a part of rice bran, contains many kinds of nutrients and biologically active components such as plant polyphenols and phytic acid, but is hard to eat. "Brown rice and rice bran fermented with *Aspergillus oryzae* (FBRA)" is a processed food that is easier for daily intake, commercially available, and rich in eating experience. During the fermentation process, dietary fibers is partially digested, and free vitamins and phenolic compounds have increased. These fermentation products are utilized for quality control to manage FBRA production. Recently, plant-derived polyphenols have shown anti-oxidative activity and biological function in various disease models. We and other research groups used raw powder FBRA to examine its biological activity through pathological and/or molecular biological analysis. Dietary administration of FBRA showed anti-tumorigenic effects in chemically induced tumors in rodents. Anti-inflammatory effects have been observed in DSS-induced colitis in rat and inflammation-mediated rodent tumor models. I will give an outline of the characteristic of FBRA, and then introduce our recently published work about "Preventive effect of FBRA on spontaneous type 1 diabetes in NOD female mice", including how to estimate the *in vivo* effect of dietary FBRA, its possible mechanisms and the limit of this study.

Keywords: Brown rice, Rice bran, Fermentation with *Aspergillus oryzae*, Animal disease model, Anti-inflammatory effect, Type 1 diabetes

1. Introduction

Lifestyle-related diseases, obesity, metabolic syndrome and cancer are now a global health problem, and the countermeasures are a global issue. Changes of our dietary habits such as increasing intake of refined sugar and polished rice largely account for the high prevalence of non-communicable chronic diseases, indicating the importance of minor food ingredients [1]. Then, improvement of the dietary habits is probably one of the best ways to decrease the risk of such diseases. Brown rice is a traditional food in Japan, and contains ordinary nutrients and many kinds of minor nutrients such as vitamins and minerals. It also contains biologically active components such as plant polyphenols and phytic acid, especially in its bran [1, 2]. Certain components in brown rice such as γ-oryzanol, have been demonstrated to show inhibitory effects on obesity and diabetes including the detailed mechanisms as described below. We should understand the benefits of such minor but biologically active ingredients, and incorporate it to our usual diet. Traditional foods are often produced by utilizing microorganisms-mediated fermentation. As well as minor bioactive components, fermented foods are attracting attentions as health-promoting functional foods. It is possible that new bioactive components produced during fermentation may promote human health.

In this chapter, I will first introduce biological activities of brown rice or its components. Next, I will mention about fermented brown rice, its components and biological activities, including our recent research result about "Brown rice and rice bran fermented with *Aspergillus oryzae* (FBRA)". FBRA is a processed food, rich in partially digested fiber, rice bran-derived phytic acid, and free phenolic compounds. We and other research groups have examined various biological activities of FBRA and reported its anti-tumorigenic, anti-inflammatory, and other effects in cultured cells and/or animal disease models. I will give an outline of the characteristic of this processed brown rice, and then I will introduce our recent work about "Preventive effect of FBRA on spontaneous type 1 diabetes in NOD female mice". How to estimate the *in vivo* effect of dietary FBRA on type 1 diabetes model, its possible mechanisms and the limit of this study will be described.

2. Constituents and biological activity of brown rice before and after *Aspergillus oryzae*-mediated fermentation

2.1 Nutrients and non-nutrient components in brown rice

Brown rice generally contains many kinds of nutrients, carbohydrates, lipids, proteins, and micronutrients such as minerals and vitamins, which are essential for our life [1]. As compared to white rice, it contains higher amount of lipids, potassium, phosphorus, ferrous, manganese, alpha-tocopherol, vitamin B1, vitamin B2, niacin, Vitamin B6, folic acid, pantothenic acid, and dietary fiber. Non-nutrient components such as plant polyphenols and phytic acid are also important components in brown rice, especially rich in a part of rice bran [1, 2]. These phytochemicals carry antioxidative activity [1–4], protect plants themselves from environmental oxidative stress, and are distributed in free, soluble-conjugated, and bound forms in the endosperm and bran/embryo fractions of the whole rice grain [4]. Beneficial biological functions of brown rice or rice bran have been shown in cultured cells and animal disease models by using whole rice, extracts with various solvents, or identified biologically active components as described in Section 2.2.

2.2 Biological activity of components in brown rice

Brown rice and rice bran, and several known constituents in brown rice have been reported to show beneficial functions against diseased conditions such as tumor and life style-related diseases. Rice bran extract from pigmented rice, containing phenolic compounds and antioxidant activity, showed antiproliferative properties against the human and mammalian cancer cell lines [4–6]. Antigenotoxic activity of rice bran was shown in *Salmonella* mutation assay and sister chromatid exchange assay [6–8]. Antitumor effects through a modification of immunity were also reported [9, 10]. Henderson et al. [11] reviewed chemopreventive properties of dietary rice bran in *in vivo* and *in vitro* studies, and highlighted the effective components and their mechanism of action from *in vitro* studies with various cancer cell lines. They classified chemopreventive components with literature-supported activity to two items: 1) components in rice bran oil including fatty acids, tocopherol, flavonoids, γ-oryzanol, and other phenolic compounds; 2) components in defatted rice bran including polysaccharide, phytic acid, and dietary fiber. These components in rice bran contributed to chemopreventive effects on various stages of carcinogenesis through anti-oxidative action, anti-proliferative/pro-apoptotic action, mucosal protection, and immune modulation [11].

Obesity and obesity-related diseases is another worldwide problematic issue of human health. As well as chemopreventive effects on multistage carcinogenesis, brown rice could show multifactorial functions against these diseased state. Certain food components such as phenolic compounds and antioxidants have been reported to have anti-diabetic effects in cultured cells and in model mice [12–14]. Such polyphenols have often worked to improve viability of β cells or decreased the apoptosis through modification of gene expression in the pancreas [13–15]. Pre-germinated brown rice prevented high fat diet induced hyperlipidemia [16] and showed hypocholesterolemic action [17]. A rice bran oil diet was reported to improve lipid abnormalities and hyperinsulinemic responses in type 2 diabetes model animal [18]. Acylated steryl β-glucosides in pre-germinated brown rice diet reduced oxidative stress in streptozotocin-induced diabetes [19]. Among bran-specific phenolic compounds, γ-oryzanol has been well demonstrated to be protective against diabetes and obesity [20–22]. It protected pancreatic islet β cells by directly ameliorating ER stress-induced β cells dysfunction [20, 21], and also functioned as an epigenetic controller in the brain reward system [20, 22]. Moreover, it enhances adipocyte differentiation and glucose uptake in insulin-resistant cells through cell signaling pathway [23]. Stress-induced and animal fat ingestion-induced hypoadiponectinemia have been ameliorated by γ-oryzanol and γ-aminobutyric acid [24, 25]. Sakai et al. [26] reported the importance of Glutathione peroxidase 4 (GPx4) against oxidative stress in the pathologies of vascular diseases such as athelosclerosis and diabetes, and suggested that vitamin E rich food such as brown rice, can compensate for GPx4 loss by protecting cells against lipid peroxidation.

2.3 Biologically active components in *Aspergillus oryzae*-mediated fermentation products of brown rice

Many kinds of biologically active compounds in rice bran have a potential to improve diseased condition and maintain human health, while rice bran itself is not easy for daily intake. However, plant-derived solid substances are often decomposed and utilized by various environmental lives. *Aspergillus* fungi are industrially important for food fermentation and production of biological/bioactive compounds, and *Aspergillus oryzae* is the most common mold for the fermentation of soybeans, rice, grains, and potatoes [27]. *A. oryzae* has also been used for Japanese traditional food, miso, sake, say sauce. During the fermentation process, *A. oryzae* produces amylase, protease, and β-glucosidase, and changes rice constituents to amino acids, fatty acids, organic acids, sugar and sugar alcohol, nucleotides, and various secondary metabolites [28]. Antioxidants and phenolic acid time-dependently increased during the fermentation [28, 29]. Production of natural iron chelator deferriferrichrysin from *A. oryzae* has been reported as a candidate of novel food-grade antioxidant [30]. *A. oryzae*-derived protease preparation showed beneficial effects on colonic environment in high-fat diet fed rats [31]. Neutral polysaccharide produced in fermented Korean brown rice vinegar was reported to have immunostimulatory activity [32]. *A. oryzae*-mediated fermentation of other grains wheat germ or sorghum also produced anti-adipogenic activity in cultured adipocytes [33] and anti-inflammatory effect in atherosclerotic mice model [34].

"Brown rice and rice bran fermented with *Aspergillus oryzae* (FBRA)" described in the following subsection is a processed food with *A. oryzae*-mediated fermentation (**Figure 1**). General constituents of raw powder FBRA used for our studies is shown in **Table 1** (excerpted from Kataoka et al. [36]). After the fermentation process, FBRA becomes to be taken more easily than the original material bran. Dried fermented

<Fermentation procedure>

Rice bran, misted with water
+
Steeped brown rice (1/10 of rice bran)

↓

First steaming, 100°C, 40min
Second steaming, 100°C, 70min in drum-type fermenter

↓

Innoculating with *Aspergillus oryzae*
Standing fermentation, approx. 12h

↓

Fermentation with stirring in rotating fermenter, approx. 20h at 37~43°C

↓

dried by heated air, 50°C, approx. 8h

Fermented product

Figure 1.
Raw powder of FBRA. Upper panel shows photo of brown rice and rice bran as starting material, and raw powder of FBRA as fermented product (kindly provided by Dr. Hideyuki Nemoto, Koken Co., Ltd., Japan). Lower panel illustrates the fermentation procedure previously described in Horie et al [35].

product is packaged and is commercially available and has been accumulating eating experiences. While eating quality of FBRA is not directly estimated by comparing with cooked brown rice, 36 healthy adult participants in our clinical study could consume 21.0 g/day FBRA for 2 weeks without dropout [37]. Dietary fibers are partially digested, and free vitamins and phenolic compounds have increased in FBRA [38, 39]. Increase of polyamines, phenolic acids, and ergothioneine have been demonstrated by LC/ESI-MS/MS [35, 40, 41]. Polyamines such as putrescine and spermidine are essential for cell growth, proliferation and tissue regeneration, but the expression of polyamine synthetic enzymes have been declining with aging [35]. Ergothioneine is an amino acid derivative which has a strong antioxidant activity as a scavenger of reactive oxygen species, and also has the potential to prevent central neurological disorders [40]. Interestingly, Takusagawa et al. [42] recently reported that *A. oryzae* can synthesize ergothioneine from histidine contained in the food material such as rice.

Analysis items	content
Macronutrients	(/100 g of FBRA)
Moisture	2.6 g
Protein	16.3 g
Fat	22.1 g
Ash	11.7 g
Carbohydrate	47.3 g
Saccharides	23.4 g
Dietary fiber	23.9 g
Energy	406 Kcal
Minerals	(/100 g of FBRA)
Sodium	15.2 mg
(equivalent to NaCl)	(0.0386 g)
Phosphorous	2660 mg
Iron	7.20 mg
Calcium	429 mg
Potassium	2090 mg
Magnesium	1070 mg
Copper	0.69 mg
Zinc	5.39 mg
Manganese	23.3 mg
Selenium	9 μg
Vitamins	(/100 g of FBRA)
Thiamine (Vitamin B1)	2.10 mg
Riboflavin (Vitamin B2)	0.79 mg
Vitamin B6	3.79 mg
Vitamin B12	0.03 μg
Vitamin E (α-tocopherol)	9.4 mg
Vitamin K1 (Phylloquinone)	28 μg
Folic acid	190 μg
Pantothenic acid	6.86 mg
Biotin	48.6 μg
Niacin	66.0 mg
(Niacin as Nicotinic acid)	62.3 mg
(Tryptophan)	222 mg
Superoxide elimination activity	1.0×10^3 units/g
Phytic acid	7.64 g / 100 g
Enzyme activity	(units/g)
Amylase	3700
Acid protease	350
Neutral protease	470
Alkaline protease	160
Lipase	640

Listed items were analyzed by Japan Food Research Laboratories (Tokyo, Japan).
(excerpeted from Kataoka K et al. [36], Journal of Functional Foods. 2021; 78: 104356, supplemental ***Table 1****)*

Table 1.
Ingredient of "Brown rice and rice bran fermented with Aspergillus oryzae *(FBRA)".*

Bioactive components produced through *A. oryzae*-mediated fermentation could bring us more beneficial effects and will contribute to preventing or ameliorating various diseased conditions. On the other hand, we should clarify what elements are effective and how the elements works against diseased conditions, including optimal dose and adverse effects. In manufacturers, management of fermentation process with using adequate control compounds described by Lee et al. [28] is essential to keep the products quality.

2.4 Biological activity of raw powder FBRA in *in vitro* and *in vivo* disease models

In parallel with the above component analysis, many research groups have been conducting *in vitro* and *in vivo* studies to estimate the functions of FBRA. A raw powder FBRA used in our study is provided by the manufacturer (Genmai Koso Co. Ltd., Sapporo, Japan). In *in vitro* studies, FBRA extract induced apoptosis of tumor cells by activating mitochondrial pathway in human colorectal tumor cells [43] and via death receptor pathway in human lymphoblastic leukemia cells [44]. These results are consistent with the previous results with those of nonfermented brown rice, while the results should be carefully interpreted because of direct addition of an aqueous extract to cultured cell lines.

Antitumorigenic *in vivo* effects of dietary administration of FBRA have been examined at 5 or 10% dietary concentration. In chemically induced tumorigenesis models, preventive effects of FBRA were demonstrated in colon [45], liver [46], esophagus [47], urinary bladder [48], oral cavity [49], stomach [50], lung [51], pancreas [52]. Preventive effects were also demonstrated in prostate carcinogenesis in TRAP rats [53] and spontaneous lymphomagenesis in AKR/NSlc female mice [54]. Sakurai et al. [55] reported inhibitory effect of oral FBRA on metastasis of colon tumor cells to the liver through a mechanism leading to a Th1-dominant immune state and activation of macrophages via anti-oxidative properties. Chemoprevention mechanisms associated with dietary brown rice components have been reviewed by Henderson et al. [11]. They depicted that rice bran constituents act through anti-oxidative protection against free radicals in initiation stage, anti-proliferative/pro-apoptotic action on malignant cells, modulation of immunity and inflammation in the early or late stage, and mucosal protection through altered microbiota and intestinal environment, and that complex mixture of rice bran-derived bioactive compounds cooperatively suppress many stages of carcinogenic process. Antitumorigenic components in FBRA might be basically the same as those in brown rice and rice bran, while fermentation process possibly influence the activity of FBRA.

Anti-inflammatory effects of FBRA have been observed against the development of hereditary hepatitis in Long-Evans Cinnamon rats [56], DSS-induced colitis in rats [57], and inflammation-related tumor models [58, 59]. Phutthaphadoong et al. [58] presented that DSS-induced inflammation promoted the colorectal carcinogenesis in *Apc*$^{Min/+}$ mice, but the increased severity was ameliorated by FBRA feeding. Onuma et al. [59] have demonstrated that FBRA prevents inflammation-related carcinogenesis in mice through inhibition of inflammatory cell infiltration.

Modifying effects of FBRA feeding on intestinal environment was investigated in rats and healthy human adults. Dietary FBRA increased resident *Lactobacillus* species in rat [60]. In healthy adults, significant effect of FBRA intake was not detected, but no adverse phenomenon was found in this clinical study at the used dose [37].

2.5 Preventive effect of FBRA on spontaneous type 1 diabetes in NOD female mice

Based on the previous findings of anti-oxidative and anti-inflammatory effect *in vivo*, and the presence of antidiabetic components in FBRA, we planned to examine suppressive effects of FBRA on autoimmune-mediated type 1 diabetes. By using spontaneously occurring model in non-obese diabetic (NOD) female mice, we have recently reported that dietary administration of 0.5% FBRA delayed the spontaneous onset of diabetes [36]. How to estimate the *in vivo* effect, its possible mechanisms and the limit of this study are introduced below.

Diabetes in NOD mice shares many features with human type 1 diabetes [61]. In this model mice, autoreactive T cells are primed in the pancreatic lymph nodes and a disequilibrium between regulatory and effector T cells occurred at around 12 weeks of age triggers β cell destruction, resulting in diabetes onset [61]. Cyclophosphamide, an immune system disturbing agent, has often used to promote an onset of diabetes in immunological studies, whereas a spontaneous onset model is often used to examine effects of food-derived components or probiotics [13, 62]. While genetic and immunologic factors are important factors in the pathogenesis of type 1 diabetes, environmental factors such as diet and microbiota can also correlate to it [62, 63]. As mentioned above, certain rice bran components such as plant polyphenols and antioxidants have been shown to be anti-diabetic in mice models and in cultured cells. Those components have worked through improved viability or decreased apoptosis of β cells in the pancreas, or through regulating expression of related genes [12–15, 19–26].

Dietary concentration of FBRA was set to 0.5% based on daily food intake of NOD mice and our preliminary result, in which intragastric administration of FBRA at 0.33 mg/g body weight/day could delay an appearance of diabetes in NOD mice. Dose-dependent effect of FBRA was examined at the concentration of 0.25% – 1.0%. The highest concentration 5% was selected since the other animal studies have been using 5% or 10% FBRA containing diet with no harmful effect.

The criteria for diabetes onset and severity of insulitis in NOD mice were decided according to previous studies. Glucosuria was weekly monitored with test paper, and the onset of diabetes was further confirmed by measuring blood glucose levels as described by Lian et al. [64]. Mice showing 2.5 mg/ml or higher blood glucose were diagnosed as diabetic. To compare insulitis levels among the groups, pancreas was resected at the end of the experimental periods and HE-stained. Level of insulitis was assessed based on the level of lymphocyte infiltration, and the islets were graded scores 0, 1, 2, 3 or 4 as described by Serreze et al. [65]. The insulitis score of each mouse was calculated as follows: accumulated score of observed islets/number of observed islets.

Dietary administration of 0.5% FBRA significantly delayed the appearance of diabetes in mice and lowered the level of insulitis score. Glucosuria and hyperglycemia appeared at around 20 week of age and the ratio of diabetic mice increased age-dependently in control diet-fed mice, but the ratio did not increase in the 0.5% FBRA -fed group. The percentage of diabetic mice was significantly lower at 24 weeks of age as compared to the control group ($p = 0.01$, log rank test). On HE-stained sections of mice pancreas, lymphocyte infiltration into pancreatic islets has already been observed at the age of 12 weeks. However, the 0.5% FBRA-fed group frequently had small intact islets and the ratio of intact islets was significantly higher than that of control-diet-fed mice (**Figure 2**, excerpted from Kataoka et al. [36]). Insulitis score of FBRA-fed group was also significantly lower compared to the control diet group. From these results, the suppressive effects of dietary FBRA is probably achieved by maintaining the number of intact islets (**Figure 3**).

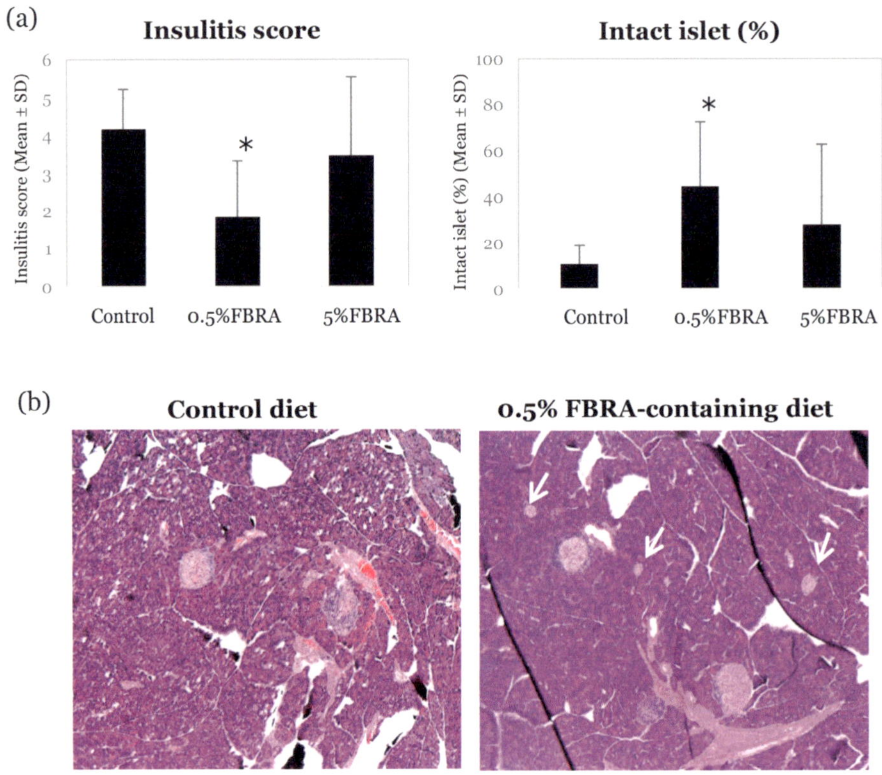

Figure 2.
Suppressive effect of FBRA on insulitis in NOD female mice. (a) Insulitis score and percentage of intact islets in NOD mice fed with control diet or FBRA-containing diet at 30 weeks of age (n = 7). Asterisk means a statistical difference between the two groups (Mann–Whitney's U test, P < 0.05). (b) Representative insulitis in HE-stained pancreatic section of NOD mice fed control diet or 0.5% FBRA-containing diet at the age of 30 weeks (HE stain, ×50). Small intact islets often observed in 0.5% FBRA-fed mice are shown with white arrows. (excerpeted from Kataoka K et al. [36], Journal of Functional Foods. 2021; 78: 104356, ***Figure 1****, with slight modification).*

Possible targets of dietary FBRA in this type 1 diabetes model include: 1) islet-specific T lymphocyte activation; 2) islet-targeting lymphocyte infiltration; 3) cytokine-mediated inflammation or ROS production; 4) regeneration of damaged islets or apoptotic cell death of damaged islets. An Inflammatory cytokine IFNγ, released from activated T cells, has an important role as a trigger of inflammation and β-cell dysfunction in autoimmune-mediated insulitis [62, 66, 67].

However, in our experiment, the percentage and number of $CD4^+$ and $CD4^+$ $IFN\gamma^+$ T cells in the spleens and pancreatic lymph nodes at 12 weeks of age were not significantly different between control diet-fed and 0.5% FBRA-fed mice. Additionally, in adoptive transfer experiments, recipient mice who received a T cell fraction from spleen of 0.5% FBRA-treated NOD mice, could not keep the ratio of intact islets and rather increased the ratio of severely damaged islets, while the number and ratio of intact islets tended to increase in 0.5%FBRA-treated recipient mice who received a T cell fraction derived from control diet-fed donor mice. These results supported that FBRA or its components might suppress the onset of diabetes through keeping an enough number of intact islets in pancreas of NOD mice, not through inhibiting the step of islet-specific T lymphocyte activation.

Pancreas has been known to have regenerative potential for autoimmune- or other factor-mediated damage to islet β cells even in adult rodents [68–70]. Pdx1 and related molecules Foxo1, Reg2, Pdcd4 have important roles in islet function and the fate of injured islet cells [15, 71–73]. Pdx1 and Foxo1 are involved in pancreatic

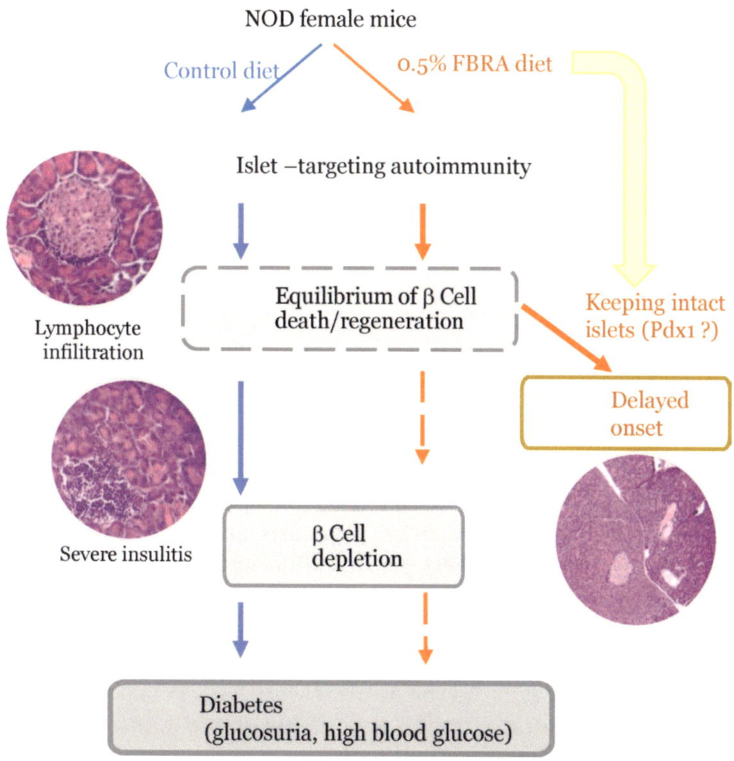

Figure 3.
Hypothetical schema of suppressive effect of FBRA on type 1 diabetes in NOD mice.

development and β cell functional regulation through changes of their intracellular translocation [71]. The inflammatory cytokine IFNγ is reported to decrease nuclear localization of Pdx1 and to trigger β cell dysfunction [67]. Oppositely, phenolic compounds in plant-derived food have recently been reported to show anti-diabetic actions via various mechanisms, including increased expression of pdx1 or restoration of nuclear localization of Pdx1 [13, 15, 67].

Then, we examined effects of FBRA on expression of Pdx1 and related molecules. In our study, mRNA levels of *Pdx1* and related molecule genes were similar in whole pancreases of 19- and 22- week-old mice between the control diet-fed group and the 0.5% FBRA-fed group. However, immuno-histochemical analyses of pancreatic sections showed a tendency for more Pdx1 in the cell nuclei in the 0.5% FBRA-fed group. Then, intracellular localization of Pdx1/Foxo1 and their phosphorylation level at appropriate ages should be further examined in NOD mice with/without 0.5% FBRA treatment.

Our study showed that consumption of FBRA throughout the experimental period suppressed the spontaneously occurring diabetes in female NOD mouse with 0.5% dietary concentration as optimal. But several limitations are still present. At the first, ameliorating effects after onset of type 1 diabetes should be examined in animal model, because autoimmune-mediated diabetes suddenly occur and is generally found as glucosuria. The second is which component (s) and how the component(s) work against pathogenesis of type 1 diabetes. Although FBRA is commercially available processed food and no harmful phenomenon has not been observed in human including clinical trial Volunteers [37], to clarify suppressive mechanisms of FBRA including optimum dose of active component(s) is very important for using it as functional food. Finally, the suppressive effect of FBRA on type 1 diabetes was observed only at lower concentrations, but not at higher concentrations (1% and 5%). At lower dietary concentration, anti-diabetic components in FBRA probably suppressed the development of diabetes through enhanced β cell viability and proliferation, but at higher dietary concentration, the other functional effect of FBRA might appear. For example, 5% dietary FBRA could increase resident *Lactobacillus* species in rat intestine [60], and certain *Lactobacillus* species has been reported to activate Th1 immunity [74–76].

3. Conclusions and perspectives

Brown rice and rice bran contains many kinds of biologically active components and their beneficial effects against diseased conditions have been demonstrated. "Brown rice and rice bran fermented with *Aspergillus oryzae* (FBRA)" is a processed food in which free vitamins and phenolic compounds have increased during *A. oryzae*-mediated fermentation. Dietary administration of FBRA showed anti-tumorigenic and anti-metastatic effects in various tumor model animals, and anti-inflammatory effects in rat hepatitis, rat colitis, and inflammation-mediated rodent tumor models. Based on these previous findings of FBRA, considering antidiabetic components in brown rice, we examined suppressive effects of FBRA on autoimmune-mediated type 1 diabetes. In non-obese diabetic (NOD) female mice, dietary administration of 0.5% FBRA delayed the spontaneous onset of diabetes and significantly reduced the insulitis score [36]. While relation of the rice variety and these anti-disease activities was not investigated, phytochemical profile is varied among different kinds of rice [2–4].

Brown rice and fermented brown rice showed very attractive beneficial functions in *in vitro* and *in vivo* studies. However, for clinical application of FBRA, its action mechanisms including determination of active ingredients and its optimal dose should be clarified in future studies.

Acknowledgements

Keiko Kataoka received financial support for FBRA relating research (Ref No. 55, 58, 59, 60) from Genmai Koso Co. Ltd. (Japan), and was supported in part by JSPS KAKENHI Grant Number 15 K00820. These studies were also technically supported by Support Center for Advanced Medical Sciences, Graduate School of Biomedical Sciences, Tokushima University.

Conflict of interest

The author do not have any conflicts of interest regarding this manuscript.

Author details

Keiko Kataoka
Department of Microbiology and Genetic Analysis, Graduate School of Biomedical Sciences, Tokushima University, Tokushima, Japan

*Address all correspondence to: kataokakeiko@tokushima-u.ac.jp

IntechOpen

© 2021 The Author(s). Licensee IntechOpen. This chapter is distributed under the terms of the Creative Commons Attribution License (http://creativecommons.org/licenses/by/3.0), which permits unrestricted use, distribution, and reproduction in any medium, provided the original work is properly cited.

References

[1] Watanabe S, Hirakawa A, Nishijima CK, Nakamura K, Beppu S, Tungtrakul P, Quin SJ, Tee E-S, Tsuno T, Ohigashi H. Food as medicine: The new concept of "medical rice". Advances in Food Technology and Nutritional Sciences Open J. 2016; 2: 38-50. DOI: 10. 17140/AFTNSOJ-2-129

[2] Ravichanthiran K, Ma ZF, Zhang H, Cao Y, Wang CW, Muhammad S, Aglago EK, Zhang Y, Jin Y, Pan B. Phytochemical profile of brown rice and its nutrigenomic implications. Antioxidants. 2018; 7: 71. DOI: 10.3390/ antiox7060071.

[3] Guofo P, Trindade H. Rice anti-oxidants: phenolic acids, flavonoids, anthocyanins, prothocyanidins, tocopherols, tocotrienols, gamma-oryzanol, and phytic acid. Food Science & Nutrition. 2014; 2: 75-104, DOI: 10.1002/fsn3.86.

[4] Ghasemzadeh A, Karbalaii MT, Jaafar HZE, Rahmat A. Phytochemical constituents, antioxidant activity, and antiproliferative properties of black, red, and brown rice bran. Chemistry Central Journal. 2018; 12: 17. DOI: 10.1186/s13065-018-0382-9.

[5] Quagliariello V, Iaffaioli RV, Falcone M, Ferrari G, Pataro G, Donsi F. Effect of pulsed electric fields-assisted extraction on anti-inflammatory and cytotoxic activity of brown rice bioactive compounds. Food Research International. 2016; 87: 115-124. DOI: 10.1016/j.foodres.2016.07.005

[6] Nam SH, Choi SP, Kang MY, Kozukue N, Friedman M. Antioxidative, antimutagenic, and anticarcinogenic activities of rice bran extracts in chemical and cell assays. Journal of Agricultural and Food Chemistry. 2005; 53: 816-822. DOI: 10.1021/jf0490293.

[7] Insuan O, Chariyakornkul A, Rungrote Y, Wongpoomchai R. Antimutagenic and anoxidant activities of Thai rice bran. Journal of Cancer Prevention. 2017; 22: 89-97. DOI: 10.15430/JCP.2017.22. 2.89.

[8] Ratanavalachai T, Thitiorul S, Tanuchit S, Jansom C, Uttama S, Itharat A. Antigenotoxic activity of Thai Sangyod red rice extracts against a chemotherapeutic agent, doxorubicin, in human lymphocytes by sister chromatid exchange (SCE) assay in vitro. J Med Assoc Thai. 2012; 95 (Suppl 1): S109-114.

[9] Choi SP, Kim SP, Nam SH, Friedman M. Antitumor effects of dietary black and brown rice brans in tumor-bearing mice: relationship to composition. Mol Nutr Food Res. 2013; 57: 390-400. DOI: 10.1002/mnfr.201200515.

[10] Liao HF, Chen YY, Yang YC, Wang CS, Chen YJ. Rice (*Oryza sativa* L) inhibits growth and induces differentiation of human leukemic U937 cells through activation of peripheral blood mononuclear cells. Food and Chemical Toxicology. 2006; 44: 1724-1729. DOI: 10.1016/j.fct.2006.05.015.

[11] Henderson AJ, Ollila CA, Kumar A, Borresen EC, Raina K, Agarwal R. Chemopreventive properties of dietary rice bran: current status and future prospects. Adv Nutr. 2012; 3:643-653. DOI: 10.3945/an.112.002303.

[12] Kaneto H, Kajimoto Y, Miyagawa J, Matsuoka T, Fujitani Y, Umayahara Y, Hanafusa T, Matsuzawa Y, Yamasaki Y, Hori M. Beneficial effects of antioxidants in diabetes, possible protection of pancreaticβ-cells against glucose toxicity. Diabetes. 1999; 48: 2398-2406.

[13] Babu PVA, Liu D, Gilbert ER. Recent advances in understanding the anti-diabetic actions of dietary flavonoids. Journal of Nutritional Biochemistry. 2013; 24: 1777-1789. DOI:10. 1016/j. jnutrbio.2013.06.003.

[14] Zhang Y, Liu D. Flavonol kaempferol improves chronic hyperglycemia-impaired pancreatic beta-cell viability and insulin secretary function. European Journal of Pharmacology. 2011; 670: 325-332. DOI: 10.1016/j.ejphar.2011.08.011.

[15] Zhang Y, Zhen W, Maechler P, Liu D. Small molecule kaempferol modulates PDX-1 protein expression and subsequently promotes pancreatic β-cell survival and function via CREB. Journal of Nutritional Biochemistry. 2013; 24: 638-646. http://dx.doi.org/10.1016/j.jnutbio.2012.03.008.

[16] Shen K-P, Hao C-L, Yen C-Y, Chen C-Y, Chen J-H, Chen F-C, Lin H-L. Pre-germinated brown rice prevented high fat diet induced hyperlipidemia through ameliorating lipid synthesis and metabolism in C57BL/6J mice. J Clin Biochem Nutr. 2016; 59: 39-44. DOI: 10.3164/jcbn.15-117.

[17] Miura D, Ito Y, Mizukuchi A, Kise M, Aoto H, Yagasaki K. Hypocholesterolemic action of pre-germinated brown rice in hepatoma-bearing rats. Life Sciences. 2006; 79: 259-264. DOI: 10.1016/j.lfs.2006.01.001.

[18] Chou TW, Ma CY, Cheng HH, Chen YY, Lai MH. A rice bran oil diet improves lipid abnormalities and suppress hyperinsulinemic responses in rats with streptozotocin/nicotinamide-induced type 2 diabetes. J Clin Biochem Nutr. 2009; 45: 29-36. DOI: 10.3164/jcbn.08-257.

[19] Usuki S, Tsai YY, Morikawa K, Nonaka S, Okuhara Y, Kise M, Yu RK. IGF-1 induction by acylated steryl β-glucosides found in a pre-germinated brown rice diet reduces oxidative stress in streptozotocin-induced diabetes. PLoS One. 2011; 6: e28693. DOI:10.1371/journal.pone.0028693.

[20] Masuzaki H, Kozuka C, Okamoto S, Yonamine M, Tanaka H, Shimabukuro M. Brown rice specific γ-oryzanol as a promising prophylactic avenue to protect against diabetes mellitus and obesity in humans. Journal of Diabetes Investigation. 2019; 10: 18-25. DOI: 10.1111/jdi.12892.

[21] Kozuka C, Sunagawa S, Ueda R, Higa M, Tanaka H, Shimizu-Okabe C, Ishiuchi S, Takayama C, Matsushita M, Tsutsui M, Miyazaki J, Oyadomari S, Shimabukuro M, Masuzaki H. γ-Oryzanol protects pancreatic βcells against endoplasmic reticulum stress in male mice. Endocrinology. 2015; 156: 1242-1250. DOI: 10.1210/en.2014-1748.

[22] Kozuka C, Kaname T, Shimizu-Okabe C, Takayama C, Tsutsui M, Matsushita M, Abe K, Matsuzaki H. Impact of brown rice specific γ-oryzanol on epigenetic modulation of dopamine D2 receptors in brain striatum in high-fat-diet-induced obecity in mice. Diabetologia. 2017; 60: 1502-1511. DOI: 10.1007/s00125-017-4305-4.

[23] Jung CH, Lee D-H, Ahn J, Lee H, Choi WH, Jang YJ, Ha T-Y. γ-Oryzanol enhances adipocyte differentiation and glucose uptake. Nutrients. 2015; 7: 4851-4861. DOI: 10.3390/nu7064851.

[24] Ohara K, Kiyotani Y, Uchida A, Nagasaka R, Maehara H, Kanemoto S, Hori M, Ushio H. Oral administration of γ-aminobutyric acid and γ-oryzanol prevents stress-induced hypo-adiponectinemia. Phytomedicine. 2011; 18: 655-660. DOI: 10.1016/j.phymed.2011.01.003.

[25] Nagasaka R, Yamasaki R, Uchida A, Ohara K, Ushio H. γ-Oryzanol recovers mouse hypoadiponectinemia induced by animal fat ingestion. Phytomedicine. 2011; 18: 669-671. DOI: 10.1016/j.phymed.2011.01.004.

[26] Sakai O, Yasuzawa T, Sumikawa Y, Ueta T, Imai H, Sawabe A, Ueshima S. Role of GPx4 in human vascular

endothelial cells, and the compensatory activity of brown rice on GPx4 ablation condition. Pathophysiology. 2017; 24: 9-15. DOI: 10.1016/j.pathophys. 2016.11.002.

[27] Park H-S, Jun S-C, Han K-H, Hong S-B, Yu J-H. Diversity, application, and systemic biology of industrially important *Aspergillus fungi*. Adv Appl Microbiol. 2017; 100: 161-202. DOI: 10.1016/bs.aambs.2017.03.001.

[28] Lee S, Lee DE, Singh D, Lee CH. Metabolomics reveal optimal grain preprocessing (milling) toward rice *Koji* fermentation. Journal of Agricultural and Food Chemistry. 2018; 66: 2694-2703. DOI: 10.1021/acs.jafc.7b05131.

[29] Shin H-Y, Kim S-M, Lee JH, Lim S-T. Solid-state fermentation of black rice bran with *Aspergillus awamori* and *Aspergillus oryzae*: effects on phenolic acid composition and antioxidant activity of bran extracts. Food chemistry. 2019; 272: 235-241. DOI: 10.1016/j.foodchem.2018.07.174.

[30] Todokoro T, Fukuda K, Matsumura K, Irie M, Hata Y. Production of the natural iron chelator deferri-ferrichrysin from *Aspergillus oryzae* and evaluation as a novel food-grade antioxidant. J Sci Food Agric. 2016; 96: 2998-3006. DOI: 10.1002/jsfa.7469.

[31] Yang Y, Sitanggang NV, Kato N, Inoue J, Murakami T, Watanabe T, Iguchi T, Okazaki Y. Beneficial effects of protease preparations derived from *Aspergillus* on the colonic luminal environment in rats consuming a high-fat diet. Biomedical Report. 2015; 3: 715-720. DOI: 10.3892/br.2015.490.

[32] Kim H, Lee H, Shin KS. Intestinal immunostimulatory activity of neutral polysaccharide isolated from traditionally fermented Korean brown rice vinegar. Biosci Biotechnol Biochem. 2016; 80: 2383-2390. DOI: 10.1080/09168451.2016.1217149.

[33] Park E, Kim HO, Kim G-N, Song J-H. Anti-oxidant and anti-adipogenic effects of ethanol extracts from wheat germ and wheat germ fermented with *Aspergillus oryzae*. Prev Nutr Food Sci. 2015; 20: 29-37. DOI: 10.3746/pnf.2015.20.1.29.

[34] Ham YM, Song HS, Kwon JE, Jeon H, Baek HJ, Kim CW, Yoon W-J, Choung ES, Kang SC. Effects of fermented *Sorghum bicolor* L.Moench extract on inflammation and thickness in a vascular cell and atherosclerotic mice model. Journal of Natural Medicines. 2019; 73: 34-46. DOI: 10.1007/s11418-018-1231-9.

[35] Horie Y, Goto A, Tsubuku S, Itoh M, Ikegawa S, Ogawa S, Higashi T. Changes in polyamine content in rice bran due to fermentation with *Aspergillus oryzae* analysed by LC/ESI-MS/MS combined with derivatization. Analytical Sciences. 2019; 35: 427-432.

[36] Kataoka K, Nemoto H, Sakurai A, Yasutomo K, Shikanai M. Preventive effect of fermented brown rice and rice bran on spontaneous type 1 diabetes in NOD female mice. Journal of Functional Foods. 2021; 78: 104356. https://doi.org/10. 1016/j.jff.2021.104356.

[37] Nemoto H, Ikata K, Arimochi H, Iwasaki T, Ohnishi Y, Kuwahara T, Kataoka K. Effects of fermented brown rice on the intestinal environments in healthy adult. Journal of Medical Investigation. 2011; 58: 235-245.

[38] Tanaka K, Horie Y, Nemoto H, Kosaka H, Jo M, Tezuka Y. Analysis of water-soluble constituents in fermented brown rice and rice bran by *Aspergillus oryzae* (FBRA). Journal of Computer Aided Chemistry. 2017; 18: 46-50.

[39] Tanaka K, Horie Y, Nemoto H, Kosaka H, Jo M, Tezuka Y. Analysis of volatile constituents in fermented brown rice and rice bran by *Aspergillus oryzae* (FBRA). Journal of Computer Aided Chemistry. 2017; 18: 42-45.

[40] Horie Y, Goto A, Imamura R, Itoh M, Ikegawa S, Ogawa S, Higashi T. Quantification of ergothioneine in *Aspergillus oryzae*-fermented rice bran by a newly-developed LC/ESI-MS/MS method. Food Science and Technology, 2020; 118: 108812. https://doi.org/10.1016/j.lwt.2019. 108812.

[41] Ogawa S, Takafuji K, Tsubuku S, Horie, Ikegawa S, Higashi T. Isotope-coded derivatization based LC/ESI-MS/MS methods using a pair of novel reagents for quantification of hydroxycinnamic acids and hydroxybenzoic acids in fermented brown rice product. Journal of Pharmaceutical and Biomedical Analysis. 2017; 142: 162-170. http://dx.doi.org/10.1016/j.jpba.2017.04.035.

[42] Takusagawa S, Satoh Y, Ohtsu I, Dairi T. Ergothioneine production with *Aspergillus oryzae*. Bioscience, Biotechnology, and Biochemistry. 2019; 83: 181-184. https://doi.org/10.1080/09168451.2018.1527210

[43] Itoh M, Nishibori N, Sagara T, Horie Y, Motojima A, Morita K. Extract of fermented brown rice induces apoptosis of human colorectal tumor cells by activating mitochondrial pathway. Phytotherapy Research. 2012; 26: 1661-1666.

[44] Horie Y, Nemoto H, Itoh M, Kosaka H, Morita K. Fermented brown rice extract causes apoptotic death of human acute lymphoblastic leukemia cells via death receptor pathway. Appl Biochem Biotechnol. 2016; 178: 1599-1611. DOI: 10.1007/s12010-015-1970-y.

[45] Katyama M, Yoshimi N, Yamada Y, Sakata K, Kuno T, Yoshida K, Qiao Z, Vihn PQ, Iwasaki T, Kobayashi H, Mori H. Preventive effect of fermented brown rice and rice bran against colon carcinogenesis in male F344 rats. Oncology Report. 2002; 9: 817-822.

[46] Katayama M, Sugie S, Yoshimi N, Yamada Y, Sakata K, Qiao Z, Iwasaki T, Kobayashi H, Mori H. Preventive effect of fermented brown rice and rice bran on diethylnitrosoamine and phenobarbital-induced hepatocarcinogenesis in male F344 rats. Oncology Report. 2003; 10: 875-880.

[47] Kuno T, Hirose Y, Hata K, Kato K, Qiang SH, Kitaori N, Hara A, Iwasaki T, Yoshimura T, Wada K, Kobayashi H, Mori H. Preventive effect of fermented brown rice and rice bran on N-nitroso-methylbenzylamine-induced esophageal tumorigenesis in rats. International Journal of Oncology. 2004; 25: 1809-1815.

[48] Kuno T, Hirose Y, Yamada Y, Hata K, Qiang SH, Asano N, Oyama T, Zhi H, Iwasaki T, Kobayashi H, Mori H. Chemoprevention of mouse urinary bladder carcinogenesis by fermented brown rice and rice bran. Oncology Report. 2006; 15: 533-538.

[49] Long NK, Makita H, Yamashita T, Toida M, Kato K, Hatakeyama D, Shibata T. Chemopreventive effect of fermented brown rice and rice bran on 4-nitroquinoline 1-oxide-induced oral carcinogenesis in rats. Oncology Report. 2007; 17: 879-885.

[50] Tomita H, Kuno T, Yamada Y, Oyama T, Asano N, Miyazaki Y, Baba S, Taguchi A, Hara A, Iwasaki T, Kobayashi H, Mori H. Preventive effect of fermented brown rice and rice bran on N-methyl-N'-nitro-N-nitrosoguanidine-induced gastric carcinogenesis in rats. Oncology Report. 2008; 19: 11-15.

[51] Phutthaphadoong S, Yamada Y, Hirata A, Tomita H, Taguchi A, Hara A, Limtrakul PN, Iwasaki T, Kobayashi H, Mori H. Chemopreventive effect of fermented brown rice and rice bran against 4-(methylnitrosamino)-1-(3-pyridyl)-1-butanone-induced lung tumorigenesis in *female* A/J mice. Oncology Reports. 2009; 21: 321-327.

[52] Kuno T, Takahashi S, Tomita H, Hisamatsu K, Hara A, Kobayashi H,

Mori H. Preventive effects of fermented brown rice and rice bran against *N*-nitrosobis(2-oxopropyl) amine-induced pancreatic tumorigenesis in male hamsters. Oncology Letter. 2015; 10: 3377-3384. DOI: 10.3892/ol.2015.3809.

[53] Kuno T, Nagano A, Mori Y, Kato H, Nagayasu Y, Naiki-Ito A, Suzuki S, Mori H, Takahashi S. Preventive effects of fermented brown rice and rice bran against prostate carcinogenesis in TRAP rats. Nutrients. 2016; 8: 421. DOI: 10.3390/nu8070421.

[54] Kuno T, Kato H, Naiki-Ito A, Suzuki S, Tanaka T, Takahashi S, Mori H. Preventive effects of fermented brown rice and rice bran on spontaneous lymphomagenesis in AKR/NSlc female mice. Asian Pacific Journal of Cancer Prevention. 2018; 19: 3217-3223. DOI: 10. 31557/APJCP.2018. 19. 11. 3217.

[55] Sakurai H, Choo M-K, Chino A, Tega E, Iwasaki T, Kobayashi H, Saiki I. Antimetastatic and immunostimulatory properties of fermented brown rice. Journal of Traditional Medicines. 2006; 23: 112-116.

[56] Shibata T, Nagayasu H, Kitajo H, Arisue M, Yamashita T, Hatakeyama D, Iwasaki T, Kobayashi H. Inhibitory effects of fermented brown rice and rice bran on the development of acute hepatitis in Long-Evans Cinnamon rats. Oncology Report. 2006; 15: 869-874.

[57] Kataoka K, Ogasa S, Kuwahara T, Bando Y, Hagiwara M, Arimochi H, Nakanishi S, Iwasaki T, Ohnishi Y. Inhibitory effects of fermented brown rice on induction of acute colitis by dextran sulfate sodium in rats. Digestive Disease and Sciences. 2008; 53: 1601-1608. DOI: 10. 1007/s10620-007-0063-3.

[58] Phutthaphadoong S, Yamada Y, Hirata A, Tomita H, Hara A, Limtrakul P, Iwasaki T, Kobayashi H, Mori H. Chemopreventive effect of fermented brown rice and rice bran (FBRA) on the inflammation-related colorectal carcinogenesis in $Apc^{Min/+}$ mice. Oncology Reports. 2010; 23: 53-59. DOI: 10.3892/or_00000605.

[59] Onuma K, Kanda Y, Suzuki-Ikeda S, Sakaki R, Nonomura T, Kobayashi M, Osaki M, Shikanai M, Kobayashi H, Okada F. Fermented brown rice and rice bran with *Aspergillus oryzae* (FBRA) prevents inflammation-related carcinogenesis in mice, through inhibition of inflammatory cell infiltration. Nutrients. 2015; 7: 10237-10250. DOI: 10.3390/nu7125531.

[60] Kataoka K, Kibe R, Kuwahara T, Hagiwara M, Arimochi H, Iwasaki T, Benno Y, Ohnishi Y. Modifying effects of fermented brown rice on fecal microbiota in rats. Anaerobe. 2007; 13: 220-227. DOI: 10.1016/j. anaerobe.2007.07.001.

[61] Gagnerault M-C, Luan JJ, Lotton C, Lepault F. Pancreatic lymph nodes are required for priming of β-cell reactive T cells in NOD mice. Journal of Experimental Medicine. 2002; 196: 69-377. http://www.jem.org/cgi/doi/10.1084/jem.20011353.

[62] Mishra S, Wang S, Nagpal R, Miller B, Singh R, Taraphder S, Yadav H. Probiotics and prebiotics for the amelioration of type 1 diabetes: Present and future perspectives. Microorganisms. 2019; 67: doi:10.3390/microorganisms7030067.

[63] Paschou SA, Paradoupoulou-Marketou N, Chrousos GP, Kanaka-Gantenbein C. On type 1 diabetes mellitus pathogenesis. Endocrine connections. 2018; 7: R38-R46. https://doi.org/10.1530/EC-17-0347.

[64] Lian G, Arimochi H, Kitamura A, Nishida J, Li S, Kishihara K, Maekawa Y, Yasutomo K. Manipulation of CD98 resolves type 1 diabetes in NOD mice. The Journal of Immunology. 2012; 188: 2227-2234. doi:10.4049/jimmunol. 1102586.

[65] Serreze DV, Fleming SA, Chapman HD, Richard SD, Leiter EH, Tisch RM. B lymphocytes are critical antigen-presenting cells for the initiation of T cell-mediated autoimmune diabetes in nonobese diabetic mice. Journal of Immunology. 1998; 61: 912-3918. http://www.jimmunol.org/content/161/8/3912.

[66] Kahaly GJ, Hansen MP. Type 1 diabetes associated autoimmunity. Autoimmunity Reviews. 2016; 15: 644-648. DOI: 10. 1016/j.autrev. 2016.02.017. http://dx.doi.org/10.1016/j.autrev.2016.02.017.

[67] Pondugala PK, Sasikala M, Guduru VR, Rebala P, Nageshwar RD. Interferon-γ decreases nuclear localization of Pdx-1 and triggers β-cell dysfunction in chronic pancreatitis. Journal of Interferon & Cytokine Research. 2015; *35,* 523-529. https://doi.org/10.1089/jir.2014.0082.

[68] Nir T, Melton DA, Dor Y. Recovery from diabetes in mice by beta cell regeneration. Journal of Clinical Investigation. 2007; 117: 2553-2561.

[69] Cano DA, Rulifson IC, Heiser PW, Swigart LB, Pelengaris S, German M, Evan GI, Bluestone JA, Hebrok M. Regulated beta-cell regeneration in the adult mouse pancreas. Diabetes. 2008; 57: 958-966.

[70] Yi P, Park J-S, Melton DA. Betatrophin: a hormone that controls pancreaticβ-cell proliferation. Cell. 2013; 153: 747-758. http://dx.doi.org/10.1016/j.cell.2013.04.008.

[71] Meng Z, Lv J, Luo Y, Lin Y, Zhu Y, Nie J, Yang T, Sun Y, Han X. Forkhead box O1/pancreatic and duodenal homeobox 1 intracellular translocation is regulated by c-jun N-terminal kinase and involved in prostaglandin E_2-induced pancreatic β-cell dysfunction. Endocrinology, 2009; 150: 5284-5293. DOI: 10.1210/en.2009-0671.

[72] Ruan Q, Wang T, Kameswaran V, Wei Q, Johnson DS, Matschinsky F, Shi W, Chen YH. The microRNA-21-PDCD4 axis prevents type 1 diabetes by blocking pancreatic β cell death. PNAS. 2011; 108: 12030-12035. DOI/10.1073/pnas.1101450108.

[73] Hill T, Krougly O, Nikoopour S, Bellemore S, Lee-Chan E, Fouser LA, Hill DJ, Singh B. The involvement of interleukin-22 in the expression of pancreatic beta cell regenerative *Reg* genes. Cell Regeneration. 2013; 2: 2. http://www.cellregenerationjournal.com/content/2/1/2.

[74] Segawa S, Nakakita Y, Takata Y, Wakita Y, Kaneko T, Kaneda H, Watari J, Yasui H. Effect of oral administration of heat-killed *Lactobacillus brevis* SBC8803 on total and ovalbumin-specific immunoglobulin E production through the improvement of Th1/Th2 balance. International Journal of Food Microbiology. 2008; 121: 1-10. DOI: 10.1016/j-ijfoodmicro.2007.10.004.

[75] Castanheira LG, Castro JM, Martins-Filho OA, Nicoli JR, Vieira LQ, Afonso LC. *Lactobacillus delbrueckii* as a potential skin adjuvant for induction of type 1 immune responses. Front Biosci. 2007; 12: 1300-1307. DOI: 10.2741/2148.

[76] Wen K, Tin C, Wang H, Yang X, Li G, Giri-Rachman E, Kocher J, Bui T, Clark-Deener S, Yuan L. Probiotic *Lactobacillus rhamnosus* GG enhanced Th1 cellular immunity but did not affect antibody responses in a human gut microbiota transplanted neonatal gnotobiotic pig model. PLOS ONE. 2014; 9: e94504.

Chapter 14

Golden Rice, VAD, Covid and Public Health: Saving Lives and Money

Adrian C. Dubock, Justus Wesseler, Robert M. Russell, Chen Chen and David Zilberman

Abstract

On July 21, 2021, Golden Rice was registered in the Philippines allowing cultivation and consumption. Research, as an intervention to combat vitamin A deficiency (VAD), started in 1991, and proof of concept for what was to become Golden Rice, was achieved in 1999. In the 1990s, 23–34% deaths globally of children less than 5 years old were caused by VAD, and in developing countries, the percentage was even higher. By 2013, progress against the Millennium Development Goals had reduced <5-y child deaths globally from VAD to about 2% of all such deaths. The progress included significant vaccination programs against measles, and better access to clean water, as well as vitamin A supplementation, all delivered through community health programs. Economic development and education about diet reduced food insecurity. In contrast to continuing VAD deaths, the Covid-19 pandemic has attracted huge political attention, including in low- and middle-income countries. Community health programs have been adversely affected by the pandemic. There is a danger that as a result VAD rates, child and maternal mortality climbs again toward 1990's levels. Adoption of Golden Rice provides a safe, culturally simple amelioration and is costless. Other countries should seize the opportunity. Bangladesh is first in line, possibly followed by Indonesia and India.

Keywords: vitamin A deficiency, micronutrient fortification, public health, costs of community health, Covid effects on VAD, vaccination, vitamin A supplementation, sustainability

1. Introduction

Vitamin A deficiency is almost non-existent in high-income country populations. Vitamin A deficiency (VAD) occurs in human populations of low- and middle-income countries and is associated with lack of dietary diversity, often associated with poverty. Staple food grains, such as rice, are readily available, easy to prepare and tasty, and an excellent source of energy, but polished white rice contains no micronutrients. Conversely, animal products (many of which contain vitamin A) and colored fruits and vegetables (which contain beta-carotene, which the human body converts to vitamin A) foods are expensive or unavailable.

From 1991 to 2013, the VAD rate among children in low- and middle-income countries declined from 39 to 29%, with notable improvements among children in East and Southeast Asia [1]. Children in sub-Saharan Africa and South and Southeast Asia continue to suffer disproportionately from VAD and its associated risks: infectious and diarrheal diseases, irreversible blindness and other sensory losses, and premature death [2, 3].

Golden Rice is any variety of rice containing the GR2E[1] genetics [4]. In addition to the carbohydrate of white rice, Golden Rice also contains organically produced beta-carotene, imparting its color, which, following its consumption, the human body converts to vitamin A. The beta-carotene content is the only difference from white rice [5].

Beta-carotene is ubiquitous in nature—all colored plant parts contain it—and in a varied human diet. Vitamin A is not found in plants, but is present in animal products. Beta-carotene from food is non-toxic [6] and the human body excretes what it does not need. It is, therefore, impossible to induce vitamin A toxicity by consuming beta-carotene, so overdosing with Golden Rice is impossible [7].

On January 25, 2001 Professor Ingo Potrykus, one of the co-creators of Golden Rice, signed a license agreement with the Philippines Rice Research Institute ("Phil Rice") to develop Golden Rice. Twenty years later, on July 21, 2021 Phil Rice received the Philippine Governments final regulatory clearance allowing cultivation and consumption of Golden Rice in the Philippines [8].

The causes of the long delay are described elsewhere [7, 9–17]. Undoubtedly, the delay has caused unnecessary human suffering and lost lives, mostly of young children and mothers.

2. Vitamin A deficiency and other nutritional deficiencies

The latest figures available estimate that of the world's population about 2 billion people are macronutrient deficient, and about 800 million people suffer from "hidden hunger" another name for micronutrient deficiency [18]. Iron, vitamin A and zinc deficiencies are the most common micronutrient deficiencies. Folate deficiencies are also widespread. Where all these deficiencies occur is strongly correlated with the global burden of poverty and disease [19], and so the distribution of them is remarkably similar to the vitamin A deficiency map (**Figure 1**).

For many years, VAD was principally associated with childhood blindness. During the early 1970s, programs of vitamin A supplementation were started in India, Indonesia and Bangladesh. In Indonesia, a development specialist with Helen Keller International noted that the true public health weight of the problem is obscured because its victims often die before they can be reported as blind [21]. Indonesian data analysis demonstrated that children with "mild" vitamin A deficiency were at a high risk of dying [22]. Subsequently, a series of seminal studies demonstrated that a universal source of vitamin A would save 23–34% of global under 5 years, child mortality [23, 24] and also, later, [2, 25–27].

These findings [23, 24] gave huge impetus to expanding vitamin A supplementation programs from the 1990s [28], which involved significant costs [29] and at the time was highly controversial [21]. The Millennium and Sustainable Development Goals made significant progress from the base year of 1990, including in combatting VAD, and with major advances in vitamin A supplementation, as well as

[1] Many transformation events were produced once in ~2004 from which event GR2E has been selected on the basis of molecular structure and insertion in the rice genome, together with agronomic performance when introduced to different rice varieties. It is the basis of the regulatory data generated and is the only form of Golden Rice, which is offered for approval and use.

vaccination programs against measles and other diseases, and improved sanitation and clean water access, in turn reducing diarrhea incidence.

Thus, from 1990 simultaneous progress was made in reducing VAD, thereby improving the immunity of populations of vulnerable children to common diseases, and at the same time reducing the incidence of those diseases.

Nevertheless, macronutrient deficiency is being reduced at a faster rate than micronutrient deficiency (**Figure 2**).

If greater attention is not paid to reducing micronutrient deficiencies, they will have a bigger impact on productive human life than macronutrient deficiencies [7].

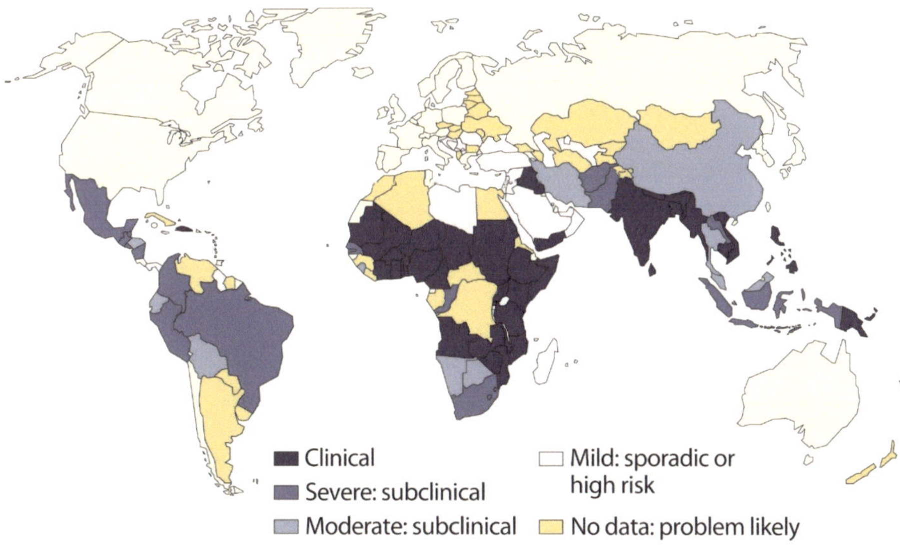

Figure 1.
Public health importance for vitamin A deficiency, by country. Source [20]. Redrawn by & courtesy of Banson.

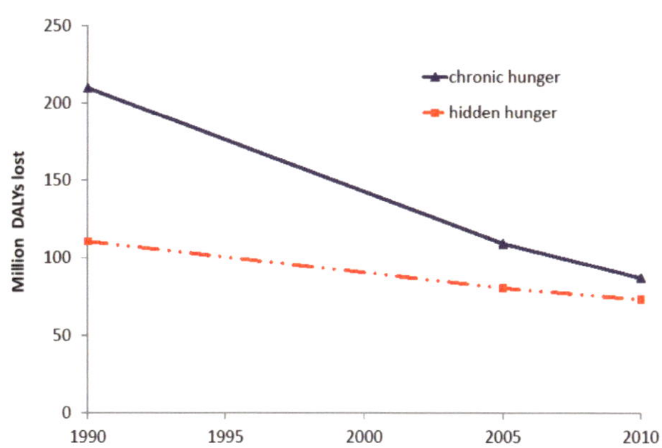

Figure 2.
DALYs lost due to chronic hunger and hidden hunger between 1990 and 2010 (part of ***Figure 2*** *from [30]).*

3. Vitamin A deficiency in Bangladesh and the Philippines

Bangladesh and the Philippines are so far the only two countries where regulatory applications been made to cultivate and consume GR. It is clear that progress in combatting VAD has been significant, a reduction from 23 to 34% of under-five child deaths in the 1990s to *circa* 2% in these two countries in 2013, the latest data available [31]. The 2013 data are useful as a proxy for comparison between countries, but has limitations [32] including that the "total VAD" deaths reported are actually solely deaths where measles or diarrhea are reported on a death certificate. But these are not the only causes of death due to immune system insufficiency due to VAD, and reporting may anyway be patchy in remote, poor districts.

In any event, a 2019 publication, reporting pre-Covid-19 pandemic data, determined that large-scale food fortification against VAD could protect nearly three million children annually [19].

In July 2021, Golden Rice was approved for cultivation and, previously for consumption, in the Philippines. Golden Rice is now awaiting final approval in Bangladesh. The Bangladesh Rice Research Institute submitted an application for consumption and cultivation in Bangladesh in 2017, which has stalled in the Department of Environment [33]; hence, no regulatory decision has been taken.

It was estimated that 3.08% of children of the 14.3 million children under age 5 in Bangladesh will die in 2019 [34, 35]. Two percent of these deaths may be attributable to VAD [31], resulting in an estimated 8826 deaths in 2019. The modified version of Golden Rice was available in 2004, and regulatory delay is the main reason why only one country as of now, The Philippines, has approved Golden Rice for cultivation. Every year of delay of approval of Golden Rice may cause at least 8500 child deaths in Bangladesh. Ten years of delay will result in over 85,000 deaths, at least some of which might have been avoided.

A recent study has estimated that substituting Golden Rice for white rice could provide 57–99% and 89–113% and of the recommended vitamin A requirement for preschool children in the Philippines and Bangladesh, respectively [5]. Such a boost to dietary beta-carotene could do much to combat VAD and is highly sustainable especially when the Golden Rice is grown by the communities which need it.

Currently, Bangladesh spends annually between USD24 million and USD47 million (a 5–6 percent increase in the cost of rice at US$480–US$783 per metric ton), on chemically fortifying 1 million metric tonnes (4% of the country's annual rice production) with at least vitamin A and zinc [36, 37]. This is sufficient rice to feed about 7% of Bangladesh's population of 163 million people. The aim of the Bangladesh Department of Woman's Affairs is to "make fortified rice available to all" [36]. The 2020 Bangladesh rice harvest was 25 million metric tonnes. The fortification program is presumably limited by the budget available.

Large-scale food fortification with chemicals represents an unusual confluence of commercial and public health interests, with useful focus on inputs and outcomes arising [19, 36, 38]. Similar focus needs to be applied to integrate public sector developed, and free, biofortified crops—conventionally bred and genetically engineered—for the same purpose, and integrates all approaches—chemical, biological, educational and cultural—to alleviate micronutrient deficiencies in populations, for least cost and maximum coverage.

In 2013, when the Bangladesh program started, there were no alternatives to industrial fortification with chemicals. Since then, high-zinc rices have been introduced by Bangladeshi rice breeders as part of the Harvest Plus program [39]. In November 2017, Golden Rice registration was applied for by the Bangladesh Rice Research Institute (BRRI) and they have been multiplying different varieties of Golden Rice seed since then. Adoption of both high-zinc rice and Golden Rice would reward Bangladesh science investment, save foreign exchange currently being spent on importing chemical fortificants, and allow the reach of biofortified rice to a greater proportion of Bangladesh society than the currently industrial fortification alone, which depends on distributive infrastructure.

The above can be achieved in steps; initially, the high-zinc rice varieties could be chemically fortified with vitamin A. Subsequently, now or after registration of Golden Rice, this and the high-zinc varieties could be introgressed (bred together) in two to three rice-growing seasons by Bangladesh's public sector rice breeders. It is very curious to understand, in light of the Bangladesh regulatory data submitted for official consideration in November 2017, 47 months ago at the time of writing, why very few regulatory meetings have been called and no regulatory decision about Golden Rice has been taken. Particularly curious, in light of the recent Philippine cultivation decision (taken 10 months after data submission) when the agro-environmental conditions for rice cultivation are so similar between the two countries. (Governments have approved Golden Rice as safe for consumption in Australia, Canada, New Zealand, the Philippines and USA [40]. Cultivation permission has not been sought in these countries, except for the Philippines, where it has been granted.)

Registration of Golden Rice for cultivation in Bangladesh would allow a refocus of the huge cost of chemical fortification of rice (currently USD24–48 m annually to chemically fortify only 4% of Bangladesh's rice production) by the Department of Women's Affairs to allow a much greater proportion of Bangladesh's population to be reached than will ever be possible if dependent on industrial fortification only. The Golden Rice option has zero cost increment, compared with white rice, to governments, growers or consumers.

What could be standing in the way of the Bangladesh's National Committee on Biosafety under the Ministry of Environment, Forest, and Climate Change, meeting and taking a positive decision, to benefit hugely VAD intervention in Bangladesh?

Delay is expensive: Delay of the use of Golden Rice in India cost USD199 per annum for the decade preceding 2014 [41, 42]. Even if all 25 million metric tonnes of Bangladesh's 2020 rice production were Golden Rice, the extra beta-carotene nutrition is free, saving a large proportion of USD600 million to USD1.17 billion if the same was achieved by chemical fortification with zinc and vitamin A. In practical terms, it appears that the Bangladesh Governments objective "to make fortified rice available to all" is unobtainable without fully embracing the results of the work of the Bangladesh governments own rice breeders, in producing high zinc as well as Golden Rice varieties.

4. Integration of fortification and biofortification

Chemical food fortification has been used to combat micronutrient deficiencies for 100 years in high-income countries and there are good data on positive impact. Conversely, there are few data, except for iodine fortification, concerning large-scale food fortification (LSFF) with vitamin A, iodine, iron and folic acid in low- and middle-income countries (LMIC), but what data there are is positive [19].

LSFF, with chemicals, is especially useful in LMIC where micronutrient deficiency is evidenced at a population level, and where rapid urbanization is accompanied by increased household purchasing power, leading to reliance on centrally processed foods [19]. One of the complex of issues in such settings, however, is to ensure that chemical fortification of different processed foods does not result in excessive intake, resulting in delivery of a tolerable upper intake level for the population (which is acceptable) and not toxicity (which is unacceptable) [19].

The Golden Rice project has been designed from its initiation principally to assist resource-poor growers and communities who do not rely on processed foods, but largely grow their own. And that remains the objective. Regarding toxicity, Golden Rice provides beta-carotene a non-toxic source of vitamin A, and not vitamin A itself (which is toxic when consumed in excess.)

A 1992 UN Conference on Nutrition confirmed that for VAD alleviation, locally available food-based strategies are the first priority, with vitamin A capsules only an interim measure [43]. However, the bioavailability of beta-carotene from commonly available fruits and vegetables is very low. Even when they are available, a young child between ages 1 year and 3 would need to eat eight servings of dark green leafy vegetables per day in order to meet the Recommended Dietary Allowance for vitamin A. This results in "the virtual impossibility for most poor, young children to meet their vitamin A requirements through vegetable and fruit intake alone" [21]. The low bioavailability of vitamin A from plant foods explains, in part, the presence of vitamin A deficiency among children living amid ample supplies of dark green leafy vegetables and other plant sources of vitamin A [21]. Conversely, a recent study has estimated that substituting Golden Rice for white rice could provide 57–99% and 89–113% of the recommended vitamin A requirement for preschool children in the Philippines and Bangladesh, respectively [5]. Such a boost to dietary beta-carotene could do much to combat VAD and is highly sustainable.

A perfect food fortificant has been described [21] as one which exhibits the following characteristics:

1. It must be a dietary staple eaten daily, with little or no variation.
2. The fortified food must reach the whole population.
3. There must be minimal effect on the cost of the staple food.
4. The micronutrient must be chemically stable.
5. The micronutrient must be undetectable by the consumers.

Golden Rice, being consumed as a staple food, matches the requirements perfectly, except for the color imparted by the beta-carotene content.

However, the golden color imparts advantages. Golden Rice is easily recognizable, so consumers—even illiterate consumers—can exercise choice. And the color is also advantageous for government programs: Each grain is naturally labeled, so "passing off"[2] as biofortified rice is not possible. With the golden color as a marker, Golden Rice can also be used—after the traits are introgressed (e.g., bred together into one variety) as a carrier for invisible micronutrient traits of rice, such as high zinc, high iron and high folate (the natural form of the folic acid used for chemical food fortification.)

Multifunctional cooperation, including between different government departments not used to working together, will be beneficial for effective use of Golden Rice [44]. (This is also the case for LSFF with chemicals [19].)

Within this requirement for multifunctional cooperation, there is clearly a role for synergistic reinforcement of what may be termed urban and rural improvement of staple foods with additional micronutrients. Social marketing research has determined that consumers of rice growing communities are interested to try Golden Rice if they can afford it and if it is good for their family's health [45], both being applicable to Golden Rice. However, although a small cultural change,

[2] Rice is fortified with chemicals by mixing rice powder with chemical fortificants, extruding and drying the result to resemble rice grains, and mixing the result in required proportion with polished white rice. If fortification is done badly, children may pick out and discard the fortified pellets, if done well unfortified white rice can be "passed off" as fortified rice by unscrupulous people.

changing from eating white rice to Golden Rice, even partially, is significant and will benefit from encouragement. There is an additional challenge to be overcome. The agronomic characteristics, such as yield, pest resistance and days to maturation, of any variety of Golden Rice are determined by the isogenic variety the beta-carotene-generating genetics have been bred into. So, there is little, except consumer demand, to encourage a grower to plant Golden Rice seed, rather than the isogenic variety. Demand may arise from the local community, if they know of the potential health benefits.

Another, more concrete demand generation, requiring cooperation between sectors, is for Government departments responsible for providing biofortified rice to urban populations to establish and communicate a buying price for Golden Rice sufficient to encourage growers to grow Golden Rice.

Another attractive program of demand generation is for school feeding programs to specify that Golden Rice must be used: simultaneously creating demand so that growers grow Golden Rice, children benefit from it nutritionally and learn about it, and inform their parents of it, generating demand also at home.

Such programs can assist Golden Rice's adoption in rural areas, as well as in urban areas, and save money compared with alternatives, at the same time as transferring wealth to growers for productive work. Such programs require cooperation between agriculture, education, women and children's affairs and public health functions of government with their own accountabilities, and should not be held back by narrow, unsubstantiated technology suspicions, which have been disproved [46–48] or for any other reason: the available health, welfare and economic benefits are too great.

Large-scale food fortification against VAD could protect nearly three million children annually by only a minimal 0.5% reduction in VAD prevalence, in a little over a year, "an effect that, importantly, would plausibly be compounded with increasing program maturity, and better intervention coverage and reach" [19].

Vitamin A capsules are only recommended for children of 6 months and older [49], and very young children do not consume solid food. These children are the most vulnerable to vitamin A deficiency: Neonate deaths in 2011 accounted for 43 percent (increased from 36 percent in 1990) of all deaths among under 5-year-olds [50]. Can a good source of vitamin A, such as Golden Rice, when part of the staple diet, improve the mother's vitamin A status, benefiting her health, and simultaneously *via* the placenta and breast milk increase the baby's resistance to disease, and reduce neonate and child mortality? [14].

For the first time since the UN's International Conference on Nutrition three decades ago [43], there is a beta-carotene-rich staple food—Golden Rice—with excellent bioavailability [51], and at no greater cost than white rice, capable of delivering a significant improvement, 57–99% and 89–113% of the recommended vitamin A requirement for preschool children in the Philippines and Bangladesh, respectively, when substituted for white rice [5]. Even partial substitution, for example, through school lunches, would contribute positively to health outcomes, especially for children from more disadvantaged households.

5. How sustainable are the reductions in VAD incidence achieved since the 1990s?

The discovery of the huge hidden mortality due to VAD, from the 1990s, focused the attention of international communities, and national governments, on the excellent cost benefit of avoiding the preventable deaths and other morbidities associated with the deficiency. This included UN meetings in 1990, 1992 and 2004 [43, 52, 53], as well as prominence in the Millennium Development Goals (MDGs) 1990–2015[3]. The huge benefit of addressing micronutrient deficiencies, compared with costs involved, was also endorsed by four rounds of the Copenhagen Consensus [54–57]. Good progress was made, although several MDG goals were missed [28]. The direction has been maintained by the sustainable development goals (SDGs) 2015–2030: It has been argued that staple biofortification with micronutrients can benefit SDGs: 1 (no poverty), 2 (zero hunger), 3 (good health), 4 (quality education), 5 (gender equality), and 7 (decent work and economic growth) [7].

Given the progress achieved in combatting VAD reducing from in excess of 23–34% of child mortality ("in excess of" because these are global percentages but VAD does not occur in industrialized countries) to circa 2% in Bangladesh and the Philippines, it is perhaps unsurprising that relatively little attention is given to VAD caused mortality currently. Much of the reduction is due to annual cycles of costly vaccination programs, including against measles, and expensive vitamin A supplementation, and community health and education as well as general economic development together allowing more food security. However, the sustainability of the reduction in VAD has to be questionable when it requires repeat annual expenditures on materials and labor. Additionally, vitamin A supplementation is not only about preventing mortality [58].

The year before the Covid-19 pandemic struck in Bangladesh and the Philippines was probably the year when child mortality due to VAD was at its lowest, as a result of the community health programs in place. Nevertheless, in 2019 nearly 15,000 children died from VAD-related illness (**Table 1**).

	Bangladesh	Philippines	Source
Number of children age 5 years and under (millions)	14.3	10.6	UNICEF [34] & PSA [59]
Child mortality rate under age 5 years (per 1000)	30.8	27.3	UNICEF [60, 61]
Child mortality under age 5 years	441,302	301,256	Calculated from above
VAD-attributed deaths in 2013 (% of child deaths)	2.0	1.8	Stevens et al., supplementary information [31]
Estimated VAD-related child mortality cases in 2019	8826	5886	Calculated from above

Table 1.
Statistics on vitamin a deficiency (VAD) among children age 5 years and under, and child mortality in Bangladesh and the Philippines in 2019.

Table 1 provides estimates of VAD, and all-cause and VAD-related mortality rates, among children age 5 years and under in Bangladesh and the Philippines in 2019. Despite a decrease in VAD in some parts of the world [31], child VAD rates in both Bangladesh and the Philippines remain high, leading to preventable mortalities due to diarrheal and infectious diseases, among other sequelae. Hence, despite VAD interventions such as food fortification and vitamin A supplementation, additional

[3] Goal 1: Eradicate extreme poverty and hunger

Target 2 Halve, between 1990 and 2015, the proportion of people who suffer from hunger.

Goal 4: Reduce child mortality

Target 5 Reduce by two thirds, between 1990 and 2015, the under-five mortality rate.

Goal 5: Improve maternal health

Target 6 Reduce by three quarters, between 1990 and 2015, the maternal mortality ratio

public health interventions to combat VAD are needed, even in normal, pre-pandemic, circumstances. We estimate that in 2019, VAD led to 8826 preventable deaths in Bangladesh and 5886 preventable deaths in the Philippines of children age five and under a total of 14,712.

For comparison, all ages COVID-19 deaths in calendar year 2020 in these two countries are recorded as nearly 19,000 (Bangladesh: 8127 and Philippines: 10,749) [62].

Thus, the scale of annual child deaths from VAD, pre-pandemic in 2019 and all-ages deaths from COVID-19 in 2020, the first calendar year of the pandemic, are of the same order of magnitude.

Even, at the time of writing[4], in the two countries, the cumulative total of Covid-19 all-age deaths 41,585 (the Philippines), and 27,814 (Bangladesh), in total 69,299, is of the same order of magnitude as VAD child-deaths circa 19,000 pre-pandemic in 2019 [63].

Each death, from whatever cause is a family tragedy. And all these numbers are a vulnerable to reporting errors and therefore approximate. What is important it that whereas no political interest was expressed in the 2019 VAD deaths, all politicians in both countries, as in all other countries in the world, were totally focused on Covid-19, and all economies of the world were brought to a stop by the arrival of the pandemic.

The Covid-19 pandemic has, severely, impacted the social structure and economies of all countries, including, directly and indirectly, low- and middle-income nations. In stark contrast to the global media interest in Covid-19 in all countries, the VAD deaths, which only occur in developing countries, have been seldom reported for the previous 30 years and were probably the lowest ever in 2019, pre-pandemic.

Regretfully, the Covid-19 pandemic has increased poverty and increased food insecurity through job losses and food price increases [64, 65]. Covid-19 has also decreased the effectiveness of community health programs, including reducing dramatically the number of measle vaccinations [66, 67]. It is to be expected that vitamin A supplementation programs have also been negatively affected: they often share resources with measle vaccination programs. Indeed, in 2020, in the first year of the Covid-19 pandemic, despite the potential benefits of this key child survival intervention, vitamin A supplementation programs only reached 41% of the target child population globally, and below 50% in all regions [68], compared with much higher coverage previously: circa 70% [69] to higher than 90% [68].

Often food price shocks lead to social instability, including riots, in LMICs [70] where food costs can be as high as 70% of family income. Such effects would further exacerbate delivery of community health programs.

As a result of the Covid-19-induced disruption of health service provision in South Asia, child mortality could potentially increase by 18–40% and maternal mortality by 14–52% over the next year [71]. Globally, the effect will be an additional 1,157,000 child deaths, and 12,200–56,700 maternal deaths [71]. As an indirect result of the pandemic, a reversal of the progress against the Millennium and Sustainable Development Goals from 1990 to 2019 reported above is to be expected.

Pre-Covid from the 1990s, simultaneous progress was made in reducing VAD, thereby improving the immunity of populations of vulnerable children to common diseases, and at the same time reducing the incidence of those diseases.

Post-Covid from 2020, simultaneously the immunity of populations of vulnerable children to common diseases could well decrease, at the same time as the incidence of those diseases increases.

Thus, it is likely VAD child deaths will increase, in Bangladesh and the Philippines only, as a result of Covid-19-induced conditions, from 2% of all

[4] October 24, 2021.

<5-y child deaths (~15,000 annually) in the direction of the previously normal 23–34% of all <5-y child deaths (170,000–250,000 annually).

We cannot know how long these second-order effects of the Covid-19 pandemic will continue, so cannot know how bad it will get. At the time of writing, 12.43% of the Bangladesh population are fully vaccinated against SARS-Cov-2 and 23.22% of the Philippine population [63], which are not indicative of a rapid return to pre-Covid normality. The VAD death figures could be even more startlingly bad if the post-Covid annual all causes child deaths in these two countries increases from the 2019 figure of 742,558 (**Table 1**).

The sustainability of VAD mitigation would be increased, and the dangers of the explosion of VAD child deaths could be significantly avoided if effective interventions appropriate to the current circumstances are quickly adopted in all relevant countries.

Practically speaking, Golden Rice is an excellent fit to the circumstances and is available.

6. Biofortification: Pioneers and the future

The creation of what became known as Golden Rice was announced by Ingo Potrykus at the XVI International Botanical Congress in St. Louis in early August 1999, a very large meeting involving 20,000 room nights and 4700 delegates from 85 countries [72], and published in "Science" in January 14, 2000 [73]. Golden Rice was widely reported, including on the front covers of the American and Asian (but not European) editions of Time Magazine on July 31, 2000.

The Second CGIAR-wide conference on Nutrition was held at the International Rice Research Institute in October 1999, organized by Howarth Bouis. On January 1, 2000 "Food and Nutrition Bulletin" ("intended for healthcare professionals") published 41 papers of this conference: "Improving human nutrition through agriculture: the role of international agricultural research", many of them anticipating feeding trials to be started soon [74].

The conference summary and recommendations were written by Dr. Bouis, subsequently Director, and then Emeritus Fellow, of Harvest Plus and a World Food Prize Laureate 2016. In his Abstract of the conference proceedings, Dr. Bouis recorded "The need for a shift in emphasis from protein-energy malnutrition to micronutrient malnutrition was recognized" [75].

The summary included comments by the then First Lady of the Philippines (a medical doctor), reporting her, and President Estrada's commitment to medical and relief missions, particularly to poor communities that are not reached by regular public health and medical centers. The "Wheat flour Fortification with vitamin A Project" was one of the first major activities of the Estrada administration in its first 100 days. She encouraged the development of more nutrient-dense crops especially rice, corn and root crops. She also encouraged the production of micronutrient-rich food products, including livestock, poultry, fish and certain vegetables and fruits, especially those that can be easily raised in backyards and community gardens [75].

Also included in the summary were comments by Muhiuddin Khan Alamgir, the then State Minister for Planning, Bangladesh. He commented that Bangladesh's Constitution recognizes "raising the level of nutrition and improvement of public health" as "among primary duties" of the state. He called for improvement in food grain quality and listed genetic engineering and technology as of special importance [75].

In 2002, the term "Biofortification" was first used [76] and in 2004, it was first defined as "a word coined to refer to increasing the bioavailable micronutrient content of food crops through genetic selection via plant breeding" [77]. In the 2004 paper, it is made clear that the human nutrition definition of "micronutrients" will apply encompassing both minerals and vitamins.

Incidentally, in crop breeding for minerals such as "high iron" or "high-zinc" varieties, what is selected are plants that have the capacity to accumulate these minerals from suitable soils. The crop varieties cannot synthesize the minerals. In the case of Golden Rice, beta-carotene is organically synthesized within the plant, independent of the soil type. The same is true of folate rice [78].

It is clear that a lot of thinking was being applied to nutritional improvement of crops at the beginning of this century, and the high public profile of Golden Rice put staple crop biofortification with micronutrients on the donor map in 2000. Harvest Plus, starting in 2003, has now tested or released 400 biofortified staple crop varieties in 63 countries as a result. They are being grown by more than 10 million households globally [79]. All have been produced through conventional, selective breeding improving existing crop varieties.

For those crops where conventional breeding cannot biofortify sufficiently, genetic engineering is necessary, and progress has been slower. Not only Golden Rice, but GMO-biofortified rice with iron and zinc [80], and with folate [78] (eventually it is hoped they will be combined in one multi-biofortified Golden Rice). In 2005, the Bill and Melinda Gates Foundation created Grand Challenge #9 and, following competitive grant allocation, funded further research into genetically engineered biofortified rice (with Peter Beyer—Golden Rice's co-creator—as Principal Investigator) as well as genetically engineered biofortified plantain/banana and cassava, and sorghum. All or some will be successfully and beneficially adopted with huge welfare and economic benefits to poor societies.

All of the successes of Harvest Plus are with single nutrients in each case—all so far conventionally bred. In the case of iron and zinc, biofortification of rice Harvest Plus has found that genetic engineering can achieve levels unattainable by conventional breeding [80]. As proposed above already for Bangladesh, the combination of delivery mechanisms—conventional and existing transgenic crops being conventionally bred together—can quite easily produce, for example, "High Zinc Golden Rice" identified by its color.

Genetic engineering can also produce combination traits: rice with beta-carotene, and simultaneously, the ability to accumulate high iron and high zinc has been developed experimentally [81]. However, with current regulatory constraints and costs it would be preferable to first register and then introgress, the different traits individually.

Gene editing has been used to construct beta-carotene rice [82], but as the construct introduced foreign genes, it was anyway a "GMO." As Beyer and Potrykus have commented, gene editing may be useful to delete function in crop plants, but with current levels of genetic knowledge, to add function requires adding genes, which makes GMO crops [83], with associated regulatory challenges under current rules.

The safety for consumption of Golden Rice has been confirmed by the regulatory authorities of Australia, Canada, New Zealand, the Philippines and the USA [40]. That cultivation is also safe has received official endorsement by regulators in the Philippines. On a separate occasion, the Philippine Secretary of Agriculture, Dr. William D Dar said of Golden Rice: "It smells and tastes the same as ordinary rice, except it is colored yellow. But I will choose 'Golden Rice' over white rice, because it has more health benefits." The Golden Rice-tasting event was part of the inauguration of the Philippines Department of Agriculture Crops and Biotechnology Center, and launch of Golden Rice, on September 30, 2021 ([84], Video 1). In an accompanying press release, Dr. Dar commented that "The recent *[September 2021]* UN Food Systems

Summit held in New York, USA, underscored the important role of biotechnology and other scientific innovations in attaining food security by all countries" [85].

GMO-produced insulin was commercialized from 1979 with no opposition, and genetic modification techniques are commonly employed in discovery and manufacture of pharmaceuticals, and beer, wine, cheese and bread are manufactured using genetically modified enzymes. Hundreds of millions of people in the United States and elsewhere and billions of farm animals have been consuming since 1997 products from genetically modified crops using the same techniques employed by Beyer and Potrykus to create Golden Rice. The European production of pork and chicken, the whole market in Europe, would collapse if it were not for imported GMO-maize and GMO-soy meal. Imported because, with very small exceptions, the Europeans will not allow cultivation. Yet, not a single case of any disease or other difficulty associated with genetically modified crops has been verifiably recorded in any human or other animal.

Every single academy of science in the world has attested to the fact that there is no scientifically valid reason for assuming that GMOs could cause harm [47]. Additionally, the European Food Safety Authority, stated, in 2010, that "The main conclusion to be drawn from the efforts of more than 130 research projects, covering a period of more than 25 years of research and involving more than 500 independent research groups, is that biotechnology, and in particular GMOs, are not per se more risky than, for example, conventional plant breeding technologies" [86].

The technology—of conventionally bred as well as genetically engineered biofortified crops—is in the seed and breeds true season to season. The biofortified traits are only introduced into modern, high-yielding crop varieties and can be easily transferred by plant breeders to new varieties as they become popular.

It is time to embrace all available tools, both forms of biofortification as well as chemical fortification, to improve the nutritional quality of staple foods by the incorporation of micronutrients, together with the macronutrients that have been the focus of plant breeding for the previous millennia since humans stopped being hunter gatherers.

These tools are complementary to other public health interventions, education, vaccination, supplementation, home gardens, breast feeding and economic development all important to population welfare.

Not only are all these tools required. Also required are all functions of the private, public, NGO and especially government sectors, working across silos of expertise to support each other's objective of improving societal public health.

Other countries should follow the Philippines example. Bangladesh is poised to do so. India has a huge VAD problem [2], equivalent to the total of the VAD of the 28 sub-Saharan African countries [21]. India has been held back from vitamin A supplementation, because "the issue of vitamin A has commercial overtones": "[W]e must look to our farmers, not to pharmaceutical companies, to protect the health of our children. The main solution to vitamin A deficiency should not be drug-based, but food-based." [21].

Golden Rice is "food based" and there are no "commercial overtones." Golden Rice does depend on farmers, first and foremost to grow it before it can be consumed, especially by their communities, as an additional intervention for vitamin A deficiency.

All the biofortification tools, and related biofortified crops described here and developed by the public sector are available without cost for use of governments, growers and consumers, as by the time of introduction the development costs have been paid.

On World Food Day, October 16, 2020—during the Covid-19 pandemic—Indian Prime Minister Narendra Modi gave a strong endorsement to staple crop biofortification as a sustainable and cost-effective solution to alleviate malnutrition [87].

The World Bank has recommended that micronutrient biofortification of staple crops, including specifically Golden Rice, should be the norm and not the exception in crop breeding [88].

The movement to common sense and reality has now become unstoppable.

7. Conclusions

In the 1990's, when vitamin A deficiency's importance was recognized not only as the principal cause of irreversible blindness in children, but also the principal cause of child mortality, VAD killed in excess of 2.0 million young children and mothers annually. At that time, VAD was responsible for between 23 and 34% of all deaths of young children globally (and a greater proportion in developing countries), and a greater cause of mortality globally than HIV or TB or Malaria [7].

A combination of successful community health programs, including vaccinations and vitamin A supplementation, as well as economic development resulting *inter alia*, in better access to clean water and sanitation, had reduced the death toll from 23 to 34% in the 1990s to circa 2% in 2019, of young children in Bangladesh and the Philippines.

The Covid pandemic, which started in 2020, has reversed the progress of community health programs achieved during the past three decades. Covid has also increased food insecurity. We cannot know for how long these conditions will last. There is an acute danger that they will result also in a reversal of VAD induced mortality from circa 2% toward in excess of 23–34% of child deaths in all LMICs.

In Bangladesh and the Philippines, in 2020 the first year of the pandemic, Covid killed as many people as VAD killed children only, in the previous year, 2019. Yet far greater attention was paid to Covid's arrival than children's deaths from VAD, which had been continuing for decades. It is long past time to pay more attention to alleviating VAD.

In 2021, for the first time since the 1992 UN International Conference on Nutrition, which recommended locally available food-based strategies are the first priority to combat vitamin A deficiency, such a staple food source with sufficient quantity and bioavailability of beta-carotene (a human source of vitamin A) is available: Golden Rice.

Golden Rice has been proven as safe to consume by Government regulators of four high-income countries and as safe to consume and cultivate in the Philippines. In only one other country has registration for Golden Rice on the same basis as in the Philippines been applied for, in late 2017, and with, at the time of writing[5], no regulatory decision: Bangladesh.

As the technology is in the seed, Golden Rice adoption requires no use of foreign exchange or industrial infrastructure. It is designed to be useful to resource poor rural communities that grow their own rice staple for consumption. And governments can pay growers to grow the Golden Rice supply necessary for urban use. The color of Golden Rice reduces the opportunities for "passing off" of normal white rice, as micronutrient-fortified rice. And Golden Rice introgressed with, for example, high-zinc and or high-iron rice and or folate rice, using conventional plant breeding will be a multi-micronutrient rice and a golden color.

All departments of government have a responsibility to work together, also with those supranational institutions supporting government public health programs, to use newly available Golden Rice.

[5] October 2021

There is a huge potential for saving lives and money—multi-millions of US dollars annually—by adopting Golden Rice, not only in the Philippines, but also Bangladesh and other countries where VAD continues to be problematic.

Video Materials

https://www.facebook.com/DAPhilRice/videos/6274026146003384/

(You can skip to minute 3:08 to see the dignitaries and congresspersons eating Golden Rice and the Philippine Secretary of Agriculture, Dr. William Dar, saying Golden Rice should be favored over white rice. Or look at 2:02 for Philippines President Duterte's address in favor of agricultural biotechnology.)

Appendix

What support is available to countries which are interested in introducing Golden Rice as an additional intervention for vitamin A deficiency?

Especially as most rice is consumed close to where it is grown, and Golden Rice will cost no more than white rice, Golden Rice should be increasingly useful, including in post pandemic circumstances, as an additional intervention to combat VAD, in all countries where rice is the staple crop of the resource poor and VAD endemic.

For such countries, and where a public sector rice breeding institute is available (in the country or a neighbouring country) to introgress the GR2E transformation event into locally adapted and preferred rice varieties, the following is required and available without cost:

1. A Golden Rice license defining responsibilities and obligations, for humanitarian development of the technology in locally adapted and preferred rice varieties, including the obligation not to charge for the extra beta-carotene in Golden Rice.

2. Breeding parent physical rice seed containing the GR2E trait from another Golden Rice Licensee.

3. Advice on establishing analytical procedures for monitoring the progress of the introgression of GR2E trait.

4. Membership of the Golden Rice licensee network for ongoing support.

5. The regulatory data package for event GR2E. Locally generated environmental impact data may be required, or regulators may agree to use data from a similar agro-ecological habitat. (The 'food, feed and processing' data package alone developed for Golden Rice GR2E is extensive, 42 megabytes of data). In due course, these regulatory studies may be published, as for example in a series of papers with Golden Rice data [5, 89–91].

Most countries make their regulatory deliberations and decisions publicly available, for example Australia and New Zealand, Canada and USA. Included in this openness are inputs from the various Government department involved, including in the Philippines.

It is a pity that In the case of Golden Rice this is all necessary even though the only difference in comparison with white rice is that the normally white endosperm, contains beta-carotene [5], a source of vitamin A for the human body [51].

6. Finance is not available, and must be sought by the potential new, or new licensee from normal sources.

It is normally a requirement that a Golden Rice licensee country has relevant laws in place governing the development and deployment of transgenic crops. If this is not the case, nevertheless discussion is encouraged.

It is to be expected that national rice breeding institutions can introgress the GR2E beta-carotene inducing trait into any public sector owned rice variety, within 2 to 4 growing seasons taking perhaps two or three years.

Other crops than rice, are also potentially able to benefit from the same technology to introduce beta-carotene synthesis to the edible parts. (Peter Beyer has already advised on this for a number of crops.)

Author details

Adrian C. Dubock[1*], Justus Wesseler[2], Robert M. Russell[3], Chen Chen[4,5] and David Zilberman[6]

1 Golden Rice Humanitarian Board, Binningen, Switzerland

2 Agricultural Economics and Rural Policy Group, Wageningen University, Netherlands

3 Tufts University School of Medicine and Tufts University School of Nutrition, USA

4 Department of Food Science and Human Nutrition, Michigan State University, USA

5 School of Public Health, Shandong University, Jinan, Shandong Province, China

6 Department of Agricultural and Resource Economics, University of California, Berkeley, USA

*Address all correspondence to: contact@goldenrice.org

IntechOpen

© 2021 The Author(s). Licensee IntechOpen. This chapter is distributed under the terms of the Creative Commons Attribution License (http://creativecommons.org/licenses/by/3.0), which permits unrestricted use, distribution, and reproduction in any medium, provided the original work is properly cited. CC BY

References

[1] Stevens GA, Bennett JE, Hennocq Q, Lu Y, De-Regil LM, Rogers L, et al. Trends and mortality effects of vitamin A deficiency in children in 138 low-income and middle-income countries between 1991 and 2013: A pooled analysis of population-based surveys. Lancet Glob Heal. 2015;3(9):e528–e536.

[2] West KP, Klemm RDW, Sommer A. Sound science, sound policy. J World Public Heal Nutr Assoc. 2010;1(5):211-229.

[3] Schmitz J, West KP, Khatry SK, Wu L, LeClerq SC, Karna SL, et al. Vitamin A supplementation in preschool children and risk of hearing loss as adolescents and young adults in rural Nepal: Randomised trial cohort follow-up study. BMJ 2012; 344(7841). d7962

[4] Paine JA, Shipton CA, Chaggar S, Howells RM, Kennedy MJ, Vernon G, et al. Improving the nutritional value of Golden Rice through increased pro-vitamin A content. Nature Biotechnology 2005;23(4):482-487.

[5] Swamy BPM, Samia M, Boncodin R, Marundan S, Rebong DB, Ordonio RL, et al. Compositional analysis of genetically engineered GR2E "golden rice" in comparison to that of conventional rice. Journal of Agricultural and Food Chemistry 2019;67(28):7986-7994.

[6] U.S. Food and Drug Administration, Grune T, Lietz G, Palou A, Ross AC, Stahl W, et al. Select committee on GRAS substances (SCOGS) opinion: Carotene (beta-carotene). 140, The Journal of Nutrition. 1979. p. 2268S-2285S.

[7] Dubock A. Golden rice: To combat vitamin A deficiency for public health. In: Queiroz ZL, de Rosso VV, Eduardo J-L, editors. Vitamin A. London, UK: IntechOpen; 2019. p. 1-21

[8] Status of Application for Propagation [Internet]. The Goverment of the Philippines, Department of Agriculture. 2021. [cited 2021 Oct 14] Available from: http://biotech.da.gov.ph/Decision_docs_jdc_propa.php

[9] Potrykus I. Regulation must be revolutionized. 466, Nature Publishing Group; 2010. p. 561.

[10] Potrykus I. "Golden rice", a GMO-product for public good, and the consequences of GE-regulation. Journal of Plant Biochemistry and Biotechnology. 2012;21(S1):68-75.

[11] Giddings LV, Potrykus I, Ammann K, Fedoroff NV. Confronting the gordian knot. Nature Biotechnology. 2012 30(3):208-209.

[12] Potrykus I. Unjustified regulation prevents use of GMO technology for public good. Trends in Biotechnology. 2013 31(3);131-133

[13] Potrykus I. Genetic modification and the public good. Eur Rev. 2013;21(S1):S68–S79

[14] Dubock A. Golden rice: A long-running story at the watershed of the GM debate. 2013;1-12

[15] Dubock AC. The politics of golden rice. GM Crop Food. 2014;5(3) 210-222

[16] Dubock A, Potrykus I, Beyer P. We-pioneered-a-technology-to-save-millions-of-poor-children-but-a-worldwide-smear-campaign-has-blocked-it. Leaps Magazine. 2019;7

[17] Wu F, Wesseler J, Zilberman D, Russell R, Chen C, Dubock A C., Opinion, allow Golden Rice to save lives. Proceedings of the National Academy of Sciences of the United States of America. 2021, (118), 51, 3p.

Available from: https://www.pnas.org/content/118/51/e2120901118

[18] Ministers GA. G20 Agriculture Ministers Meeting Communiqué [Internet]. 2016 [cited 2017 Mar 3]. Available from: http://www.g20.org/English/Documents/Current/201606/t20160608_2301.html

[19] Keats EC, Neufeld LM, Garrett GS, Mbuya MNN, Bhutta ZA. Improved micronutrient status and health outcomes in low-and middle-income countries following large-scale fortification: Evidence from a systematic review and meta-analysis. The American Journal of Clinical Nutrition 2019;109(6):1696-1708.

[20] WHO Global prevalence of vitamin A deficiency in populations at risk 1995-2005. 2009

[21] Semba RD. The vitamin A story: Lifting the shadow of death. Koletzko SB, editor. The Vitamin A Story: Lifting the Shadow of Death. World Review of Nutrition and Dietetics, 104; 2012. 1-207

[22] Sommer A, Hussaini G, Tarwotjo I, Susanto D. Increased mortality in children with mild vitamin A deficiency. Lancet. 1983;ii:585-588.

[23] Sommer A, Djunaedi E, Loeden AA, Tarwotjo I, West KP, Tilden R, et al. Impact of vitamin A supplementation on childhood mortality: A randomised controlled community trial. Lancet 1986;327(8491):1169-1173.

[24] West KP, Katz J, LeClerq SC, Pradhan EK, Tielsch JM, Sommer A, et al. Efficacy of vitamin A in reducing preschool child mortality in Nepal. Lancet 1991;338(8759):67-71.

[25] Sommer A. Vitamin A deficiency and clinica disease: An historical overview. The Journal of Nutrition 2008;138:1835-1839.

[26] Mayo-Wilson E, Imdad A, Herzer K, Yakoob MY, Bhutta ZA. Vitamin A supplements for preventing mortality, illness, and blindness in children aged under 5: Systematic review and meta-analysis. BMJ. 2011;343(251):d5094

[27] Thorne-Lyman AL, Parajuli K, Paudyal N, Chitekwe S, Shrestha R, Manandhar DL, et al. To see, hear, and live: 25 years of the vitamin A programme in Nepal. Maternal & Child Nutrition. 2020:1-12.

[28] United Nations. The Millennium Development Goals Report 2015 [Internet]. 2015. Available from: http://www.un.org/millenniumgoals/2015_MDG_Report/pdf/MDG 2015 rev (July 1).pdf

[29] Neidecker-Gonzales O, Nestel P, Bouis H. Estimating the global costs of vitamin A capsule supplementation: A review of the literature. Food and Nutrition Bulletin 2007;28(3):307-316.

[30] Gödecke T, Stein AJ, Qaim M. The global burden of chronic and hidden hunger: Trends and determinants. 17, Global Food Security. 2018. p. 21-29.

[31] Stevens GA, Bennett JE, Hennocq Q, et al. Trends and mortality effects of vitamin A deficiency in children in 138 low-income and middle-income countries between 1991 and 2013: A pooled analysis of population-based surveys. Supplementary appendix. Lancet Glob Heal. 2015;3(e528–e536).

[32] Hamer DH, Keusch GT. Vitamin a deficiency: Slow progress towards elimination. Lancet Glob Heal. 2015;3(9):e502–e503.

[33] Staff Correspondent. Golden rice release stuck due to DoE reluctance: minister. Daily Observer. 2020; Available from: http://www.goldenrice.org/PDFs/Golden Rice stuck - Daily Observer 2020.pdf

[34] United Nations Children's Fund (UNICEF). For Every Child, Reimagine. UNICEF Annual Report 2019. [Internet]. 2020. Available from: https://www.unicef.org/media/74016/file/UNICEF-annual-report-2019.pdf

[35] United Nations Children's Fund (UNICEF). Bangladesh: Key Demographic Indicators [Internet]. 2021 [cited 2021 Apr 11]. Available from: https://data.unicef.org/country/bgd/

[36] Das RC. Status of Rice Fortification in Bangladesh [Internet]. 2021 [cited 2021 Sep 23]. Available from: https://www.gainhealth.org/sites/default/files/event/documents/status-of-rice-fortification-in-bangladesh-ram-chandra-das.pdf

[37] Montgomery S. Rice Fortification: Quality and Cost [Internet]. 2021 [cited 2021 Sep 23]. Available from: https://www.gainhealth.org/sites/default/files/event/documents/rice-fortification-quality-and-cost-scott-montgomery.pdf

[38] Food Fortification Initiative, Global Alliance for Improved Nutrition (GAIN), Iodine Global Network, Micronutrient Forum. Providing Actionable Food Fortification Data [Internet]. Global Fortification Data Exchange. 2021 [cited 2021 Oct 22]. Available from: https://tinyurl.com/y4w4naxv

[39] Harvest Plus. Bangladesh Releases New, Improved Zinc Rice Variety [Internet]. 2015 [cited 2021 Oct 14]. Available from: https://www.harvestplus.org/knowledge-market/in-the-news/bangladesh-releases-new-improved-zinc-rice-variety-0

[40] ISAAA. GMO Approval Database [Internet]. 2021 [cited 2021 Apr 8]. Available from: https://www.isaaa.org/gmapprovaldatabase/event/default.asp?EventID=528

[41] Wesseler J, Zilberman D. The economic power of the Golden Rice opposition. Environment and Development Economics 2014;19(6):724-742.

[42] Wesseler J, Zilberman D. Golden Rice: No progress to be seen. Do we still need it? Environment and Development Economics 2017;22(2):107-109.

[43] FAO of UN. The International Conference on Nutrition [Internet]. 1992. Available from: http://www.fao.org/docrep/v7700t/v7700t02.htm

[44] Dubock A. Golden Rice: Instructions for use. Agric Food Secur. 2017;6(1).

[45] Abalajen S, Alonto II K, Bitagun K, Capareda J, Diaz J, Miranda R, et al. Attitudes and Influences relevant to Golden Rice's potential use in the Philippines [Internet]. 2009. Available from: http://www.goldenrice.org/PDFs/GR_FR_Story Board.pdf; Www.Goldenrice.Org

[46] National Academy of Sciences, Engineering and Medicine. Genetically Engineered Crops: Experiences and Prospects. Washington, DC: The National Academies Press; 2016

[47] Roberts R. Support Precision Agriculture [Internet]. 2016. [cited 2017 Mar 3]. Available from: http://supportprecisionagriculture.org/Nobel Laureates June 2016

[48] Philippines National Academy of Science and Technology. Statement on Genetically Modified Organisms (gmo) — Golden Rice and bt Eggplant [Internet]. 2021. Available from: https://www.nast.ph/images/pdf%20files/Publications/Statement/Statement%20on%20GMO.pdf

[49] WHO. Guideline: Vitamin A supplementation in infants and children 6 – 59 months of age. World Heal Organ. 2011;1-25.

[50] UN Interagency Group for Child Mortality Estimates. Levels and trends in Child Mortality: estimates developed by the United Nations Inter-Agency group for Child Mortality Estimation [Internet]. 2010. Available from: http://www.childinfo.org/files/Child_Mortality_Report_2010.pdf

[51] Tang G, Jian Q, Dolnikowski GG, Russell RM, Grusak MA. Golden rice is an effective source of vitamin A. The American Journal of Clinical Nutrition 2009;89(6):1776-1783.

[52] United Nations. UN World Summit for Children [Internet]. New York; 1990. Available from: http://www.unicef.org/wsc/

[53] UNICEF. United Nations Children's Fund: Vitamin and mineral deficiency: A global damage assessment report. [Internet]. 2004. Available from: http://www.eldis.org/vfile/upload/1/document/0708/DOC23773.pdf

[54] Copenhagen Consensus. Copenhagen Consensus 2004. 2004;4-5.

[55] Consensus C. Copenhagen Consensus 2008. 2008;25-26.

[56] Copenhagen Consensus 2012 [Internet]. [cited 2018 Nov 15]. Available from: https://www.copenhagenconsensus.com/sites/default/files/cc12resultspressreleasefinal_0.pdf

[57] Copenhagen Consensus Center. Copenhagen Consensus - Post 2015. 2016; Available from: http://www.copenhagenconsensus.com/post-2015-consensus/nobel-laureates-guide-smarter-global-targets-2030

[58] Bhutta ZA, Baker SK. Premature abandonment of global vitamin A supplementation programmes is not prudent! 44, International Journal of Epidemiology. Oxford University Press; 2015. p. 297-299.

[59] Philippines Statistics Authority Government of the Philippines. Updated Population Projections Based on the Results of 2015 POPCEN [Internet]. October 4, 2019. [cited 2021 Feb 2]. Available from: https://psa.gov.ph/content/updated-population-projections-based-results-2015-popcen

[60] UNICEF. Bangladesh Key demographic indicators [Internet]. 2021 [cited 2021 Feb 2]. Available from: https://data.unicef.org/country/bgd/

[61] United Nations Children's Fund (UNICEF). Philippines: Key demographic indicators [Internet]. 2021 [cited 2021 Apr 11]. Available from: https://data.unicef.org/country/phl/

[62] Johns Hopkins University & Medicine. COVID-19 Map - Johns Hopkins Coronavirus Resource Center [Internet]. Johns Hopkins Coronavirus Resource Center. 2020. p. 1. [Cited 2021 Jan 31]. Available from: https://coronavirus.jhu.edu/map.html

[63] Johns Hopkins University & Medicine. Covid-19 Map - Johns Hopkins Coronavirus Resource Center [Internet]. 2021. [Cited 2021 Oct 24]. Available from: https://coronavirus.jhu.edu/region/

[64] Dabalen A. and Paci P. How-severe-will-poverty-impacts-covid-19-be-africa [Internet]. World Bank Blogs: Africa Can End Poverty. 2020 [cited 2020 Dec 8]. p. August 5th. Available from: https://blogs.worldbank.org/africacan/how-severe-will-poverty-impacts-covid-19-be-africa

[65] Laborde D, Martin W, Swinnen J, Vos R. COVID-19 risks to global food security. Science. 2020;369(6503):500-502.

[66] Anonymous. WHO and UNICEF warn of a decline in vaccinations during COVID-19. Saudi Medical Journal. 2020;41(8):898-899.

[67] UNICEF. Protecting the most vulnerable children from the impact of coronavirus: An agenda for action [Internet]. 2020 [cited 2020 Dec 8]. Available from: https://www.unicef.org/coronavirus/agenda-for-action

[68] United Nations Children's Fund (UNICEF). Vitamin A deficiency [Internet]. 2021. Available from: https://data.unicef.org/topic/nutrition/vitamin-a-deficiency/#:~:text=Out of the 46 countries, 71 per cent were reached.

[69] Wirth JP, Petry N, Tanumihardjo SA, Rogers LM, McLean E, Greig A, et al. Vitamin a supplementation programs and country-level evidence of vitamin A deficiency. Nutrients 2017 1;9(3);190.

[70] Fedoroff NV. Food in a future of 10 billion. Agric Food Secur. 2015;4(11).

[71] UNICEF. Direct and indirect effects of the COVID-19 pandemic and response in South Asia. 2021.

[72] Editor-Desconocido. XVI International Botanical Congress. August 1-7, 1999. Abstracts. In St. Louis, USA; 1999.

[73] Ye X, Al-Babili S, Klöti A, Zhang J, Lucca P, Beyer P, et al. Engineering the provitamin A (beta-carotene) biosynthetic pathway into (carotenoid-free) rice endosperm. Annual Review of Plant Biology [Internet]. 2000;66(5451):303-305.

[74] Anonymous. Table of contents. Food and Nutrition Bulletin. 2000;21(4):14

[75] Bouis HE. Improving human nutrition through agriculture: The role of international agricultural research. Conference summary and recommendations. Food and Nutrition Bulletin 2000;21(4):550-567.

[76] Welch RM. Breeding strategies for biofortified staple plant foods to reduce micronutrient malnutrition globally. The Journal of Nutrition 2002;132(3):495-499.

[77] Welch, Ross and Graham R. Breeding for micronutrients in staple food crops from a human nutrition perspective. Journal of Experimental Botany 2004;55(396):353-364.

[78] Storozhenko S, De Brouwer V, Volckaert M, Navarrete O, Blancquaert D, Zhang GF, et al. Folate fortification of rice by metabolic engineering. Nature Biotechnology 2007;25(11):1277-1279.

[79] Bouis H. Linking agriculture and nutrition: An overview of biofortification and harvestplus [Internet]. West African Biofortification Webinar. 2021.

[80] Trijatmiko KR, Duenãs C, Tsakirpaloglou N, Torrizo L, Arines FM, Adeva C, et al. Biofortified indica rice attains iron and zinc nutrition dietary targets in the field. Scientific Reports 2016;6(2015):1-13.

[81] Singh SP, Gruissem W, Bhullar NK. Single genetic locus improvement of iron, zinc and β-carotene content in rice grains. Scientific Reports 2017;7(1):6883.

[82] Dong OX, Yu S, Jain R, Zhang N, Duong P, Butler C, et al. Marker-free carotenoid-enriched rice generated through targeted gene insertion using CRISPR-Cas9. Nature Communications 2020;11:1-7.

[83] Beyer P, Potrykus I. Can you make a CRISPR Golden Rice? An interview with Beyer & Potrykus [Internet]. 2017. [cited 2018 Oct 11] Available from: https://mycrispr.blog/2017/04/03/can-you-make-a-crispr-golden-rice/

[84] ISAAA. Golden Rice Favored as It Tastes, Smells like Regular Rice, but More Nutritious [Internet]. Crop

Biotech Update. 2021. [cited 2021 Oct 21] Available from: https://www.isaaa.org/kc/cropbiotechupdate/article/default.asp?ID=19065

[85] Government of Philippines Department of Agriculture. New Department of Agriculture biotech center to help ensure food security [Internet]. 2021. Available from: https://www.philrice.gov.ph/new-da-biotech-center-to-help-ensure-food-security/

[86] Biotechnologies. A Decade of EU-Funded GMO Research (2001-2010) [Internet]. 2010. Available from: ftp://ftp.cordis.europa.eu/pub/fp//kbbe/docs/a-decade-of-eu-funded-gmo-research_en.pdf

[87] Harvest Plus. On World Food Day, India PM Modi Endorses Biofortification to Address Malnutrition [Internet]. 16 October 2020. [cited 2020 Dec 16]. Available from: https://www.harvestplus.org/knowledge-market/in-the-news/world-food-day-india-pm-modi-endorses-biofortification-address

[88] Htenas AM, World Bank Group. An overview of links between obesity and food systems: Implications for the agriculture GP agenda (English). [Internet]. Washington DC; 2017. Available from: http://documents.worldbank.org/curated/en/222101499437276873/An-overview-of-links-between-obesity-and-food-systems-implications-for-the-agriculture-GP-agenda

[89] Oliva N, Florida Cueto-Reaño M, Trijatmiko KR, Samia M, Welsch R, Schaub P, et al. Molecular characterization and safety assessment of biofortified provitamin A rice. Scientific Reports. 2020;10(1):1-13.

[90] Mallikarjuna Swamy BP, Marundan S, Samia M, Ordonio RL, Rebong DB, Miranda R, et al. Development and characterization of GR2E Golden rice introgression lines. Scientific Reports. 2021;11(1):2496

[91] Biswas PS, Mallikarjuna Swamy BP, Abdul Kader M, Amalgir Hossain M, Boncodin R, Samia M, et al. Development and field evaluation of near-isogenic lines of GR2-EBRRI dhan29 golden rice. Frontiers in Plant Science. 2021;12:619739.